Return to Volcano Town

REASSESSING THE 1937–1943 VOLCANIC ERUPTIONS AT RABAUL

Return to Volcano Town

REASSESSING THE 1937–1943 VOLCANIC ERUPTIONS AT RABAUL

R. WALLY JOHNSON AND
NEVILLE A. THRELFALL

Australian
National
University

ANU PRESS

PACIFIC SERIES

Australian
National
University

ANU PRESS

Published by ANU Press
The Australian National University
Canberra ACT 2600, Australia
Email: anupress@anu.edu.au

Available to download for free at press.anu.edu.au

ISBN (print): 9781760466039
ISBN (online): 9781760466046

WorldCat (print): 1394053743
WorldCat (online): 1393909427

DOI: 10.22459/RVT.2023

Cover design and layout by ANU Press

Front cover: This photograph of simultaneous eruptions at Vulcan (left) and Tavurvur volcanoes, Rabaul, was taken from Nangananga village by Torsten Blackwood overlooking Karavia Bay on 22 September 1994 (see Figure 8.3 for further details). The statue is of a *tubuan* (masked ritual figure) called 'Ia Patarai' from the Ralalar Clan who handed over this piece of land to the Catholic Church in 1882. The Nangananga Monument was built in 1982 to mark 100 years of the Catholic Church in the Rabaul area (1882–1982). The transaction from the Catholic Church to the Ralalar Clan was made using traditional Tolai shell money (*tabu*) and is acknowledged by the Christian cross on top of the *tubuan* statue. This information on the Nangananga Monument was kindly provided by Steven and Jennifer Gagau, Robin Bangin and Ia Kaum.

This book is published under the aegis of the Pacific editorial board of ANU Press.

Contents

List of Illustrations

About the Authors

The authors of this book have quite different professional backgrounds, yet they share a strong, long-lived interest in the Rabaul area of New Britain, Papua New Guinea (PNG)—its history, people and volcanoes. Both first came to New Britain in the 1960s: Neville Threlfall in 1961 as a missionary involved in pastoral ministry for the Methodist and later United Church, Wally Johnson in 1969 as a member of an Australian Government field party mapping the geology of New Britain.

Wally Johnson was born on Tyneside in the north-east of England and is a graduate of the Imperial College of Science and Technology, University of London. He is an honorary associate professor in the Department of Pacific Affairs, College of Asia and the Pacific, The Australian National University (ANU). Johnson worked for many years for the Australian Government's geoscientific agency in Canberra, Geoscience Australia, first as a volcanic geologist and research scientist and later in senior management roles. Most of his research career has focused on the volcanology of Papua New Guinea, much of which was undertaken in close association with colleagues at the Rabaul Volcanological Observatory, headquarters of the national volcanological service of Papua New Guinea. Johnson is an honorary life member of the International Association of Volcanology and Chemistry of the Earth's Interior and an honorary fellow of the International Union of Geodesy and Geophysics. His previous publications with ANU Press are *Fire Mountains of the Islands: A History of Volcanic Eruptions and Disaster Management in Papua New Guinea and the Solomon Islands* (2013) and *Roars from the Mountain: Colonial Management of the 1951 Volcanic Disaster at Mount Lamington* (2020).

Neville Threlfall was born in Western Australia and grew up on a farm in the north-east Wheatbelt. He attended Perth Modern School and the University of Western Australia, graduating with a Bachelor of Arts in history and French, and then undertook theological studies and pastoral ministry for the Methodist Church. Offering to serve in Methodist Overseas Missions, Threlfall was posted to PNG, and, while in the Rabaul region (1961–80), became heavily involved in working with the Methodist Mission Press (later Trinity Press) in the production of literature in English and several PNG languages. He also worked with the Bible Society of PNG on the Bible in Kuanua and Tok Pisin. He lived for over nine years in Rabaul town and on nearby Matupit Island, an island that plays a key role in *Return to Volcano Town*. Threlfall wrote *One Hundred Years in the Islands: The Methodist/United Church in the New Guinea Islands Region 1875–1975* (1975), and, in 1980, was invited to write a history of Rabaul. The research and writing for this history were conducted in part at The Australian National University during tenure of a visiting fellowship in 1981–82. Threlfall next returned to parish ministry in Western Australia, and later in New South Wales, where he is now retired on the Central Coast. His book *Mangroves, Coconuts and Frangipani: The Story of Rabaul* was published in 2012 by the Rabaul Historical Society. He was awarded the PNG Companion of the Order of the Star of Melanesia in 2020 for his pastoral ministry, Bible translation and historical writing.

Acknowledgements

The contents of this book draw heavily on our previous publications on the history and volcanology of the Rabaul area and we therefore must acknowledge again the many people—far too numerous to list here individually—who have helped us over the years, ranging from villagers and townspeople carrying local knowledge to international and national historians and scientists. Included in this wide range of people are the staff of the Rabaul Volcanological Observatory, as well as those of the University of Papua New Guinea Library and the National Archives of Papua New Guinea (both in Port Moresby). We thank too the helpful staff of the following agencies in Canberra: Geoscience Australia (the N.H. [Doc] Fisher Geoscience Library), National Archives of Australia, Australian War Memorial and National Library of Australia.

We also gratefully acknowledge the valuable assistance of staff at The Australian National University, Canberra, notably at the Menzies Library, Pacific Manuscripts Bureau and ANU Press, and members of the Pacific Editorial Board chaired by Stewart Firth. We extend our particular thanks to ANU colleagues Robin Hide and Bryant Allen for labouring through the first draft of the manuscript, and to two, anonymous, external reviewers who were both generous and helpful in their reports of the submitted version. Finally, we extend our thanks to our copyeditor, Rani Kerin, and to illustrators Chandra Jayasuriya and Karina Pelling for redrafting several new diagrams.

The 1937–43 photographs used in the original *Volcano Town* still form a significant part of this new book, *Return to Volcano Town*, and duplicate negatives for many of the original prints are still held by Geoscience Australia—formerly the Bureau of Mineral Resources—an agency of the Australian Government. This does not mean, however, that the photographs were taken by employees of the Australian Government. Most, in fact, were taken by local, camera-wielding, European witnesses, some of them with a good eye for photographic composition, including, for example—alphabetically and not exclusively—Sarah Chinnery, Roger Davies, E.A. Hawnt and Eric Hoskin. We acknowledge photographers in the captions of the illustrations where we know their names.

We have added in this new edition some other illustrations—both diagrams and photographs—and have omitted a few others; we have also given more attention to interpreting several of the eruption photographs that were taken in May 1937 and to presenting them more systematically with regard to points in the narrative. Many of the photographs were included as bromide prints in the albums and collections of several individuals who were not necessarily the photographers, and some seem to have been produced commercially as postcards. We nevertheless must acknowledge the generosity of the print owners in allowing us to digitally scan many of the photographs in their individual collections.

Rabaul-based photographers in 1937 were also generous in providing their photographs to Australian newspapers and magazines who used them in their publications, and not always with acknowledgements. We have tended not to favour the immediately reported text of these media publications as information sources for the events of 1937–43 because of their secondary nature. We do, however, tend to favour articles in issues of the *Rabaul Times* (editor G. Thomas) and *Pacific Islands Monthly* (editor R.W. Robson).

Introduction

More than 500 people, mainly local Tolai villagers, were killed in 1937 by a volcanic eruption on the shores of Blanche Bay in the north-eastern Gazelle Peninsula area of East New Britain (Figures 0.1 and 0.2). Both Vulcan and Tavurvur volcanoes were in simultaneous activity for a short while in 1937. This disaster took place during colonial times when Australian authorities used Rabaul town as their administrative capital, subjugating the Tolai to a different way of doing things. The disaster was the central theme of *Volcano Town: The 1937–43 Rabaul Eruptions*, a book we wrote about 40 years ago (Johnson and Threlfall 1985). The primary aim of the book was to present photographs of the 1937–43 eruptions and their damaging effects for the benefit of people throughout the Blanche Bay area in the early 1980s who were concerned that another disastrous eruption might be imminent.

This apprehension was because of the increased numbers of local earthquakes being felt by people and recorded by the Rabaul Volcanological Observatory (RVO) as well as the slow rise of part of Matupit Island (changes had begun to be noticed about 10 years earlier). The years 1983–85 became known as the 'crisis' period. The great majority of people in the Blanche Bay area in the 1980s were post–World War II newcomers or else were too young to remember the eruptions of 1937–43. Our earlier book, *Volcano Town*, thus was very much a 'public awareness' document for local consumption rather than an account suitable for a wider audience beyond the shores of Papua New Guinea (PNG), including an academic one.

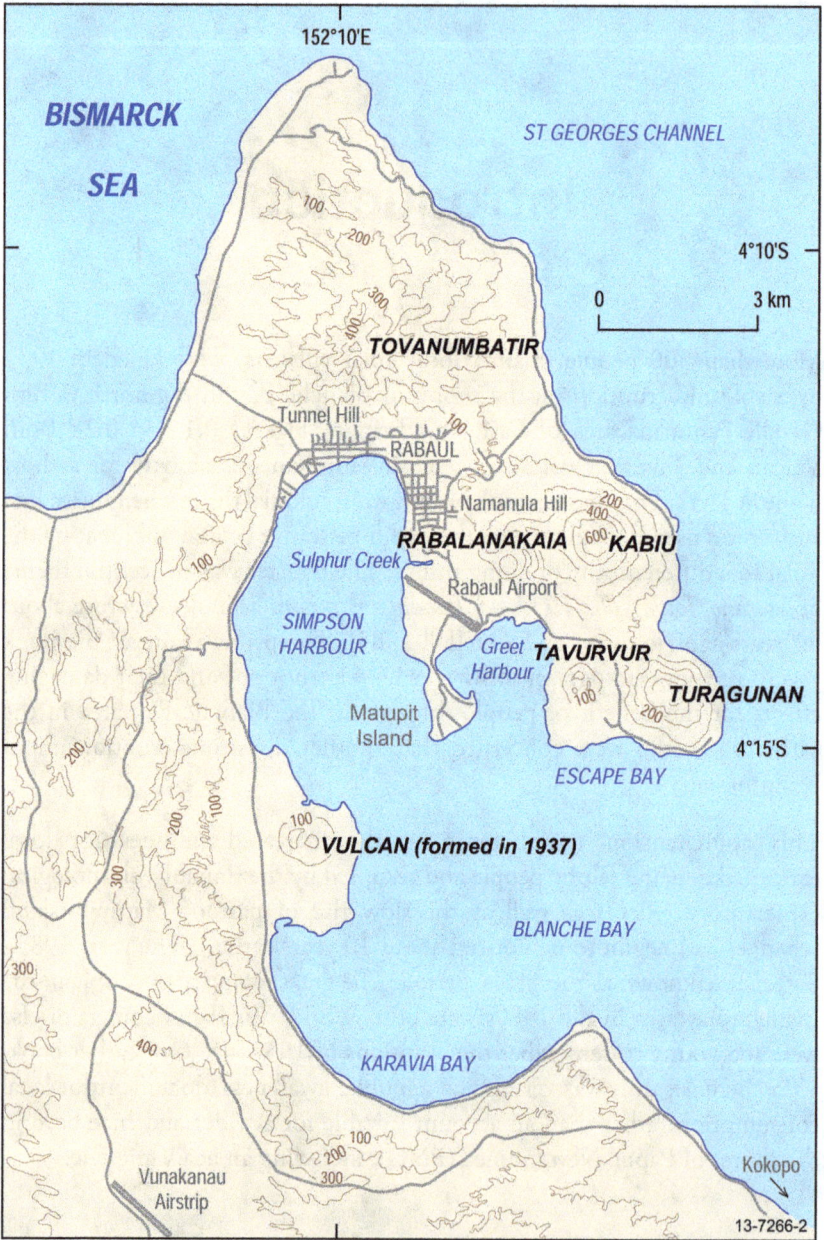

Figure 0.1. Blanche Bay area in the early 1980s.

The volcanic peaks of the Blanche Bay area, including the active volcanoes of Vulcan and Tavurvur, are shown in this map of the north-eastern Gazelle Peninsula from the early 1980s (adapted from Johnson and Threlfall 1985, 4). The street layout for Rabaul town is shown as it was before the 1994 eruption destroyed most of the town's eastern part.

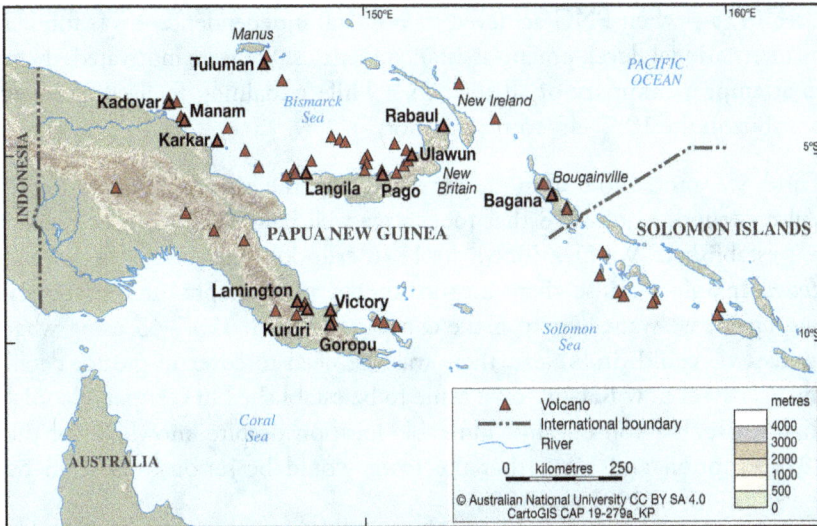

Figure 0.2. Volcano distribution in Papua New Guinea and Solomon Islands.
The triangles in this map of modern-day Papua New Guinea and Solomon Islands represent volcanoes that have known or inferred Holocene eruptions, or those with possible Holocene eruptions, together with three active geothermal fields (not named on this map) where there is no known Holocene volcanism (adapted from maps by Simkin and Siebert 1994, 58; Siebert, Simkin and Kimberly 2010, 75). Rabaul (including neighbouring Tavui) is home to the easternmost volcanoes of the zone known as the Bismarck Volcanic Arc that runs westwards through New Britain along the southern margin of the Bismarck Sea to the Schouten Islands (which include Kadovar and Bam) in the far west (Carey 1938). The Rabaul volcanic complex (Figure 0.1) is still recognised today as PNG's highest-risk volcanic area on account of the frequency and size of its past eruptions and the close proximity, and therefore vulnerability, of nearby communities (Lowenstein and Talai 1984; RVO 2014).

There were several motivations for us in returning to the 1937–43 eruptions in Blanche Bay. First, Vulcan and Tavurvur volcanoes were again in eruptive activity in 1994, damaging much of Rabaul town, with those at Tavurvur continuing intermittently until 2014. We were interested therefore in seeing how these 'twin' eruptive periods from two different times compared with one another. Second, wideranging disaster-management and geoscientific investigations have been prolific in recent decades, including ideas on how the volcanic systems in Blanche Bay 'work' and how at-risk communities can be better prepared for likely disasters. This work began when the RVO was re-established in 1950, accelerating through the following decades and including geological studies of volcanic deposits produced by eruptions many times the size of those in 1937–43. Much of this investigative work

3

after 1975—when PNG achieved its political independence—was funded by international development-assistance agencies. We were motivated, then, to attempt a summary of all this work while remaining focused as far as possible on the 1937–43 eruptive period.

Third, we wanted to deal in greater detail with the historical Tavurvur and Vulcan eruptions of 1878 that took place well before the town of Rabaul was established. We mentioned the 1878 eruption only briefly in *Volcano Town*, mainly because there are no known photographs of the activity but, again, we want here to make comparisons with 1937–43 using what sources we could find. Next, there was the need to cover in greater detail the history of how Rabaul town came to be established in German colonial times in such a volcanically vulnerable location despite knowledge of the 1878 eruption and given that the town would be seriously affected by the eruptions of 1937 and 1994.

Our final motivation was our own frustration at not having provided a suitable bibliography in *Volcano Town*. This oversight has been addressed here in our new book. Copies of much of the quoted material are available digitally in an RVO information management system, and we have made publicly available our own research papers through donation to archives in Canberra, as listed in Appendix 1. Acronyms and abbreviations are listed in Appendix 2 and a short glossary of some volcanological terms is provided in Appendix 3.

We have attempted to be consistent in our spelling of the names of geographic features in the north-eastern Gazelle Peninsula. The choice of name, however, is not so straightforward as some features have several names, ranging from origins in local dialects through to European languages, especially English and German. A good example is the naming of the active volcano known widely as 'Vulcan', also spelt 'Vulkan'—and named after the Roman God of Fire—but also known and possibly in some cases misspelled as Vulcan Crater, Kaia, Rakaia, Raluan, Baluan, Keravia, Kalamanagunan and so forth. European colonists might ask local people for the name of a place, or peak, or river, but the final spelling they recorded would depend on the language and dialect being spoken (Linggood, Fellmann and Rickard 1940). For example, the settlement and river now known and spelt most commonly as Kerevat would also be spoken as Keravat. Another example is Rapindik, or Rapidik, both of which are used by the Tolai. Rapidik is the spelling used by speakers of the Raluana dialect, meaning 'the secret place' and associated with the meeting of secret societies such as the *tubuan*. Rapindik, however, is used most commonly as the spelling in official volcanological reports written in English. The common-use spelling of placenames in modern published maps of the area has been influential in our choices of spelling, rather than any consistent usage that may have existed in 1937–43. However, in the case of 'Dawapia Rocks'—for the steep-sided volcanic pinnacles in Simpson Harbour that are also known as the 'Beehives' or 'Beehive Rocks' or simply Dawapia or Davapia—we can be criticised for mixing the Tolai and English origins of the name.

1

Towards an Unsanctioned Capital

1.1. The Tolai

The people known as the Tolai live on a large active volcanic complex in the north-eastern part of East New Britain in modern-day Papua New Guinea (Figure 1.1). They are a populous, Austronesian-language group made up of different matrilineal descent lines or *vunatarai*. The Tolai and other Melanesian people such as the non-Austronesian ('Papuan') Baining to the west and the Taulil and Butam in the south have lived in these volcanic surrounds for centuries.

The Tolai themselves are thought to have migrated to the area centuries ago from New Ireland to the east, some through the Duke of York Islands. They settled successfully by taking horticultural advantage of the rich soils of the area's volcanoes and are thought to have displaced the Baining people, who moved into the mountains to the west. The Tolai coastline is long, so providing ready access also to fisheries and other marine resources (Figure 1.2). One well-known name for the Tolai language is Kuanua, a word from the Duke of York Islands meaning 'over there'. Another name, however, is *tinata tuna*, 'our own language'. European linguists, anthropologists, ethnographers, social historians and museum artefact collectors in later times would take considerable research interest in how the Tolai lived and thought (Epstein 1968; Epstein 1969, 1992; Salisbury 1970; Sack 1973; Neumann 1992; Threlfall 2012; Martin 2013).

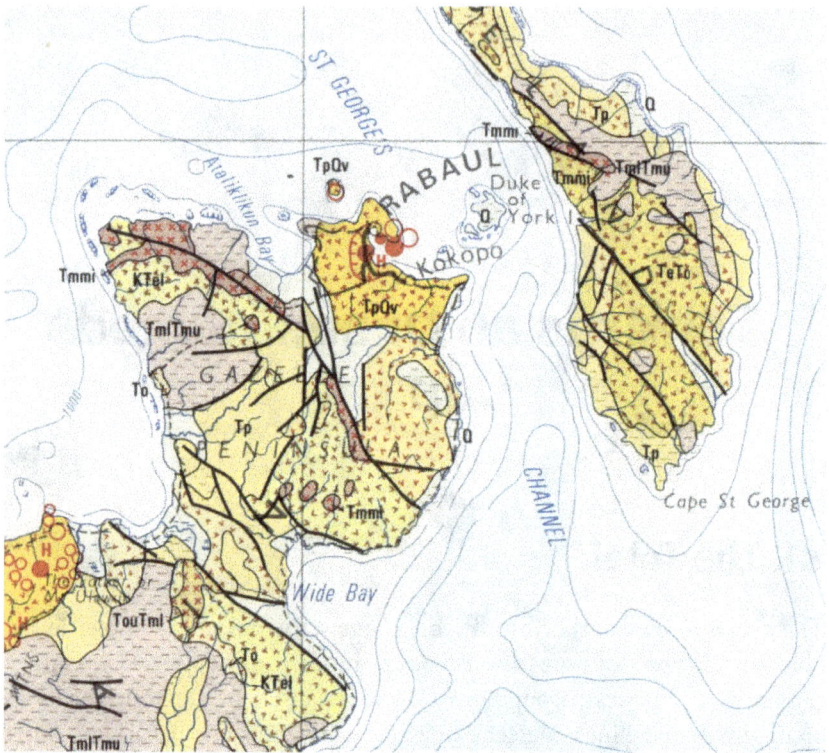

Figure 1.1. Geological map of north-eastern New Britain and southern New Ireland.

The approximate extent of the large active volcanic complex at present-day Rabaul and Kokopo is shown in this detail from a geological map by the orange area (plus v-pattern) and labelled 'TpQv' (after D'Addario, Dow and Swodoba 1975). Most Tolai people live within this same area — that is, mainly north of the Warangoi River (southern boundary) and east of the Kerevat River (western boundary). Latitude 4°S and longitude 152°E are shown by the straight lines intersecting just to the north-east of Watom Island (which is also labelled TpQv). The distance across the area is about 225 kilometres. Ulawun volcano is represented by the filled red circle near the lower left-hand margin.

The whole Tolai area is vulnerable to volcanic eruptions and earthquakes, and its coast to the impact of tsunamis. However, a particularly large eruption in the late seventh century—identified and dated (677–79 CE) by modern geological methods (McKee, Baillie and Reimer 2015)—would have made much of the area uninhabitable, including Watom Island, where Lapita pottery and obsidian artefacts from earlier Austronesian settlement were buried by the eruption's deposits (Green and Anson 1991). Reoccupation of the devastated area, and re-establishment of productive gardens, presumably took place over an extended period as vegetation growth gradually recovered on new soils suitable for gardens.

Figure 1.2. Tolai fishermen working fish trap nets.

Tolai fishermen are seen here working traditional fish traps at the entrance to Blanche Bay (Mennis and Mennis 2019). The volcanoes in the background are Tovanumbatir (in the centre on the distant left) and Turagunan (on the closer right). Watom Island is just visible beneath the cloud to the left of Tovanumbatir. The photograph was taken from the south-east by the late Brian Mennis and reproduced as a postcard by Robert Brown and Associates (Port Moresby).

There appear to have been two main settlements by the 1870s on the shores of what later would be called Blanche Bay: Matupit Island, west of and close to Tavurvur volcano on the eastern side; and Malaguna (Malakuna) on the north-western corner of the bay and referred to as 'the big place' by one Tolai elder from Matupit (ToMaran 1951). There were other settlements, too, down the western side of Blanche Bay including Karavia in the south-west where there was a market for the exchange of produce. Life in those days, however, was not peaceful. There was a male warrior culture and fighting between competing clans, perhaps as payback for early misdemeanours. There were inflicted deaths, revenge killings and cannibalism of those who had been killed (ToMaran 1951). And there was sorcery.

The western side of Blanche Bay seems to have been preferred as a place for settlement, rather than the eastern side where visible volcanoes were much closer, although this did not seem to have bothered the Matupits. There are no known Tolai traditions of recent volcanic eruptions taking place on the western side such as would take place in 1878 and then 1937. The north-eastern corner of the bay and adjoining shores were not preferred for settlement because of the presence of swamps and mangroves and, therefore, likely related debilitating diseases. This place was called Rababaul, meaning 'the many mangroves'; *ra-baul* is the Tolai name for the mangrove tree. One other favoured settlement area was at Nodup on the coast overlooking St Georges Channel to the east, on the flanks of Kabiu volcano, and facing out to the Duke of York Islands.

Numerous features of Blanche Bay—volcanic peaks and ridges, gullies and rocks, prominences and islands, reefs and inlets—have been named individually by the Tolai. Even small, seemingly insignificant locations are named after particular plants, trees and so forth. These physical features are of overall importance because of their intimate association with past events, ancestors and a spirit world that is a part of the Tolai natural environment and belief system. Naming places and incorporating them in genealogies were important aspects of Tolai descent-line claims of landownership. Land still holds special meaning for the Tolai as it does for many other Melanesian groups, reflecting a fluid system of collective land tenure. Early European intruders in search of 'purchasable' land who attempted to individuate, codify and regulate land tenure using courts of law appear to have been generally slow to understand the complexities—religious, historical, social and political—lying behind this indigenous view of landownership. Tolai country in effect became a 'land with two laws' (Sack 1973).

Volcanic activity was acknowledged by the Tolai both as an agent of landscape creation and change, and as a manifestation of the workings of the spirit world. The word *kaia* in the Kuanua language of the Tolai refers to the different spirits that appear, most commonly, as giant snakes called *valvalir* or *kaliku*. The most prominent *kaia* live inside the craters of volcanoes and may cause volcanic eruptions. Rakaia—'the Spirit'—was the name given by the Tolai to the volcanic island that was created in Blanche Bay in 1878 even though more than one *kaia* was thought to live there (Neumann 1992, 235–6, quoting in English the German missionaries Josef Meier and August Kleintitschen). Some *kaia* can be befriended and some may be of assistance. One story tells of two Tolai sorcerers instructing two *kaia* who were disguised as snakes and sending them to the volcano at Matupit (Tavurvur)—which was also called Rakaia—to start the 1878 eruption (Meier 1908; an English translation is given by Neumann 1996, 13). The sorcerers also made two magic sticks that were thrown into a fire and any crackling was to be the signal for the *kaia* to use their powers again and stop the eruption. Many earthquakes of volcanic origin are felt in the Rabaul area, and the Kuanua word for earthquake, *guria*, is now part of the vocabulary of modern Tok Pisin.

Figure 1.3. *Tubuan* **ceremony on Greet Harbour.**

Tolai men, including two *tubuan*, participate in a traditional ceremony on Greet Harbour on 29 October 2004. Tavurvur volcano in the right background is inactive, producing only vapour emissions. The photograph was provided courtesy of Shane Nancarrow of Geoscience Australia. The sketch of a *tubuan* on the right is from McCarthy (1963, 221).

Beings in the physical guises of *tubuan* and *dukduk* are created by the Tolai in the form of painted conical, volcano-like heads and bodies of leaves that cover all of the enclosed person's body, except for feet and calves (Figure 1.3). *Tubuan* and *dukduk* heads rise from the leafed bodies as if—to present-day volcanologists in particular—they are volcanic peaks rising from a forested landscape, and the bodies rustle and vibrate during tremor-like shaking movements, as if allegories for volcanoes in eruption and for ground-shaking by earthquakes. *Tubuan* are 'raised' in commitment to the rituals of male secret societies that are also called *tubuan*. The Tolai have complex rites at initiation, mortuary, funeral and other ceremonies that were—and to an extent still are—part of the intricate tapestry of traditional cultural activity in the Blanche Bay area.

Another important aspect of Tolai culture is the accumulation and use of 'shell money' or *tabu*. This refers to the custom of creating a currency by stringing together small *Nassarius* seashells abundantly onto lengths, or 'fathoms', of lawyer cane, to be used in small amounts for trade and exchange, and also, by coiling the lengths into large hoops, for ceremonial displays of wealth (Simet 1991; Kalua 2022). *Tabu* is still used widely today by the Tolai in association with the PNG national currency of kina and toea, as summarised in a very recent article:

> This little snail shell is much more than simply a bartering medium. It has traditionally been an intrinsic part of mourning ceremonies, bride price, initiations and other significant customary practices. It is indispensable during sacred rites, particularly so during Kinavai, a ritual which pays tribute to the origins of shell money and the oceanic lineage of the clan. Its exchange forges bonds, resolves

disputes and honours ancestors. It is not just a symbol of prestige but the representation of a profound spiritual belief and its use signifies deep respect within Tolai society. It also plays a crucial role in inter-clan networking and the reinforcement of traditional governance. (Kalua 2022, 37)

1.2. Enlightenment Years in the Pacific

European voyagers of the Enlightenment encountered the Tolai coastline for themselves in the late eighteenth century. The foreign seafarers and their sponsors from Europe were driven by a curiosity for the new lands and people of the vast Pacific Ocean that had been encountered earlier, in particular by the Spanish and was thus called the 'Spanish Lake' (Spate 1979). The French and British, however, were the two nations who jostled for ascendancy in the south-west Pacific during the Enlightenment's 'long eighteenth century'—that is, a time span including decades immediately before and after the eighteenth century proper, and more precisely defined as 1660–1832 CE using key events from just British history (Sloan 2003). Both France and Britain were driven to find new lands and navigable waters, make scientific discoveries, discover resources and generate new maps (Suarez 2004). They also gave considerable thought to the possible economic and political gains that might be made through European colonisation of these distant but potentially resource-rich lands. Germany, however, exerted a strong influence in the islands of the south-west Pacific as the nineteenth century progressed, including when the Tolai came under pressure to change their traditional ways of life and adapt to a different world.

Captain Philip Carteret in 1767 was on a British circumnavigation of the world when he discovered what he called St Georges Channel and became the first known Enlightenment voyager to get close to Tolai country and to see volcanic activity from within it (Carteret [1767] 1965; see also Hawkesworth 1773). The English explorer and former buccaneer William Dampier, 67 years earlier, had used this name of the English patron saint for a stretch of water that he mistakenly identified as a bay, and indeed had called it St Georges Bay, but Carteret demonstrated that this body of water was, in fact, an important navigable waterway separating Nova Britannia or New Britain (named by Dampier in 1700) from Nova Hibernia or New Ireland (named by Carteret). Both islands, together with the Admiralty

Islands to the north-west, thus comprised a large archipelago north of the previously named and main island of New Guinea and north-west of Solomon Islands. Carteret also identified a large, flat island in the middle of St Georges Channel to which he gave the name 'Duke of York's Island', later to be called the Duke of York Islands, or just 'the Duke of Yorks', when smaller islands adjacent the main one were included in the name for convenience.

Philip Carteret was also the first European traveller to identify and name volcanoes on the western shore of the channel, chronicling the following for Thursday 10 September 1767:

> [O]n the main or New Britain side are some very high land and 3 remarkable hills close to one and other which We named the Mother and daughters, the mother is the middlemost and biggest, behind these we saw great smoke and imagined it most [sic] have been from a Vulcano. (Carteret [1767] 1965, 342; see also the edited version given by Hawkesworth 1773, 376).

An exaggerated sketch of the volcanic view was made from a point about 28 kilometres to the north-east (Figure 1.4). Carteret was evidently unaware of the large bay that existed behind the three volcanoes and of the smaller, active volcanoes found on its eastern side.

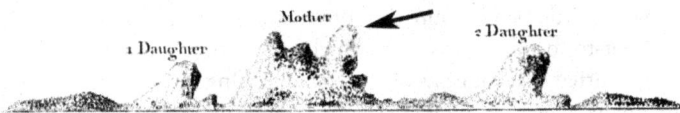

The Mother *and* two Daughters, *bearing S.W. diſtant 5 leagues.*

1 Daughter   Mother   2 Daughter

Figure 1.4. Philip Carteret sketch of Rabaul volcanoes in 1767.

The volcanoes of the Rabaul area are shown highly exaggerated, and even overhanging, in this sketch made from the north-east by an unknown artist on Philip Carteret's voyage to the area in 1767. The sketch is part of an engraved nautical chart entitled *Nova Hibernia* that was published on 1 November 1772. This chart also appeared in the first published record of the Carteret voyage (Hawkesworth 1773) but not in a more definitive, two-volume version edited by Wallis (1965). The 'great smoke' reported by Carteret may be represented by the small, lighter-coloured area — indicated by the arrow — which rises slightly higher than the summit of the Mother. If so, the volcano in eruption is most likely to have been Rabalanakaia, lying on the south-western side of the Mother. Tavurvur volcano sits behind the Mother on this bearing and therefore does not appear in the profile.

The French voyager Louis-Antoine de Bougainville was also on a circumnavigation of the globe when, in 1768, he came to the southern end of New Ireland, having followed the north-eastern coast of the large Solomons island that was named after him (Dunmore 2005). He gave the name Port Praslin to his anchorage on the southern tip of New Ireland, an anchorage that would become a place of French tragedy more than a hundred years later. Bougainville was evidently unaware of Carteret's discovery of St Georges Channel the year before and continued his journey north-westwards along the east coast of New Ireland rather than through the strait. Neither did he record any volcano sightings.

Captain James Cook raised the Union Jack on the east coast of Australia in 1770, leading eventually to the beginning of British settlement on the continent in 1788. This was when the First Fleet, under the leadership of Captain Arthur Phillip, and Second Captain John Hunter, arrived at what is now known as Sydney Harbour. A French expedition under the leadership of La Pérouse visited the settlement shortly after this first arrival, but then sailed north into Melanesia where they vanished.

John Hunter was obliged to return to England in 1791 and travelled on a Dutch transport, northwards and through St Georges Channel, stopping off at Duke of York Island where the vessel took on fresh water. He noted:

> The hills mentioned by Captain Carteret, on the coast of New Britain, by the name of the Mother and Daughters, are very remarkable. A little way within the south-eastern Daughter there is a small flat-top'd hill, or volcano, which all the time we were in sight of it, emitted vast columns of black smoke. On this coast there appeared many extensive spots of cleared and apparently cultivated land. (Hunter [1793] 1968, 155; see Figure 1.5).

The interaction of the ship's crew with the local islanders was not peaceful. It ended with the islanders submitting to the authority of the guns and cannon fired by the crew. Some islanders were probably killed by grapeshot.

Figure 1.5. John Hunter sketch of Rabaul volcanoes in 1791.
The Mother volcano dominates the central background in this idealised and disproportional view (Hunter [1793] 1968, Plate 11). Its caption is simply 'Canoe of the Duke of York's Island'. North Daughter is shown to the right of the Mother, and a volcanic plume can be seen drifting from the 'flat-top'd hill' to the right of South Daughter. The steep background escarpment to the left of South Daughter in the top-left corner is the large bay — still unnamed by the Europeans — whose northern and eastern margins are rimmed by the four volcanoes shown here. The small island left of centre may represent one of the Credner Islands.

Whether later voyagers of the Enlightenment identified the active volcano and even entered the bay, shown in the top-left corner of Figure 1.5, is unknown. Presumably many of them, like Captain Hunter, would have seen the bay's entrance. Two French voyagers are known to have transited St Georges Channel—Bruni D'Entrecasteaux and Louis Duperrey—although neither recorded anything significant volcanologically. D'Entrecasteaux came through the strait in 1792 on one of his two voyages in New Guinea waters searching for, but not discovering the whereabouts of, the lost vessels of the La Pérouse expedition (Rossel 1808; D'Entrecasteaux 2001). Louis Duperrey, known for his explorations and mapping of the Pacific Ocean (Suarez 2004), sailed through the channel on *La Coquille* in 1823 after visiting Port Praslin on the southern end of New Ireland (Figure 1.6). The naturalist on board, René Lesson, recorded the animosity of the local people to their enemies in New Britain—probably the Tolai—on the western side of St Georges Channel:

> [H]atreds are never more alive and more fierce than when they occur between two tribes descended from the same family; and yet the hatred which divides them is so great that the name of Birare (native name for Dampier's New Britain) [see Figure 1.6] pronounced in front of a native of Port Praslin, is sufficient to provoke a terrible anger and to make him utter imprecations in his own tongue which, to judge from the violence of the accompanying actions, must be terribly vigorous and drastic. (Whittaker 1975, 229, translated from French)

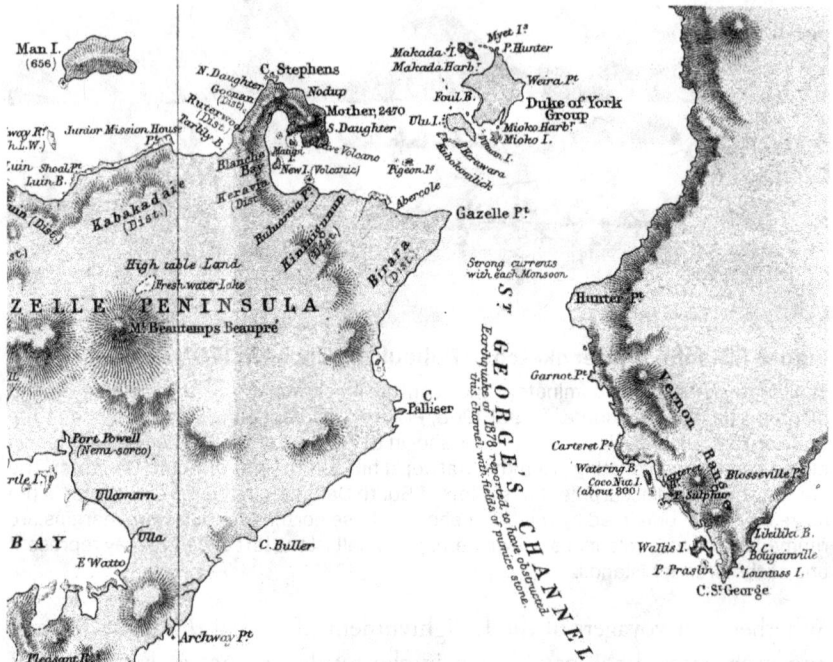

Figure 1.6. Wilfred Powell map of the St Georges Channel area in 1878–79.
Many of the coastal places named by early European voyagers are shown in this detail from a sketch map of the St Georges Channel area published by Wilfred Powell (1883) as a result of his travels in 1878–79. New Ireland is on the right. The distance across the map is about 75 kilometres. Note that Powell has distinguished two districts ('Dist.') on New Britain, including Kininigunan west of 'Gazelle Point', which in the years ahead would become a centre of European settlement and development. Note also that Powell mistakenly attributes the floating pumice obstructing the channel to an 'Earthquake'.

1.3. Traders, Missionaries and the First Maps of Blanche Bay

St Georges Channel and its nearby coastlines became known even further during the course of the nineteenth century when transiting European voyagers stopped at the Duke of York Islands and elsewhere on the adjacent shorelines for replenishment of water and fuel supplies, thus encountering local people. By the 1870s, the ambitions of European seafarers had changed and many now were more interested in commercial opportunities than they were in Enlightenment-style discoveries.

Mercantile shipping companies were established, particularly by the Germans, and island trading began in commodities such as copra, pearl shell, bêche-de-mer and turtle shell throughout the Western Pacific. Traders were left on islands, where they lived solitary lives dependent on the goodwill of local people who were interested in trading for Western goods and materials such as cloth, tobacco and, later, firearms. Mostly itinerant, these lone traders were also, and necessarily, dependent on the not-always frequent arrivals of company ships. Whalers and the warships of European navies, notably those of Britain and Germany, also prowled New Guinea waters during the mid-nineteenth century.

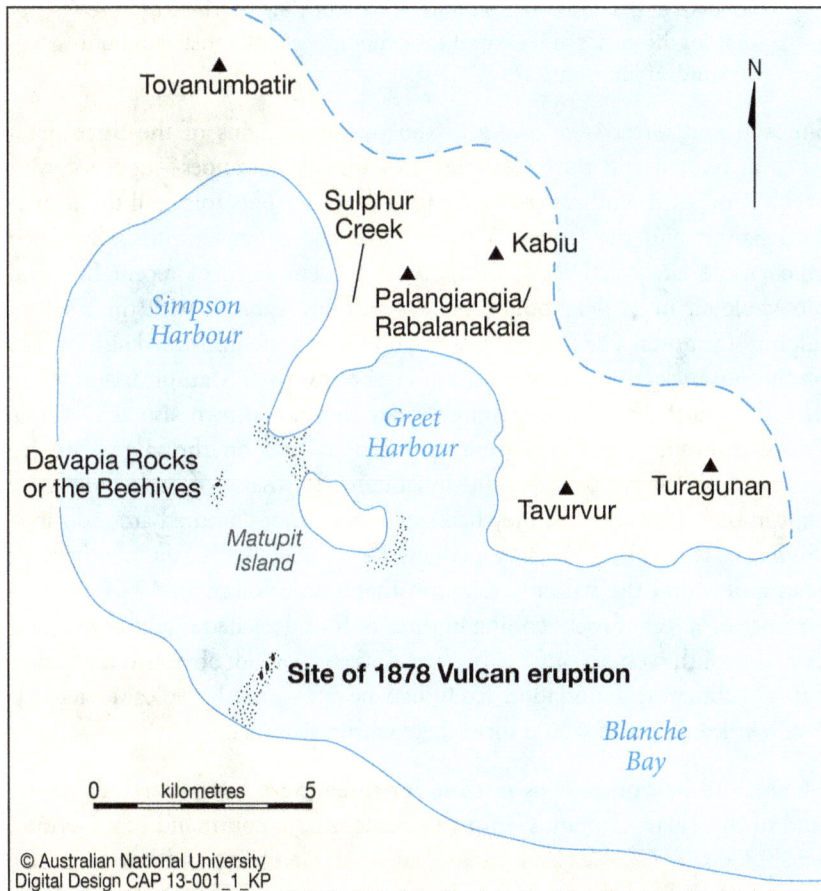

Figure 1.7. Sketch map of Simpson and Greet harbours in 1872.

This illustration has been adapted from the sketch map made by Simpson and Greet in 1872, which was referred to as 'annexed plan 4' by Simpson (1873). The triangles represent mapped peaks, but the names shown are those used today. Sulphur Creek was not identified by Simpson and Greet, and Matupit Island was called Henderson Island.

The British naval surveyors Captain C.H. Simpson and Lieutenant W.F.A. Greet came to St Georges Channel and then entered the large bay west of the Mother on 17 July 1872, naming it after their vessel, HMS *Blanche*. They produced the first map of Blanche Bay, naming waters in the north of the bay after themselves: Simpson Harbour and Greet Harbour (Figure 1.7). Their map is somewhat sketchy and distorted, but it does give a first impression of the imposing scale of Blanche Bay and the prominence of the peaks of North Daughter, Mother and South Daughter. The beauty of the bay impressed them, but Simpson wrote, more pragmatically, that:

> [O]ne cannot look at it a moment without being struck at the natural strength of the position in a military point of view ... [there is] water in it for the navies of the world to anchor in, perfectly sheltered from all winds. (Simpson 1873, 4–5)

Simpson and Greet were aware of the volcanic origins of the three main peaks at Rabaul and they identified two further volcanoes—one just west of the Mother, the other west of South Daughter. The Tolai call the former Palangiangia and the latter is Tavurvur, which, wrote Simpson, 'has been much more lately active ... [and] shows evident signs of recent fire, and the whole air in its neighbourhood was still impregnated with the smell of sulphur' (Simpson 1873, 4). Simpson and Greet were impressed also by two precipitous rocks rising from Simpson Harbour west of Matupit Island. They named them the Beehives, although today they are known also as Dawapia Rocks. Simpson and Greet were surprised to find on the larger Beehive 'a village containing perhaps 200 inhabitants ... many of their houses are built in the water on piles, they had numerous canoes moored around them' (Simpson 1873, 5). However, perhaps their most significant observation, bearing in mind the volcanic eruption that would follow in 1878, was the presence of 'a reef of rocks ending in three or four detached islands' extending from the south-western shore of the bay. This reef was not obviously a volcano, but its submarine foundation likely had been formed by volcanic activity. A new volcanic island would form there within six years.

The second half of 1875 is notable for three more foreign arrivals to the land of the Tolai: Captain Georg von Schleinitz in command of a German warship, SMS *Gazelle*; a German maritime merchant and entrepreneur, Eduard Hernsheim; and a Methodist missionary, Rev. George Brown. Each left their mark on the developmental history of the St Georges Channel area. Schleinitz entered Blanche Bay, surveying and describing its volcanic and marine nature (Schleinitz 1876, 1889). He produced a more accurate map of the bay's features, compared with the one prepared by Captain Simpson, and included bathymetric measurements of the two harbours (Figure 1.8). The elongated reef on the western shore is better defined and is shown running towards the small, young volcano—now known as Tavurvur—on the other side of the bay and just west of South Daughter. The island on the south-western side of Greet-Hafen (Harbour) was named 'Matupi Henderson' by Captain Schleinitz (Figure 1.8), thus incorporating the local Tolai name, Matupit, for this now well-known geographic feature.

Eduard Hernsheim first came to Micronesia in 1874, eventually establishing a major German shipping and trading network throughout the wider Western Pacific region (Firth 1978b; Hernsheim 1983; Anderhandt 2012). Hernsheim there came up against competition, particularly from the German company J.C. Godeffroy und Sohn of Hamburg, which had established an economic centre in Samoa as early as 1857 (Firth 1983). Hernsheim then tried his luck to the south, in Melanesia south of the equator, starting in the Admiralty Islands area, but then, in October 1875, visiting St Georges Channel and the Duke of York Islands, where, again, there were Godeffroy traders in place. He noted too that Rev. George Brown (Figure 1.9) had recently established a mission centre at Port Hunter on the northern end of the main Duke of York Island. Hernsheim, assisted by his traders, would remain an influence in the economic development of the area until the 1890s. This initial German trading influence represents the start of a gradual, de facto colonisation by the Germans—but through an economic process rather than by military invasion and occupation, yet nevertheless taking advantage, where opportunities arose, of the presence of warships and the advantages of pistol and rifle firepower. Economic development also, presumably, benefited now from the pacifying presence of a Christian mission.

Rev. George Brown, accompanied by a team of Fijian and Samoan missionary teachers, had arrived at Port Hunter on 15 August 1875 at a location strategically placed for missionisation in both north-eastern New Britain to the west and New Ireland to the east. Their aim was to establish centres of missionisation throughout this previously un-Christianised area including the Duke of York Islands (Brown 1877, 1908; Threlfall 1975). Local people offered them land nearby at Kinavanua from where they began exploring the two large islands to the west and east. Brown and others, after just two weeks and travelling on two mission vessels, entered Blanche Bay and anchored off Matupit Island where they established good relations with the islanders. One of the Fijian pastor-teachers, Penijimani Caumea, and his wife were transferred to the island on 17 November, representing the first of the Christian denominations to establish mission work on Matupit. Catholic and Seventh-day Adventist missions would be established later.

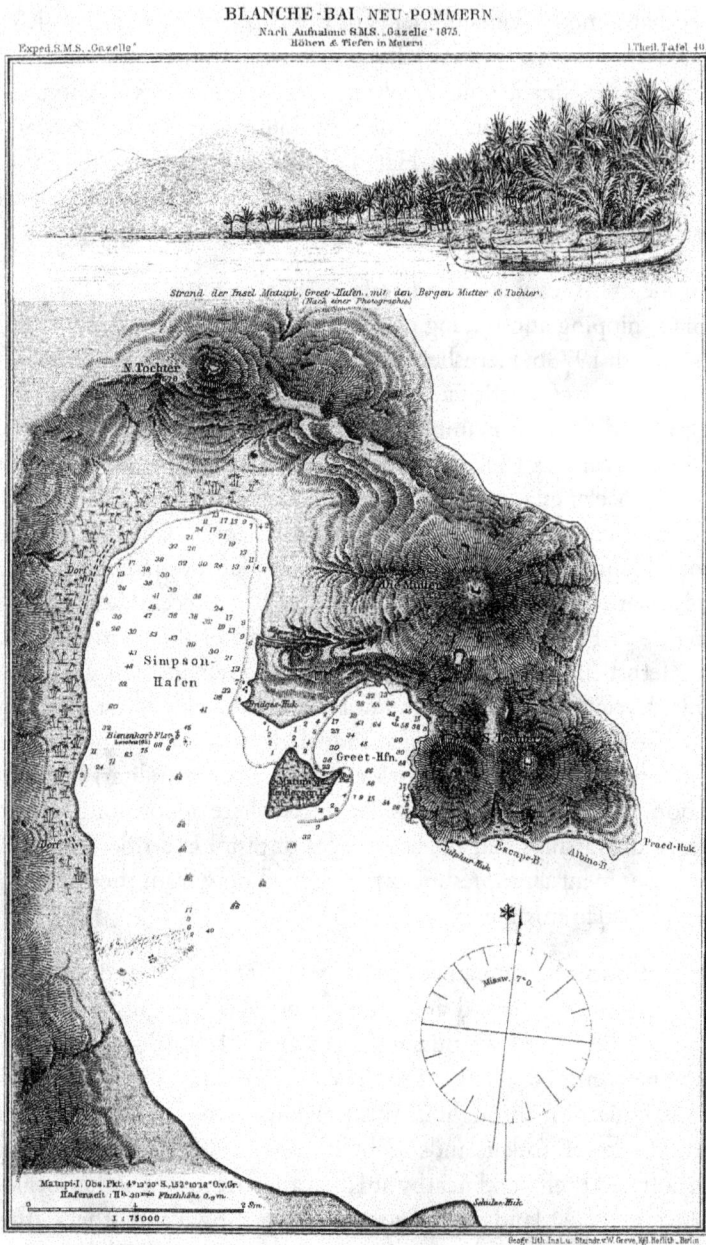

Figure 1.8. Bathymetry of the two named harbours in Blanche Bay mapped in 1875.

This map of Blanche Bay was published by Schleinitz (1889, tafel 4, opposite p. 240) after his surveying of 1875 in the *Gazelle*. The sketch at the top is of Greet Harbour and Matupit Island and was adapted from a photograph. The Mother volcano, Kabiu, is the high peak in the background, and Palangiangia/Rabalanakaia is down to its left.

A good deal of interest was paid by many Europeans to the origin and age of Matupit on account of its apparently recent emergence from the sea and the establishment of its well-settled population (Brown 1908; Laufer 1956; Epstein 1969; Mennis 1972; Sack 1987). Soil cover was poor and soil had to be brought onto the island by the settlers to develop sustainable gardens. George Brown reported that:

> Matupit was thickly populated and seemed to be a very healthy island. Captain Ferguson [a trader] told me that it was a comparatively recent upheaval, as some of the oldest men remembered a time when no such island existed. This was, I think, confirmed by the size of the cocoanut [sic] trees, as they were all young trees, and appeared to have been planted about the same time. There had recently been a considerable sinking along the shore, amounting to more than six feet in some places, so that it could scarcely be considered to be a very safe dwelling place. (Brown 1908, 93–4)

One interpretation of the name 'Matupit' is that it derives from *a mata*, meaning a hole or depression, and from *pit*, meaning 'stopped short' (Epstein 1969, 105). A traditional 'origin' story is that people crossing the bay would come to the 'hole', which teamed with turtle and fish, and be stopped short. This led to European speculation that the hole was, in fact, a volcanic crater and that the island must be of recent volcanic origin. The island in this case would be just like the other young nearby volcanoes in Blanche Bay (Heming 1974). The first settler on the island was said to be a man called Diararat and some Matupit people said they could trace their heritage back to him. Genealogy, therefore, played a part in estimating that the island's emergence may have taken place as recently as the eighteenth century and been connected in some way to the volcanic eruptions noted by Carteret in 1767 and Hunter in 1791. Volcanic geologists, however, would later argue that Matupit is not a volcano even though it is made up of pumiceous volcanic materials. Nevertheless, the island would play an important part much later in understanding how the young volcanoes of Blanche Bay came into existence.

Later in 1875, Brown climbed the volcano on the other side of Greet Harbour from Matupit Island, west of South Daughter volcano, noting:

> The crater was of great depth, with almost perpendicular sides, and was still smoking in many places; but the most recent eruption seemed to have been on the lower land near the beach, where we found the ground quite loose and very hot, with a good deal of sulphur on the surface. The sides of the crater [volcano] were full

of the nests of the megapodes. This bird does not build mounds here, as in other places, but deposits its eggs in the warm, loose ashes of which the hill is formed. (Brown 1908, 129)

Brown here is referring to a quasi-permanent geothermal area on the eastern shores of Greet Harbour and to the megapodes that still nest there today, providing an egg food resource for landowners. Foreshadowing comprehensive geological results obtained a century later, Brown later speculated that:

[M]any years ago the whole of Blanche Bay was itself an active volcano, and I think the soundings which have been recently taken by one of H.I.G.M. ships of war [i.e. SMS *Gazelle*] will probably confirm this opinion. The land on all sides is of pumice formation. My opinion is that the present entrance to Blanche Bay has been formed during some terrible eruption, which burst up the land there, and admitted the sea into the crater, and that this now forms the bay itself. (Brown 1908, 238)

1.4. Volcanic Eruptions in 1878

There is no single, definitive historical account of the volcanic eruptions in Blanche Bay in 1878. Rather, information has to be assembled, at least in the first place, from fragmentary records left by Europeans, none of whom was in the bay when the first outbreaks took place. A notable compilation, however, was made by the late R.J.S. Cooke for a paper on submarine eruptions in Papua New Guinea (Johnson, Everingham and Cooke 1981). Further, four Europeans who reported on the 1878 eruption, including Rev. Brown, appear to have benefited from verbal information provided by Mr William Hicks, a trader of biracial heritage from Fiji who lived on Matupit Island and who witnessed the start of the eruption.

George Brown was in the Duke of York Islands (Brown 1878, 1903, 1908). He recorded in his diary for 30 January 1878 that he had heard on this day that the volcano 'at Matupit' was active but that he could not leave to investigate because of ill health (Brown 1878). Matupit Island itself is not a volcano, so Brown apparently was referring to the volcanic cone on the east side of Greet Harbour—known today as Tavurvur—which is out of view from the Duke of York Islands unless eruption clouds rise above South Daughter volcano. Also, the volcanic unrest that apparently took place on 30 January may not necessarily have been a true explosive eruption of magma, such as happened later, but rather some sort of preliminary disturbance—perhaps increased vapour emission from the volcano, vent-clearing explosions or even geyser activity in the geothermal area at Greet Harbour. There is also some confusion in that the name Matupit, or Matupi, is given in some reports, including Brown's, to the active volcano itself (Tavurvur) and to Greet Harbour–Matupit Harbour.

An eruption in Blanche Bay was apparently well underway by 5 February, as Brown (1878) recorded in his diary that he saw on that day 'Matupit in full eruption, a thick heavy cloud of smoke hanging over it and hiding the coast for miles'. He had not mentioned the eruption in entries for the two previous days, but over the next 11 days he was able to collect further information. Frequent and locally strong earthquakes had preceded the eruption (Brown, diary entry, 16 February). The earthquakes were particularly violent in Blanche Bay overnight on Sunday 3 February, although they were not felt by Brown in the Duke of York Islands. The next morning, 4 February, two tsunamis eroded shorelines in the bay, and soon afterwards 'clouds of steam were observed rising from the Bay in a direct line' between Tavurvur and the south-west shore of the Blanche Bay. A submarine volcano developed into an island about 1.5 kilometres from the shore at the south-western end of this line where previously there had been only the reef and islets mapped by Simpson and then Schleinitz. The other vapour clouds disappeared and then Tavurvur 'burst out with terrific power' a few hours later. People on Matupit Island fled to safer parts of the bay's shoreline, probably mainly to the north-west, so avoiding any ash fallout from Tavurvur. February is in the monsoon or 'north-west' season when winds blow mainly to the south-east.

The explosive eruptions from both eruption centres in Blanche Bay were also described by Wilfred Powell in his spectacularly titled *Wanderings in a Wild Country: Or Three Years amongst the Cannibals of New Britain* (Powell 1883). Powell was an adventurous Englishman, one of the 'landed gentry' back in Britain, and perhaps best described as a 'gentleman-explorer'. His uncle was the father of Lord Baden Powell, founder of the Boy Scout movement. Powell spent a good deal of time in 1878–79 exploring New Britain in the ketch *Star of the East* (see Figure 1.6). He was imprecise about the times of the volcanic events at Rabaul in 1878, including making the error that the eruption was in May. He reported that he had been unaware of the eruption until 'huge blocks' of floating 'pumice stone' surrounded his ketch anchored at Makada in the Duke of York Islands, but that he could see 'smoke and fire' plainly. The next day he set out to visit Blanche Bay but found he had to sail 'a long way round to the northward, to avoid the enormous fields of pumice-stone that had drifted down the channel' (Powell 1883, 112). The pumice had been blown out of Blanche Bay by the north-west winds, thus interfering with the movements of ships, whaleboats and canoes. Powell also produced a dramatic description of Tavurvur's activity that he saw upwind having climbed to the summit of the Mother:

> From our situation we could gaze down into the fiery crater beneath. In the evening the sight became more than grand—it was awful; every few moments there would come a huge convulsion, and then the very bowels of the earth seemed to be vomited from the crater into the air; enormous stones, red hot, the size of an ordinary house, would be thrown up, almost out of sight, when they would burst like a rocket, and fall hissing into the sea. At the same time angry flames would dart up, almost to the altitude on which we stood, and of the most dazzling brightness. Then all would die down to a low, sulphureous breathing, spreading a blue flame all over the mouth of the crater, whilst over us and all the country near hung a panoply of thick black smoke, broken only by the falling of red-hot stones in showers, which destroyed all the vegetation to leeward to a distance of about two miles. (Powell 1883, 113)

Eduard Hernsheim encountered the floating pumice in St Georges Channel on 9 February on his return to his base at Makada in the Duke of York Islands (Hernsheim [1878a] 2015, 1878b). He had retired to his cabin for the night after rounding Cape St George on the southern tip of New Ireland, but was woken by a commotion on deck. The engines had been stopped, and the ship's captain came up to him

> in great anxiety and said that we must be on the wrong course, for there was no passage here of any kind—we were heading straight for land. And in fact a low-lying coast appeared to extend far and wide before us. When we drew closer we found that what we had taken for land was drifting lava [sic] through which we could force our way only with great difficulty. (Hernsheim 1983, 38)

Hernsheim 'found the entire northeast coast of New Britain shrouded in thick smoke':

> Thick layers of pumice covered the ocean as far as the eye could see … The passage between the Duke of York Islands and New Britain was closed by a thick blanket of pumice up to 1.5 and 1.8 m thick. (Hernsheim 1878b, 372–3, translated from German)

Similarly, Rev. Brown (1878) recorded in his diary on 13 February that: 'The whole channel in front of our house [in the Duke of York Islands] and seemingly for miles to the southward is full of enormous fields of pumice stone from the volcano.'

Another European author of an account of the 1878 eruption was Captain H.W. Wendt, in charge of the vessel *Peter Godeffroy* (Wendt 1879). He arrived in the Duke of Yorks area in mid-year, well after the end of the eruptions at Blanche Bay. He was, therefore, not an eyewitness and had to depend on information provided by an unnamed 'agent' on Matupit Island. Wendt dated the Blanche Bay events one day later than did Brown: a strong earthquake overnight on 4 February and another the next morning. This second earthquake, he wrote, was accompanied by six tsunamis, initially 4 metres high, beginning at 6 am. Wendt also, as well as Brown, referred to a 'line of fire' (Wendt 1879, 179, translated from German) that made its way across the bay from the volcano (Tavurvur) on the mainland south-westwards towards the site of the submarine eruptive centre. The word 'fire' here presumably referred to visible water vapour rather than to flames from any combustible material emerging from the sea. Wendt concluded that the two eruptive centres were probably 'closely linked with one another'.

George Brown and Wilfred Powell made separate visits to the newly formed island in Blanche Bay shortly after the eruption had ceased but while Tavurvur volcano across the bay was continuing its explosive activity. Brown made his visit to the island on Friday 15 February together with Mr Hicks in his whaleboat, leaving from Malaguna (Malakuna) on the north-west side of the bay. Hicks had evacuated there from Matupit Island. Brown had had to avoid floating pumice masses in St Georges Channel on the way to visit a teacher-missionary at Malaguna. He noted on the southward trip from Malaguna past Dawapia Rocks (the 'Beehives') that they 'were gradually sinking as the Houses which were some feet above high water mark on my previous visit are now quite flooded at high water' (diary entry, 16 February). This was in contrast to the new volcanic island further south that Brown said had been 'upheaved' and the former reef 'raised up' several feet; in reality, the new island had been formed mainly by the subaerial accumulation of newly erupted pumice and ash.

The surface of the new island was still hot and care had to be taken while walking on it, especially near a water-filled crater where the water was of scalding temperature. The heat, however, had not prevented Mr Hicks on an earlier visit from planting coconuts on part of the new island, thus claiming possession of the land. Who actually owned the new land would be a recurrent topic of future discussion and dispute. The hot waters in the bay also had killed and cooked fish and turtles, the valuable but now damaged shells of the turtles falling off easily as a result. Powell noted that the island was semicircular in form and gave its dimensions as about 2 miles in extent and 70 feet high, noting also that a short spur on the north-eastern side was terminated by 'a small island covered with bushes' (Powell 1883, 113)—that is, the remains of the reef and islets mapped by Schleinitz and Simpson. Brown gave the name 'Hicks Island' to the recently formed land (Brown 1903, 466).

George Brown also described the ongoing eruptive activity at 'the old crater' on the other side of the bay towards Matupit Island:

> [I]t presented a grand and awful sight; billow after billow of thick black smoke and flame were shot out with great force, and formed a very high column, which towered up far above the surrounding mountains [including the Mother] ... For a few minutes there would be a comparative lull, then a deep rumbling sound, after which there was a loud roar, followed or accompanied by violent expulsions of ashes and pumice, and cloud after cloud of thick smoke following each other in quick succession ...

> Not a green leaf was to be seen, though all was covered with grass and trees a fortnight before. The dead and blackened trees, with almost every branch beaten off by the stones, stood like spectres on the hillsides and gave a most mournful aspect to the scene, whilst the cocoanut [sic] trees on Matupit and places far enough away to escape destruction, were so weighted by the dust and ashes that their leaves hung straight down by their stems, giving them a rather comical appearance—in fact we all agreed that they were very much like a lot of closed gigantic Chinese umbrellas ...

> As we passed down the bay on our way home [to Malaguna] we found that another point of land had been formed near Escape Bay, which was about twenty feet in height and extended seawards about 150 yards from the old shoreline. The whole of the vegetation from Point Praed [north-westwards] to the volcano was entirely destroyed, the prevailing wind having carried the pumice in that direction ... the banana plantations and cocoanuts [sic], on which the natives depended for food, not showing a sign of life. We heard that one woman who was unable to get away was killed by the first showers of stones. (Brown 1908, 243–5)

Figure 1.9. Portrait photograph of George Brown in later life.

This portrait of Rev. George Brown DD is from the frontispiece to his autobiography (1908). The photographer and the precise date of the photograph are unknown.

The date of the last eruption from the 'old crater' is unknown. Wendt (1879) was told that the volcano continued to be active for another three weeks after the formation of the new island to the south-west, and Powell (1883, 114) wrote that the eruption lasted 'upwards of a month'. Brown (1878) recorded that the eruption was still in progress on 24 February, and he visited the volcano after the activity had ceased, noting that a new crater had been formed. A duration of three to four weeks is in contrast to the much shorter duration of perhaps three to four days for the submarine, island-forming eruption that, notably, had begun its activity a few hours before that of the volcano on Matupit Harbour. The two volcanoes, therefore, were in simultaneous 'double eruption' for only a few days.

The effects of the floating pumice drifts on the open sea were apparently experienced long after the two Blanche Bay eruptions had ceased their activity, and not only in St Georges Strait. Guppy (1887) wrote that the pumice was carried eastwards by the ocean currents, deluging the shores of the Solomon Islands, impeding navigation and temporarily suspending the bêche-de-mer industry. Large quantities of pumice were washed up on the islands of the Ellice Group. Captain Harrington sailed through the pumice for four days and noted that the bottom of his ship was 'scraped clean of paint' by the abrasive rock (Harrington 1878, 373–4).

The changing seasons and wave erosion over the following years impacted on the shape of the new island and shorelines produced by the eruptions in Blanche Bay, at least until vegetation was established and the bare land surfaces became more stabilised. George Brown referred to a visit he made to the island in August 1897, noting that it was

> much reduced in size and height, and is now only about two miles in circumference, and about 30 or 40 feet high. The crater continued to emit boiling water for at least two years after the eruption, and the lagoon still exists; the pumice has consolidated, part of the shoreline has been washed away, and some of the material has been deposited in what was formerly a channel between the new island and a small rocky islet, but which channel is now quite filled up. The whole island is covered with vegetation, and there are casuarina trees on it at least 30 feet in height. (Brown 1903, 468)

The island was also mapped later by the Germans, receiving from them the name Vulkan Island, after the Roman God of Fire (see Figure 1.13). The Tolai, on the other hand, used the names Rakaia, Baluan and Kalamanagunan. The 'old crater' on Greet or Matupit Harbour became known by the Tolai name of 'Tavurvur'—meaning, appropriately, 'the hornet's nest'.

In conclusion, only an approximate account can be given of the 1878 eruption using what are incomplete European sources. The compilation is sufficient, however, for comparisons to be made with the 'double' eruption that took place in 1937, and it can be supplemented further by a valuable and nearly forgotten account of the 1878 eruption dictated by a Tolai eyewitness. The account was published in 1951, more than 70 years after the eruption, in the *Pacific Islands Monthly* magazine, and was said to have been told more than 30 years earlier—that is, about 40 years after the 1878 eruption itself. The eyewitness was an old Matupit man called ToMaran, who told his story in 'pidgin' English, or Tok Pisin, which was then translated, transcribed and edited into English—a series of steps that may have resulted in some loss of original meaning (ToMaran 1951; see also ToMaran n.d.). Tok Pisin is a creole language that today is one of the official languages of modern Papua New Guinea.

ToMaran recalled that the weeks and months prior to the eruption were 'a time of famine and hunger' for the Matupits because the ground on their island had been 'getting steadily hotter—it was this that had killed our gardens and started the famine' (ToMaran 1951, 67). Severe tremors then began to be felt 'so that at times our people were thrown down and none could remain standing'. The Matupits had shell-money wealth and one day a large group of them decided to cross the bay to Karavia to buy much needed food, but on the way 'with a noise like a great cannon, there burst forth out of the sea not far from us a great explosion which threw the sea-water high in the air'. They returned to Matupit, one canoe being caught up in tsunamis from the submarine disturbances. Then, next morning while at Matupit watching the eruption across Blanche Bay:

> [S]uddenly a new opening appeared along the beach near the sea, throwing fire high up the mountain side [at Tavurvur]. Stones were shot high into the air, also dense smoke [from which] fell much ash, so that some of the people of Talawat, nearby, were killed. (ToMaran 1951, 67)

The Matupits decided to evacuate to Malaguna, 'the big place' to the north-west:

> We embarked in our canoes, and the trader, Bell [or 'Bill', the shortened name of William Hicks], went off to Malaguna, also … The falling ash had fouled the food in our canoes. None of what we had brought to Malaguna could be eaten, and we suffered hunger again, and for that the people of the big place fought us. Yes, our men stole from their gardens so that trouble arose and they cried out that they would kill and eat us. A mary and a man were killed where Komines [a Japanese boatbuilder] now stands. (ToMaran 1951, 67)

The ToMaran account has two points of disaster significance that were not alluded to by the European chroniclers. The first of these is that 'many' people died as a result of the eruption. A significant number of fatalities might be expected for people exposed to fallout of heavy volcanic debris or to tsunami impact along shores. Brown (1908, 245), however, said he heard that only one woman had died, near Point Praed, 'but I could not find out whether this was true or not'. The second point in the ToMaran account relates to whether the 'drought' on Matupit was caused by the ground on the island becoming volcanically heated before the eruption. Again, there is no confirmatory evidence from other sources, but the claim is of some interest in comparing precursory volcanological events with those in 1937.

Eventually the 1878 eruptions ceased and the Matupits returned to their island. However, the

> ownership of the land gave us some trouble to define, for the old boundaries could not be seen. Trees were broken, rocks covered up, and the beach was not the same shape it had been. Many owners had died. (ToMaran 1951, 67)

Village discussions were held and 'the land was re-alloted [sic] and the ownership then decided has held good till this day' (ToMaran 1951, 67). Factional fighting among some Tolai also broke out over landownership of the newly created volcanic island across the bay from Matupit. The fight was between the Tavui-Liu and Valaur people on one hand and the Davuan and Matupit on the other, according to Sack (1987, 12, see footnote), who referred to another Tolai version of the 1878 eruption compiled by Johann To Vairop and published in the German mission magazine *Hiltruper Monatshefte*.

The fierce reactions of a disaffected Tolai leader, Talili, and his supporters were encountered in April 1878 south-west of Blanche Bay when four Fijian missionaries of the Methodist mission were attacked, killed and cannibalised. This led, controversially, to a forceful response by a group of traders, missionaries and friendly villagers, leading to the destruction of gardens and houses in the name of 'compensation', and to further killings (Threlfall 1975). Rev. George Brown was not part of the retaliating group and he arranged a peacemaking ceremony afterwards. However, Brown accepted responsibility for the actions of his Fijian missionaries and, in consequence, his reputation became unfairly stained as a man of violence. The Tolai placename Taliligap is a contracted form of 'Talili shed the blood'. Similar confrontations involving the Tolai would be met head-on by a new settler community that grew in the Blanche Bay area in the colonial decades ahead.

1.5. Eruption at Sulphur Creek in the Mid-Nineteenth Century

The value of Tolai 'oral history' has been somewhat controversial, questioned by some in favour of the written European word, but defended strongly by European historians of the Tolai such as Peter Sack (1987) and Klaus Neumann (1992, 1996). The question of how much volcanological and historical truth can be derived from Tolai stories can be addressed with regard to an account that was collected in 1984 and translated from the Kuanua by Lily Waisea in association with Inge de Saint Ours, and then edited and published by Klaus Neumann (1996, 18–19). The account was told by John ToVuia, then in his early eighties.

ToVuia's story is of interest because he and his ancestors knew of *six* specific eruptive events from Blanche Bay volcanoes taking place before the two eruptions in 1878. Two were from Tavurvur. A third eruption was from Palangiangia (meaning 'teethy ridge'), also known as Raliplip (meaning 'the fence'). An eruption from Rabalanakaia, within Palangiangia, was next, and the remaining two events were from Sulphur Creek, as follows:

> The eruption of Iavatirane or Tokurkurung, which is now known as Sulphur Creek, followed. Its first submarine eruption. Its second eruption took place over 130 years ago [i.e. before about 1854]. It created two craters: Tokurkurung-Tawa, featuring a swamp, near the Rabaul Golf Club [in 1984], and Tokokurung-Maga, a dry crater on the Rabaul Town side of Sulphur Creek. Iavatirane was Sulphur

Creek's first name. It originated during its first major eruption, the people began calling it Tokurkurung as they heard a thunder-like rumbling when it was building its latest craters. (Neumann 1996, 19)

Sulphur Creek volcano is clearly identifiable today as small craters at the head of the straight, canal-like inlet—also known as Matanatara or Matamatar Creek—that runs eastwards from the harbour towards Rabalanakaia. The mid-nineteenth century eruption at Sulphur Creek mentioned by John ToVuia was first recorded much earlier as a metaphoric story of battling spirit-beings from the Rabaul area. An elderly Roman Catholic missionary, Father Georg Boegershausen, in 1937 informed the *Rabaul Times* about meeting elderly villagers who remembered the volcanic activity at Sulphur Creek (Boegershausen 1906, 1937). This was during the priest's first years of living on Matupit Island, from 1900 to 1911. His main source was a prominent village leader, ToMulue, who said in the retelling by the priest:

All that land rose during a heavy earthquake. I was a young man when it happened, and now I am, but for an old woman at Davaun, the only living witness. It was in the morning about 9 o'clock, we were holding a sing-sing, when near Matanatar the earth broke in eruption. The crater is called Kururung maqe—it is close to the hot water creek in Rabaul. A big crab had a quarrel with a snake and caused the eruption. Stones were thrown inland ... A large district of new land rose on the mainland and two-thirds of Matupit from Kikila to the passage. (Boegershausen 1937, 15)

Father Boegershausen estimated that the eruption must have taken place sometime in 1845–50. He also mentioned in his 1906 report that an earthquake had been felt a few months previously in Rabaul, on 1 October 1905. It was 'only a local one', but it might have heralded a volcanic eruption (Boegershausen 1906, 112).

Rev. Brown had noted that:

The old people told us that there was a small eruption, not nearly so large as the present one [1878], some thirty or forty years before [i.e. 1838–48], but since then the volcano has been very quiet indeed. (Brown 1908, 242)

Brown, however, did not state, or perhaps did not realise, that the volcano being considered may have been the one at Sulphur Creek. This is in contrast to two other Matupit islanders who were boys at the time of the eruption and who did recall the event—namely, Tulue (Boegershausen 1937) and ToLivai (Cilento 1937a, 1937b).

The precise time, duration and character of the Sulphur Creek eruption, or eruptions, are uncertain from all these accounts. There is also a volcanological question of whether the volcano at the head of Sulphur Creek can be considered as a 'satellite' or 'adventive' eruptive centre of the nearby and larger Palangiangia. Note also that Palangiangia volcano is the prime candidate for the source of the eruption cloud seen by Carteret in 1767 (Figure 1.4).

1.6. The Colony of German New Guinea

The years of the late nineteenth century following the 1878 eruptions were eventful, not because of any notable volcanic activity but because of increasingly robust European settlement and the creation of a German New Guinea colony, as recorded by many historians (Sack 1973; Hempenstall 1978; Hahl 1980; Firth 1983; Hernsheim 1983; Neumann 1992; Hiery 1995; see also Südsee-Handbuch 1920). Land acquisition from Tolai people became a requirement for the settlers, and there was more foreign-company trade and local investment. Shipping became more frequent through St Georges Channel and steamers would stop off at the Duke of York Islands to take on coal that was stockpiled and sold there (Figure 1.10). The Tolai themselves were thrown into a new environment that required adaptation to this Western invasion of colonialism and all of its requirements and demands. Conflicts with the Tolai, attacks and retaliations are part of this history of economic development, but also pacification as Christian missions expanded their proselytisation.

The year 1879 can be said to mark the start of development of a 'planters' community in the area. The part-Samoan Emma Forsayth, together with an Australian, Captain T. Farrell, arrived from Samoa, settling first at Mioko in the Duke of York Islands and establishing a new commercial base there (Robson 1965). Emma, or 'Queen Emma' as she became known later, brought members of her extended family in Samoa to join her, including her sister Phoebe and Richard Parkinson, Phoebe's husband. Parkinson in 1882 began surveying the nearby New Britain coastline and chose Ralum on the Kinigunan, or Kininigunan, coast (Figure 1.10), about halfway between Raluana Point and Cape Gazelle, as a site where plantations might be developed. New headquarters were established at Ralum that later included not only extensive plantations but also a small port and Gunantambu, where an increasingly wealthy Emma later built a luxurious residence.

German commercial interests continued to strengthen in the St Georges Channel area. French colonists were taking more than a passing interest in settlement too. Ships began arriving from France in 1880 carrying colonists who had invested in establishment of Nouvelle France at the southern end of New Ireland. This was where Louis de Bougainville in 1768 had anchored at Port Praslin (Figure 1.6)—named after a French duke, although Bougainville did not necessarily recommend it as a place of settlement. The Marquis de Rays in France thought otherwise, and convinced people requiring a new start in life to invest and travel to the other side of the world. The proposal was a scam and the results were disastrous (e.g. Biskup 1974). There was hardly any land suitable for agriculture, diseases were rife, people died and settlement failed. Survivors were rescued and Emma Forsyth benefited by salvaging the equipment and furnishings intended for New France.

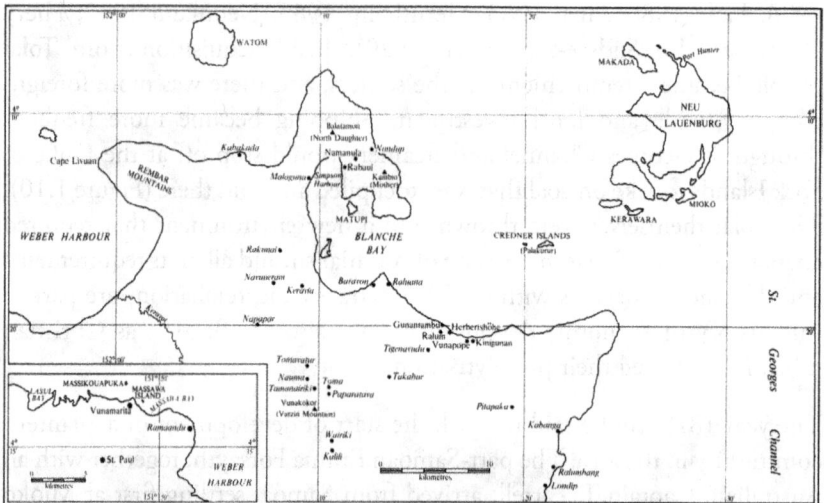

Figure 1.10. Map of the Rabaul region in German times.
The north-eastern Gazelle Peninsula and the Duke of York Islands are seen in this map from German times (Hahl 1980, Map 4).

Figure 1.11. Commemorative plaque at Nodup.

Nodup on the St Georges Channel coast plays a special role in the story of volcanic eruptions in the area. It would become a key assembly and departure point for evacuees from the town of Rabaul in 1937. Nodup was also the place where, in 1882, MSC fathers established their first mission station for work among the Tolai. The commemorative plaque shown in this figure was erected at Nodup and the photograph is from a publication celebrating 100 years of the Catholic presence in the Rabaul area (MSC 1982, n.p.).

A French Roman Catholic missionary also came out with the de Rays expedition but, in 1881, after the failure of the Port Breton settlement, he moved for a few months to north-eastern New Britain, visiting Matupit and then Nodup, or Nordup, on the shore of St Georges Channel, before returning to France (Mennis 1972; Threlfall 1975; MSC 1982). Missionary Rev. George Brown also departed from the New Britain area in 1881 owing to ill health, but Methodist missionisation continued unbroken. The Pope in that year, 1881, assigned the vast Vicariate of Melanesia and Micronesia to the Missionaries of the Sacred Heart of Jesus or MSC and, in 1882, three Roman Catholic missionaries arrived, first setting up at Nodup (Figure 1.11) and then, in 1883, at Kinigunan. They even established, in 1884, a mission at Malaguna within Simpson Harbour and then a presence on Matupit Island alongside the Methodists. More significantly, vicariate headquarters were eventually built, in 1889, near Kinigunan and named Vunapope— the 'place of the Pope' (or 'Popies'). From the early 1880s, Catholicism and Methodism were, therefore, very much in competition, no doubt providing some confusion to the unbaptised and non-aligned Tolai (Figure 1.11).

Hernsheim & Co., like other trading companies, was well established throughout the St Georges Channel area, and Eduard Hernsheim by 1883 had built his own home and trading base, including a copra factory, at Raulai on the eastern side of Matupit Island within Blanche Bay. He was evidently untroubled by any memory of the 1878 eruption or by the ongoing volcanic threat from the damaging, previously active volcano, Tavurvur, at his doorstep across Greet Harbour. Hernsheim, rather, clearly was charmed by the beauty—and economic potential—of the bay:

> [T]he site looked out over a wonderful panorama of the deep blue waters of Blanche Bay. To the left rose the pointed cones of the 'Mother' and 'South Daughter' mountains, and alongside them the perpetually smoking crater of the volcano sent up its light sulphurous vapours. To the right the coast line, fringed with coconut palms, stretched as far as the eye could see. In between lay the sea, of every imaginable shade from deepest blue to the green and white of the surf. (Hernsheim 1983, 65)

Evenings in late May 1883 and afterwards were especially impressive:

> The evening lights on these mountains, when the green gradually changed to red and then to deep violet and every single tree stood out against the background in the clear air, turned this spot into a scene of the greatest beauty, which I can never forget. These colour effects were particularly magnificent in the year 1883, and I heard later that this was due to the mighty eruption of the volcano of Krakatoa in the Sunda Straits, thousands of miles away, which filled the whole of the atmosphere with its cosmic dust and gave rise to these wonderful colour effects by the refraction of light. (Hernsheim 1983, 81)

Hernsheim here is referring to the catastrophic volcanic eruption of 20 May 1883 when Krakatau volcano in the Dutch East Indies collapsed, producing large explosive eruptions, devastating tsunamis and a large, mainly submarine, caldera (Symons 1888). What Hernsheim would not have recognised was that eruptions of this scale had also taken place in the Blanche Bay area and, indeed, that the magnificent flooded bay itself owed its existence to more than one caldera-formation event in the not-so-distant geological past.

Figure 1.12. Dawapia Rocks in 1883.
Some Tolai people had re-established themselves on this, the larger of the Dawapia Rocks (or Beehives), by the time this photograph was taken in 1883, five years after the 1878 eruption, and the year before Germany declared its protectorate. Source: Methodist Church of Australia, Department of Mission Papers, held by the State Library of New South Wales. Published courtesy of the Uniting World Mission.

There had been growing concern from European settlers in the colonies of Australia—which were all still part of the British Empire—and especially in Queensland, about the increasing German presence in the area to the north-east of Australia. This concern was expressed symbolically in April 1883 when Henry Chester, the Thursday Island magistrate, raised the Union Jack on behalf of Queensland at Port Moresby on the south coast of New Guinea. He claimed all of New Guinea island east of 141°E, together with the islands adjacent to it, as far as 155°E. The British Government did not ratify the claim, however, and in 1884, the Reich made its own claim—after discussions with Britain—on a German protectorate for the north-eastern part of mainland New Guinea and adjacent islands including New Britain and New Ireland.

In the early 1880s, Otto von Bismarck, Germany's 'Iron Chancellor' (Ludwig 1927), was not a strong supporter of German colonisation and imperial administration of distant territories in the Pacific. Other Germans, however, were committed to economically aligned foreign-policy objectives, profits and nationalistic empire building. These included wealthy investors and bankers such as Adolf von Hansemann in Berlin, as well as representatives of the other trading and shipping companies already operating in the St Georges Channel region. For example, the German naturalist Dr Otto Finsch, exploring the north-eastern coast of New Guinea on board the *Samoa* in early 1884, had encountered the active volcanoes there, but had apparently been more focused on the identification of potential harbours for German settlement and economic development.

Bismarck in 1883–84 was under some domestic political pressures to accede to a program of colonisation in the New Guinea area. Adolf von Hansemann then founded the Neu Guinea Compagnie, which, in 1885, and together with Bismarck's support, received a far-reaching imperial charter to administer a new German protectorate in the region (Firth 1983; Sack and Clark 1979). Appropriate flag-raising ceremonies were carried out in the new territory. Hansemann focused very largely on settlement of the north-eastern mainland of New Guinea, now called Kaiser-Wilhelmsland, which adjoined the south-eastern part and had been claimed by the British in 1884 as a protectorate. German names and spellings began to proliferate, some expansively, on maps, notably the Bismarck Archipelago, including Neu Pommern (New Britain), and the Bismarck Sea. Even a small volcanic peak inland from Blanche Bay and known locally as Vunakokor was named Mount Varzin, after Bismarck's country estate in Germany.

The New Guinea Company was concerned primarily with economic profit, with attracting European settlers and with developing new plantations and settlements. Asian immigration was encouraged for many tasks that the few Germans could or would not do (Cahill 2012), as was the immigration of labourers from islands such as Bougainville to work on the plantations. However, governmental administrative duties such as running a protectorate-wide judicial system and police force, formal relations with neighbouring countries and general military defence tended to be treated secondarily. Visiting naval warships of the Reich provided support where needed. There was also the hope that economically viable mineral resources might be discovered, and indeed gold was found to exist in the Waria River area of Kaiser-Wilhelmsland. This was much later, however:

in 1903, gold prospectors from British New Guinea crossed the border into the German protectorate and found colours (Nelson 1976). No goldmining industry was ever developed by the Germans.

Administrative headquarters for the New Guinea Company was established at Finschhafen on the mainland at the southern entrance to Dampier Strait, and Georg von Schleinitz—who had sailed into Blanche Bay at Rabaul in 1875—arrived at his new headquarters in June 1886 with his wife and children as the first *landeshauptmann* or administrator. Topographic and bathymetric mapping was carried out in the protectorate, including in the Blanche Bay area (Figure 1.13). However, there were serious health challenges for the Europeans living in Kaiser-Wilhelmsland. Almost 50 of them died from disease during 1887 alone, including Schleinitz's wife (Sack 1973).

Douglas Rannie, a self-described 'Sometime Government Agent for Queensland', visited Blanche Bay on 30 June 1887, anchoring off Matupit Island. He was visited by German officers and the manager of Hernsheim & Co., 'entertained most hospitably' and taken to different sites around Blanche Bay that 'struck [him] at the time' as being, for Germany, 'an ideal harbour to transform into a great naval base for future operations in the world' (Rannie 1912, 282–3). Rannie was taken to

> a mud island which had been thrown up about two weeks previously. There were springs of hot water still bubbling up from several places round it. The island stood about sixty feet high and was about three-quarters of a mile in circumference. (Rannie 1912, 282)

Here Rannie was apparently describing Vulcan Island, which had been created in January 1878 and not just 'two weeks' previously. His description of the features of the island, though very brief, if correct, is nevertheless of interest in that thermal activity (in the form of shoreline hot-water springs) was still taking place at Vulcan Island in 1887, nine years after the 1878 eruption. Rannie also wrote that at the north-western entrance to St Georges Channel, and before entering Blanche Bay, 'we could see the three active volcanoes on New Britain, named the Mother and Two Daughters, belching smoke high into the heavens' (Rannie 1912, 277). However, none of these volcanoes was geothermally active, either in 1887 or up to the present time.

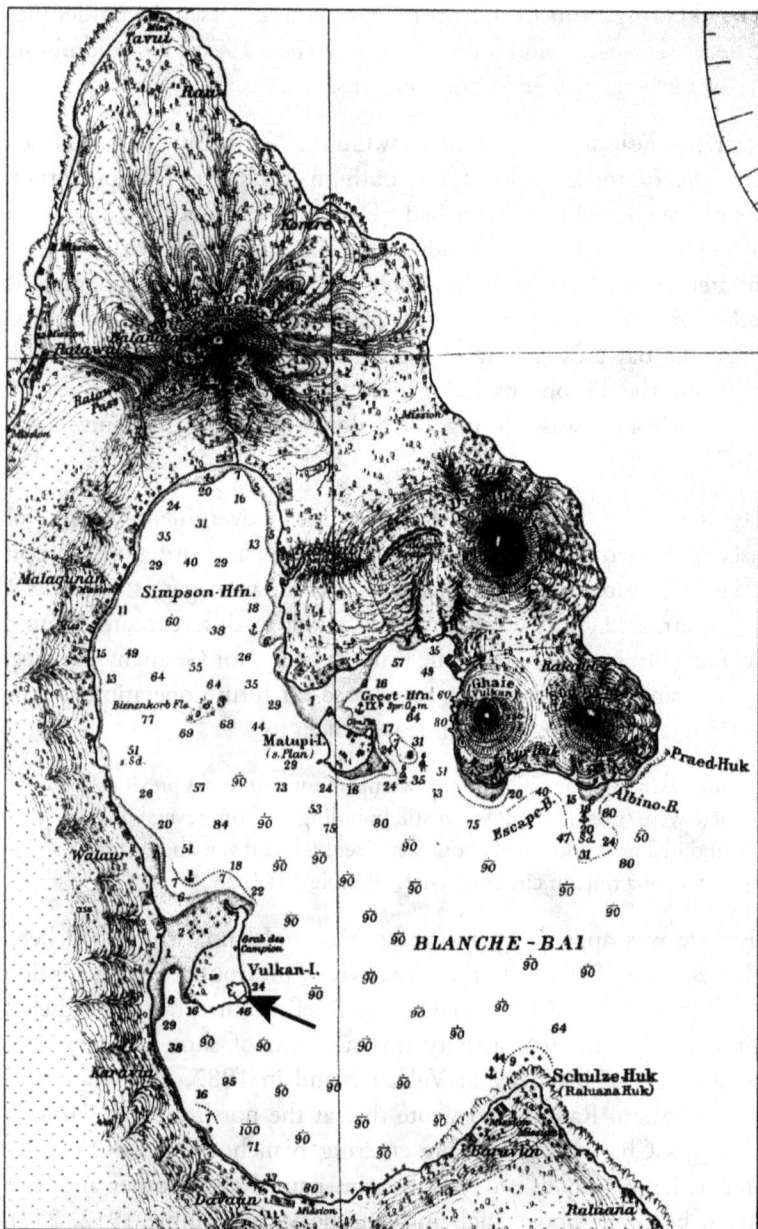

Figure 1.13. Bathymetric map of Blanche Bay, 1888.

The volcanoes of Blanche Bay are shown in this detail from a German chart drawn from surveys after the 1878 eruption of Blanche Bay, and now showing Vulkan I (Vulcan) and its 1878 crater (arrow). Note that 'Ghaie'/Matupi refers to Tavurvur volcano and that ghaie means 'kaia'. Source: Gazelle Halbinsel and Neu-Lauenberg published in *Nachrichten über Kaiser Wilhelms-Land und den Bismarck-Archipel für 1888* (Anonymous 1888). 1:100,000 scale chart.

The year 1888 was also challenging for German colonists. On 13 March 1888, a conspicuous active volcano in Dampier Strait originally known as Volcano Island but now named Ritter—after a German geographer—produced a devastating tsunami that swept onto the inhabited coasts of nearby western New Britain, the shores of adjacent islands such as Umboi and Sakar, and along the north coast of Kaiser-Wilhelmsland (Anonymous 1888a, 1888b; Cooke 1981; Johnson 1987, 2013; see also Figure 0.2). What, at Ritter, formerly had been a steep-sided conical island was reduced to an arcuate remnant and west-facing escarpment. Maximum run-up heights for the tsunami were 12–15 metres in Dampier Strait, decreasing towards more distant parts of the Bismarck Archipelago. The wave height was about 2 metres at Hatzfeldhafen on the mainland, almost 400 kilometres north-west from Ritter, but the oscillations had reached 7–8 metres by 8 am, threatening the station there.

The New Guinea Company's primitive 'capital' at Finschhafen was not directly in line with the main westward vector of the Ritter tsunami of 13 March 1888, but it was still vulnerable:

> [S]hortly after 6.30 a.m. on this day, a noise like thunder was heard, and at the same time the sea and the harbour water was strongly disturbed, so that it rolled in and out and endangered moored vessels. The current was so strong that the reef south of the islet Madang was completely dry in about 2 minutes and stood about 5–6 feet out of the water. Then the water rose again with the same violence. The time between the lowest and highest levels was only 3–4 minutes. (Anonymous 1888a, 76, translated from German)

Tsunami waves were also recorded on Matupit Island in Blanche Bay at the other end of New Britain. An anonymous observer there reported:

> [F]rom 0815 to 1100 the sea retreated from the island by 12–15 feet [4–5 metres] below its position at lowest tide and after that individual waves rose by just as much above the high water mark. The phenomenon revealed itself essentially on the southeast and northern side of the island while the western side remained untouched. The waves partially came from the south partially from the west-northwest. The water seemed to be stirred up in its depths. It looked turbid and carried dirty foam. Earth tremors or an underground roar were not noticed. The weather was clear and there was a weak southeasterly breeze. The natives by default, blamed the volcano, Alaia [probably Ghaie] opposite Matupi because the evil spirit dwelling there had been offended by the taking away pieces of

lava for building purposes. They fought the rising waves with sticks and threw stones at them with their catapults. (Anonymous 1888b, 148–9, translated from German; see also Wharton 1889)

The overall death toll from the Ritter tsunami is unknown but hundreds of coastal villagers must have lost their lives west of, and on the western coast of, New Britain; two Germans are also known to have perished. Ritter volcano for many years after 1888 was thought to have produced a major volcanic eruption and collapse like that of Krakatau in 1883, in the Dutch East Indies. However, any explosive volcanic activity at Ritter was comparatively minimal, and the cause of the tsunami was actually gravitational sliding of a huge segment of the volcanic cone westwards into the sea rather than collapse directly into an underlying magma reservoir. Tsunamis resulting from large volcanic landslides, as well as from sea floor earthquakes, can travel great distances and may become trapped in sheltered harbours; the name 'tsunami' derives from the Japanese for 'harbour wave'. The waves entering in Blanche Bay and observed at Matupit Island may have interfered with one another, possibly in places producing amplification of wave heights. Volcano-related tsunamis had already been experienced 10 years earlier in Blanche Bay during the 1878 eruptions, so now the Germans could add 'Ritter 1888' to their colonial experience of volcanoes and what volcanic hazards to expect.

1.7. Creation of the New Colonial Capital

Administrator Georg Schleinitz in Kaiser-Wilhelmsland and Adolf Hansemann, the owner of the New Guinea Company in distant Germany, had conflicting opinions about development strategies for the struggling protectorate. However, Schleinitz, now a widower, resigned and left Finschhafen on 19 March, just five days after the Ritter tsunami disaster. During his time as administrator, Schleinitz had arranged for an administrative station for the New Guinea Company to be developed on Kerawara Island in the Duke of York Islands, but it was abandoned. His successor, Rudolf von Kraetke, had decided by 1889 to purchase a large area of land near Ralum in the Kinigunan district on mainland New Britain. Richard Parkinson began work there on the new company station located at Kokopo—meaning 'cliff' or 'landslide'. This terrace-like area was renamed Herbertshöhe, meaning 'Herbert's heights', after Bismarck's younger brother.

Eduard Hernsheim on Matupit Island had recommended to Kraetke that picturesque Simpson Harbour within Blanche Bay was a far more suitable location for development, as the anchorages at Herbertshöhe were unprotected and large-scale harbour works would be required for ships to anchor there during the north-west monsoon. German administration in the Bismarck Archipelago—but not yet the entire protectorate—was nevertheless transferred to Herbertshöhe in 1891. A French priest, Louis Couppé, in 1889 had become the first bishop of the Catholic Vicariate, the headquarters of which was built at Vunapope near Kinigunan. The ever-opportunistic Emma Forsayth at Ralum made her home at Gunantambu—a social centre where her young female relatives from Samoa were a magnet for unmarried male employees of the company. This entire area, including Ralum Plantation, Herbertshöhe and Vunapope, was now becoming established as an important colonial centre for the protectorate.

Eduard Hernsheim on Matupit Island in 1891–92 became anxious about his declining state of health, which

> was made even worse by nerve-wracking, continual earth tremors and mysterious rumblings, which appeared to indicate an imminent eruption of the volcano [Tavurvur] just opposite my house. Towards the end of the year these phenomena became so imminent that my last remaining employees gave notice and left me, so that for some months I myself plus one clerk and a few Chinese remained the only inhabitants of Matupi apart from the natives … [T]he menacing and unsettling natural phenomena were affecting my nervous system. (Hernsheim 1983, 114)

An eruption did not take place, but Hernsheim, aged 45, left German New Guinea on 14 May 1892 and returned to Germany, where 'he could be confident that the income from his Pacific enterprises would allow him to lead a very comfortable life at home' in Hamburg (Sack and Clark, cited in Hernsheim 1983, v). Hernsheim left the company's business based on Matupit in charge of his new partner and nephew, Max Thiel.

A key person in this story of 'Volcano Town' is lawyer Albert Hahl (Firth 1978a; Hahl 1980). Hahl first came to the Neu Guinea Compagnie Protectorate in early January 1896 as imperial judge on a three-year term, arriving first at Friedrich Wilhelmshafen—locally also called 'Madang'—in Kaiser-Wilhelmsland, now the capital of the protectorate. He soon moved on and, by 14 January, had reached his base in Herbertshöhe, New Britain, headquarters of the company in the Bismarck Archipelago. Hahl there was to be 'the sole official of the Reich in the protectorate … independent of the influence of the Neu Guinea Compagnie' whose obligations were, as summarised by Hahl, to 'attempt the economic development of the country by means of trade with the natives and by regular plantation agriculture on its own account' (Hahl 1980, 3, 7). Hahl's home afforded a magnificent view over the wide expanse of sea sparkling in the sunlight, as far as the mountains of the volcanic peninsula at Matupi and past the islands of the Neu Lauenburg Group to the mountain ranges of Neu Mecklenburg which loomed dark on the horizon. (Hahl 1980, 9)

Herbertshöhe itself had become recognised as the main colonial centre. Hahl returned to Germany in 1899, where he transferred to the Colonial Service of the Foreign Office. His duties on his return to Herbertshöhe later in the year would have to take into account the recent purchasing by Germany from Spain of the Mariana and Caroline islands. This produced an 'Island Territory', followed in 1906 by the addition of the German Protectorate of the Marshall Islands—all adjacent to what became known as the 'Old Protectorate' (Firth 1983; see also Figure 1.14). There had also been agreement between the German Reich and the Neu Guinea Compagnie relating to transfer of sovereignty to the Reich. Rudolf von Bennigsen became the new imperial administrator, or governor, of the two-part German New Guinea. Hahl, who became vice-governor, was based in Ponape, but he returned to Herbertshöhe by June 1901, impressed by developments there, including construction of two new hotels: 'I would not have known the quiet little Kokopo of former days' (Hahl 1980, 83).

The year 1902 was critical in the 'Volcano Town' story: later that year, Hahl travelled back to Berlin for discussions with the Colonial Section of the Foreign Office, which 'knew that a good shipping service was essential' in the Pacific (Hahl 1980, 94), and the need to hand over the construction and operation of a new ship, the *Stephan*, to Norddeutscher Lloyd was identified. Hahl next travelled to Bremen for discussions of future plans in German New Guinea with the general manager of Norddeutscher Lloyd. 'These plans envisaged the creation by the Lloyd on Simpson Harbour of a permanent base for coastal shipping, including the construction of a wharf and the necessary warehouses.' Agreement was eventually reached that the company would 'establish a commercial base and an entrepôt port on Simpson Harbour' (Hahl 1980, 95). In other words, Herbertshöhe would not be that port. There is no known record of whether the geological risks involved in building new facilities in Simpson Harbour were discussed in Bremen, nor whether any attempt was made to prepare comparative costs for the construction of suitable wharf facilities at Herbertshöhe. Similarly, there is no known record of any conversations Albert Hahl and Eduard Hernsheim may have had when they met in Colombo in 1898 about the advantages of Simpson Harbour for shipping—a view that Hernsheim had always favoured (Figure 1.15).

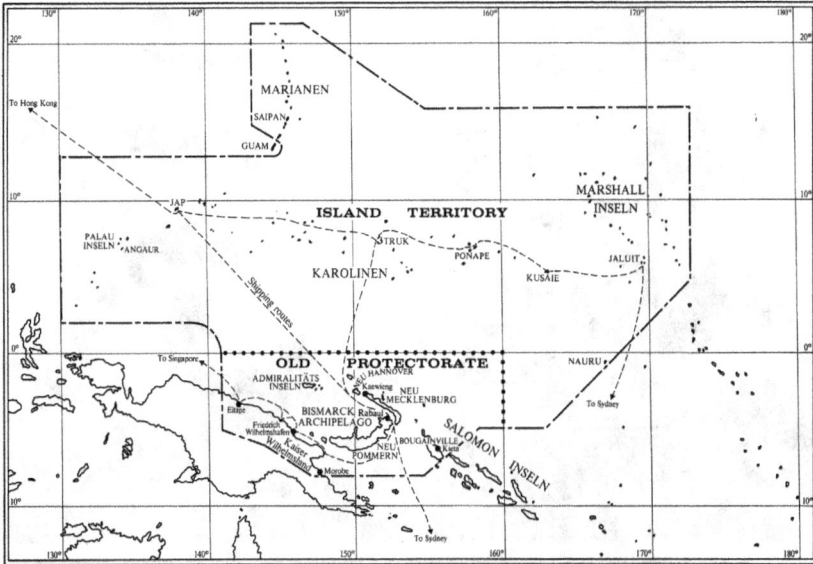

Figure 1.14. Geography of the Old Protectorate and Island Territory in German times.

The two constituent territories of German New Guinea are shown in this map (taken from Hahl 1980, Map 1). Principal shipping lanes are shown passing through St Georges Channel, connecting with Hong Kong, Sydney and Singapore. Guam in the north-west was and still is US-held.

Hahl returned to German New Guinea and, in November 1902, was appointed governor of the protectorate to succeed Herr von Bennigsen. He later reported:

> The Norddeutscher Lloyd had lost no time in translating into fact the promise given in Bremen [in 1902]. The first experts arrived as early as 1903; soundings and surveys were carried out in the harbour of Rabaul; the steamers brought building materials; a great wharf was completed and on it were erected warehouses and water tanks; the foreshore was cleared and houses and office buildings went up. By the end of 1904 the installations were opened for shipping. (Hahl 1980, 115; see also Figure 1.15)

Also, Norddeutscher Lloyd was granted a monopoly for shipping goods to and from Rabaul, which 'dealt a severe blow to the Australian firm Burns Philp & Co.' (Hahl 1980, 115).

Figure 1.15. Portrait photographs of Eduard Hernsheim and Albert Hahl in later life.

The photograph on the left of Eduard Hernsheim in later life is from the *Hamburger Fremdenblatt* for 24 April 1917 (no. 112B, evening edition, p. 9). The photograph of Albert Hahl on the right is from the frontispiece of Hahl's 1980 book; an original print of this portrait was supplied to editor Peter Sack and reproduced courtesy of the Hahl family.

English writer Bessie Pullen-Burry travelled widely in 1906–07, visited New Britain probably for four weeks, and was hosted and guided by German residents in the Herbertshöhe and Blanche Bay area (Pullen-Burry 1909). She met there the Austrian anthropologist Richard Thurnwald, who was undertaking linguistic and anthropological studies in the German colony. Pullen-Burry also experienced an earthquake whose strength, some of her hosts declared, would prove to be the severest felt, so far, during German times. 'Evidently the Mother [volcano] was responsible for the quakes', Pullen-Burry wrote, noting also that the waters of Blanche Bay were 'abnormally agitated between the shore and the volcano' (Pullen-Burry 1909, 228–9). Further: 'Looking westward [sic] across the bay, I recognised the all-powerful Mother, whose eruptive playfulness I had already experienced several times' (Pullen-Burry 1909, 220). However, Pullen-Burry is almost certainly incorrect that the extinct Mother volcano was responsible for the geophysical unrest at this time. Other visitors to the Herbertshöhe area around this time were Lily and Karl Rechinger, botanists from Vienna. They were able to use the *Seestern* (Figure 1.16) to explore the Blanche Bay area and to visit Matupit Island (Figure 1.17) as well as Tavurvur, commenting on its botany (Rechinger and Rechinger 1908).

Figure 1.16. The German steamer *Seestern* on Greet Harbour.

Governor Albert Hahl wrote that the arrival of the government steamer, the *Seestern*—for use by the governor and officials throughout German New Guinea waters—'was the highlight of the year 1903' for the colony (Hahl 1980, 101). The date of its arrival in New Britain was 21 August 1903 (*Sydney Morning Herald* 1904). The *Seestern* is seen here left of centre anchored in Greet Harbour. The vessel is bedecked with flags on the occasion of an official jubilee ceremony (foreground) on 4 November 1909 commemorating the raising of the German flag on Matupit Island in 1884 (Overlack 1972–73, image on p. 134). The photograph was taken from Matupit Island looking north-eastwards across Greet Harbour towards the peak of the Mother (Kabiu) volcano.

The severe earthquake felt by Pullen-Burry may have been the one that took place on 14 September 1906. By this time, seismographs had been installed in different countries worldwide, such that instrumentally recorded data for large earthquakes could be shared and calculations made of the size and location of each earthquake. This 1906 event was assigned the large-magnitude value of 8.1 and given an epicentre, to the nearest degree, of 7°S latitude and 149°E longitude (Sieberg 1910; Gutenberg and Richter 1954). The earthquake, notably, is also the oldest one listed in a table compiled in the early 1960s for numerous principal earthquakes in New Guinea and Papua between 1906 and 1962 (Brooks 1965, Table 1). Its poorly defined epicentre, however, represents a point in the western Solomon Sea hundreds of kilometres from Rabaul (Figures 6.5 and 6.10).

Figure 1.17. Matupit children in 1906–07.

This photograph of children on Matupit Island was published in the book by Lily and Karl Rechinger (1908, tafel 2). The children would have been born after the 1878 eruption at nearby Tavurvur just across Greet Harbour. The volcanic peak in the background is Tovanumbatir to the north.

Meanwhile, Governor Hahl was proceeding with his development of the new town on Simpson Harbour irrespective of earthquakes:

> Rabaul grew apace. The major firms were anxious to secure suitable blocks of land in this fast-growing port … The extensive plain round the bay … had to be surveyed and an appropriate townplan prepared. There was provision for a business section near the life-giving wharf; residential sections; a special quarter for the Chinese and another for the Melanesians. (Hahl 1980, 115)

The town plan covered the low, flat land around the north-eastern corner of Simpson Harbour, where swampy areas of mangrove (*ra-baul*) had to be drained. Rabaul's main business district was in the south-east, looking westwards out across the waters to Malaguna and Dawapia Rocks and to the western escarpment of Blanche Bay. A botanical garden was designed for the north-eastern part of the town on the slopes of North Daughter volcano, and lines of trees were planted along some streets, later providing welcome shade for times when the weather was hot and windless. The new town appeared in some ways idyllic, yet its southern edge was close to Sulphur Creek volcano and, from many points, there were clear views towards

Tavurvur, the active volcano that, in 1878, had erupted explosively during the monsoon season sending ash mainly to the south-east. The position of the new town of Rabaul was now in the ash fall zone from Tavurvur during the south-east trade wind season. Rabaul's generally low height above sea level (compared with the 'heights' at Herbertshöhe) also meant that the town was more susceptible to tsunamis that might become trapped in the harbour, creating seiches. Flooding of the town could be a problem too, particularly where water streamed off the escarpments near the edge of the town after torrential downpours. Poorly constructed buildings in both Herbertshöhe and Rabaul might also be susceptible to ground-shaking caused by both local and regional earthquakes.

Building of the new town and its wharves had advanced significantly by 1908. This was also the year that Professor Karl Theodor Sapper, a German scientist from the University of Tübingen, travelled to the Bismarck Archipelago and Solomon Islands as part of the Sapper–Friederici Expedition of 1908–09. The study area also included north-eastern New Britain. Sapper came to German New Guinea in 1908 with considerable volcanological expertise on the nature of explosive volcanic eruptions and their hazardous impacts (Termer 1966; McBirney and Lorenz 2003). He was 42 years old in 1908 and had spent his early years in volcanically active Central America and southern Mexico, later developing further experience through extensive fieldwork in ethnology, geography and geology, including volcanology and geomorphology. He had made a visit in 1902–03 to Mount Pelée volcano on the small island of Martinique in the Caribbean to investigate the disastrous eruption that had just taken place there. Thousands of people in the town of Saint-Pierre had been killed by a searing cloud of ash, rocks and dust that had raced down the flanks of the volcano on 8 May 1902 (Lacroix 1904).

Albert Hahl had not invited Karl Sapper to visit German New Guinea, so far as is known. Rather, the German Colonial Office had ordered Sapper to the area, almost certainly because of his specific volcanological experience (McBirney and Lorenz 2003). The implication, although speculative, is that someone in the Colonial Office had concerns about the new development going on in Simpsonhafen in the shadow of so many volcanoes and required some professional volcanological advice. In addition, at least some of the planter community, particularly of the large and influential trading companies, were not supportive of some of Hahl's colonial policies, and perhaps had expressed some concern to Berlin about his decisions. Rather than local people becoming better educated and more self-reliant in a changing world, which was Hahl's policy, the companies required a compliant labour force for their commercial plantations. However, a better port at Rabaul would provide the companies with an improved international shipping service, so they could hardly criticise Hahl for that.

Much of Sapper's and Friederici's time was spent surveying the geology, geography and ethnology of New Hanover and New Ireland, as well as the offshore volcanic islands of the Tabar-Feni group (Sapper 1910a). Albert Hahl and Karl Sapper together crossed Bougainville Island from east to west in mid-July 1908, supported by 50 Melanesians—20 soldiers and 30 carriers—and accompanied by American ethnologist G.A. Dorsey and government officer A. Doellinger (Sapper 1910b; Hahl 1980). Balbi and Bagana volcanoes, both thermally active, were seen during the crossing. Sapper also spent time in Blanche Bay examining the active volcanoes of Palangiangia/Rabalanakaia, Vulcan Island and Ghaie or 'Taburbur' (Tavurvur), as he called the volcano on Greet Harbour (Sapper 1910c). There is no known record of the conversations held between the governor and the professor about Rabaul, volcanoes or volcanic hazards, and Hahl does not refer to Sapper's Rabaul investigations in his memoirs or even to Sapper's considerable volcanological expertise. There is no doubt, however, that Sapper had serious concerns about the volcanic risks involved in developing the new capital at Rabaul. He wrote that the Rabaul eruption of 1878 was 'a warning to the inhabitants of Blanche Bay' and that there was

> the question whether it was advisable to establish the new capital of the territory at Rabaul in this endangered area of Simpson Harbour. It is of course possible that the volcanic force will lie dormant for decades, even centuries, but it is also possible that it will soon become active again; nothing could be more unpredictable. (Sapper 1910c, 191; see also Sapper 1937)

Engineer Ludwig Kohl in 1909 also warned of future volcanic activity and referred to the need for geodetic measurements:

> In all likelihood the whole area—the Mother–Matupi–Vulcan Island area—is involved in a relatively marked movement of elevation, so that frequent surveys will be necessary. Movements of elevation frequently and suddenly occur in association with earthquakes. (Cilento 1937a, 3; 1937b, 39)

Neither Sapper nor Kohl was in a position in 1908–09 to argue that the shift to Rabaul be reversed, but Sapper did offer the following practical advice for volcanic-risk reduction:

[P]rovision should be made to alleviate the effects of further volcanic eruptions or devastating earthquakes by specially constructed dwellings, which should be kept low and built of timber to offer maximum resistance to earthquakes; on the other hand, they would have to have steep roofs so as to render harmless a rain of ash or pumice [which] cannot settle on a steep roof and will slide off. (Sapper 1910c, 193)

Karl Sapper, after leaving German New Guinea, maintained an interest in Rabaul volcano and in New Guinea volcanoes generally in his later volcanological papers, including his benchmark book *Vulkankunde* (Volcanology), albeit briefly (Sapper 1927), and notably in 1937 (Sapper 1937). Other German scientists interested in volcanoes and their eruptive activity incorporated New Guinea examples in their work, including K.L. Hammer (1907), A. Wichman (1912) and, later, G. Hantke (1939).

German marine surveyors and navigation technicians also worked in the Blanche Bay area in these early century times. Earthquakes were felt on board vessels that moored in Greet Harbour in 1909, although they were referred to by the mariners as 'seaquakes' (Südsee-Handbuch 1920, 41). A notable visiting vessel was the German Navy's hydrographic surveying vessel, SMS *Planet* (Figure 1.18), which operated throughout several parts of marine German New Guinea (Intemann 2017). SMS *Planet* is famous for its deep-sea soundings in 1909–10 of the sea floor trench that runs eastwards south of New Britain, then south-eastwards down the south-west side of Bougainville Island. The greatest depth or 'deep' recorded in the south-east was 4,998 fathoms (9,140 metres), and 4,050 fathoms (7,407 metres) south of New Britain (Südsee-Handbuch 1920, 20). This major discovery would have strong implications decades later for ideas about the tectonic plate development of New Britain and its volcanoes, including those on Blanche Bay. The deepest part of the submarine trench off Bougainville Island is commonly referred to even today as the 'Planet Deep'.

'Seaquakes' were felt in Greet (or Matupi) Harbour and in the wider Blanche Bay area on 18, 19 and 20 February 1909, and again on 18, 19 and 26 November. Others were felt in the early morning of 8 December 1909, when

eight shocks of longer duration were observed. The longest shaking lasted thirty seconds. This shook the whole ship (S.M.S. *Planet*) severely, so that almost the whole crew awoke. Some of the men came on deck in the belief that danger threatened the ship … On board

S.M.S. *Kormoran*, which also was lying in Matupit harbour, almost the same phenomena were noted. During the 8th and 9th December several minor seaquakes without special phenomena were observed. (Südsee-Handbuch 1920, 41)

Residents were again reminded of the geological instability of the Blanche Bay area early the next year:

During the night of 24th–25th February, 1910, there was a severe earthquake in Matupi. It began without previous warning, with a severe shock four seconds long … About a minute later the phenomenon was repeated; the second shock was six seconds long. The first had been immediately followed by a moderately heavy swell from the south-east. The compass showed a deflection of four points—north-east to north … [Six other shocks were recorded between 12.52 am and 12.56 am.] Between 12.50 and 2 a.m. a strong sulphurous smell was noticed. This was the heaviest seaquake for many years. On shore the shocks were markedly vehement, and in houses lamps, pictures, crockery, &c., were thrown down. (Südsee-Handbuch 1920, 42)

Figure 1.18. SMS *Planet* in still waters.

The German hydrographic vessel SMS *Planet* is seen in this digitally enhanced image from a photograph kindly provided by the Universitätsbibliothek, Goethe-Universität Frankfurt.

The precise location of this distant earthquake sequence in February 1910 appears to be unknown, but another strong earthquake later that year, on 7 September, was measured instrumentally (Brooks 1965). The approximate position of this later event was beneath the south coast of New Britain (151°E, 6°S; depth 80 kilometres) about 240 kilometres south-east of Rabaul. Its magnitude was estimated to be '7 ¼' (Brooks 1965, Table 1).

Kommandant K-Kapt (Korvettenkapitän) Habernecht of SMS *Planet* published a short note on an earthquake felt in Greet Harbour on 8 September 1911, together with a tide-gauge record that showed the height of the wave from a small tsunami that followed to be only 20 centimetres (Habernecht 1912).

Governor Hahl wrote that Rabaul by 1909 'had long since outstripped in importance the town of Herbertshöhe, which had become a sleepy hollow'. He also had been 'forced to transfer some Government offices [from Herbertshöhe] and to increase the staff' in order to service the increased trade and communications at Rabaul (Hahl 1980, 131, 133). Hahl remained that year 'in lonely splendour in Herbertshöhe', but he allocated funding—without prior approval from a later disapproving Colonial Office in Berlin—to complete the move to Rabaul, which in 1910 became the new capital of German New Guinea (Hahl 1980). A neat network of tree-lined streets and roads was laid out beneath and between North Daughter and Rabalanakaia volcanoes, forming the basis for an attractive but vulnerable tropical town.

The town by 1913 was well developed (Figure 1.19). Three wharves had been constructed adjacent to the main part of the town in the south-east where there was a treasury, post office and district court. The town was home to German administration-run departments of agriculture and health, but there was no provision for a volcanologist or volcanological observatory. Government House, a home for the governor, had been built on the cooler heights of Namanula Hill together with a European hospital and school. The town was multiracial and racially stratified. White Europeans constituted the top layer, followed by Asians in the middle—including Chinese, Malays and Ambonese—followed by Melanesians at the bottom. The Japanese were readily accepted by the whites, especially successful businessmen such as boatbuilder Captain Isokide Komine. Chinese businessmen such as Lee Tam Tuck, known best as Ah Tam, were also successful. Both Komine and Ah Tam ran businesses along the foreshore in the western part of the town where the native hospital was located (Figure 1.19). Hahl's opinion of the Chinese was no-one 'wished to or was able to do without [them], but no-one wanted to have them in the country' (Hahl 1980, 145; see also Cahill 2012).

Hahl recorded that the passage between Matupit Island and the mainland to the north was in 1895 '16 feet deep and in 1913 totally dry', thus referring to sea floor uplift in this area (quoted in Fisher 1939a, 17). A causeway had been built by the Germans across the passage but it was submerged at high tide after 1913–14, implying some sea floor subsidence. Earth movements in the Matupit area would also be recorded in the years afterwards—clear indication of geological instability of this part of Blanche Bay.

No formal risk assessment was made of Rabaul town and its environs, so far as is known. However, the main part of Rabaul was vulnerable, being restricted to a coastal flat and hemmed in by steep escarpments on one hand and by the sea on the other (Figure 1.19). Timely evacuation out of Blanche Bay using vessels might not be possible if there was no forewarning of eruptive activity from the active volcanoes. Two trafficable roads led out of town, one westward towards Malaguna, then eventually out of Blanche Bay towards either Herbertshöhe to the south or northward over the ridge to Ratavul on the north coast. The other route was eastward up and over Namanula Hill and then down the eastern slopes to the shores of St Georges Channel. Evacuations triggered by volcanic eruptions were not required by the people of Rabaul during the short remaining time of what was known as German New Guinea.

People living in the Rabaul area in 1913–14 probably were becoming less apprehensive about volcanic hazards as they were also having to cope with another natural hazard: drought. They were concerned, too, by the building up of serious political tensions in Europe, where a war involving Germany seemed a real possibility. German New Guinea was a long way from its home country and even though ships of the German Navy patrolled the Western Pacific, the colony had very few men trained militarily, and the coastal towns of Rabaul and Herbertshöhe were clearly vulnerable to any major naval attack from a hostile enemy. International radio communications were not the best either, although a radio station was being built defensively at Bitapaka several kilometres inland and south-east of Herbertshöhe as part of a German telecommunications network for the region and the German Navy. A successful defence initiative of Rabaul and Herbertshöhe against a determined invading force such as from nearby Australia was, therefore, unrealistic.

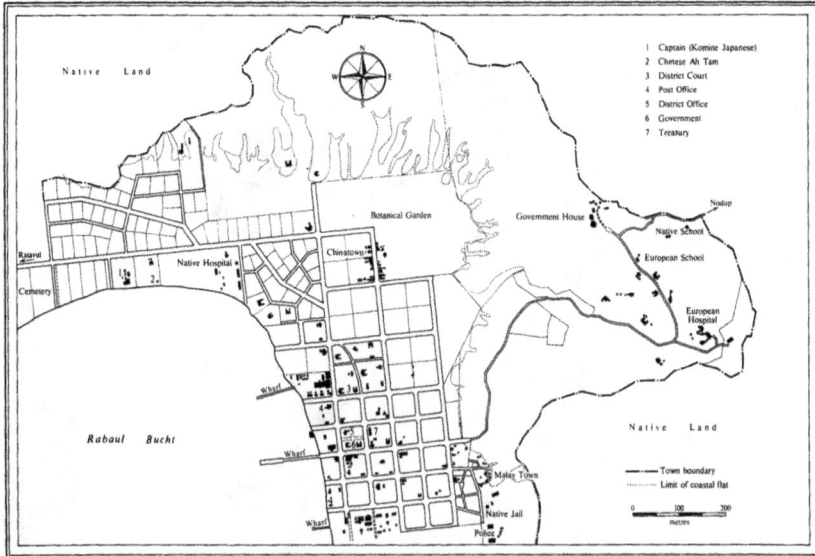

Figure 1.19. Map of Rabaul town in 1913.
The restriction in German times of Rabaul town to flat land mainly hemmed in by escarpments and the sea is shown in this town plan from 1913 (Sack and Clark 1979, Map 5). A road winds eastwards up to the cooler heights of Namanula Hill where Government House was built.

Albert Hahl did not have to wrestle with the challenge of possible invasion of the colony because he left German New Guinea in April 1914. Then, soon after reaching Germany,

> he was abruptly relieved of his post as governor, ostensibly because of ill-health. Though the facts are still unclear, he was almost certainly forced to resign, probably as a result of pressure on the German government from dissatisfied plantation companies or perhaps because the Kaiser was displeased at his failure to develop the Waria goldfields. (Firth 1978a, 44)

Another reason might have been Hahl's development of Rabaul as the colony's capital without prior sanction to do so from his Berlin masters.

2

Australia Takes Possession of the At-Risk Capital

2.1. Earthquakes without Local Seismographs

The Commonwealth of Australia was created in 1901 by federation of the Australian colonies. In 1906, Australia took over the administration of British New Guinea after its renaming as the Territory of Papua. Australia maintained strong ties with the 'mother country', Great Britain, as seen in its decision to follow Britain into conflict after it declared war on Germany on 4 August 1914, ushering in World War I. Australians had long-lived and ongoing concerns about foreign powers—particularly Germany—being on Australia's doorstep, so there was little hesitation in establishing an Australian Naval and Military Expeditionary Force for invasion purposes in 1914. This large assembly of troops invaded German New Guinea on 11–12 September, capturing and taking over the running of the capital Rabaul and thus the former Old Protectorate as a whole. Australian troops met military resistance from some German-led forces on the road to the Bitapaka wireless station and there were casualties on both sides (Mackenzie [1927] 1987).

Rabaul was occupied by an Australian garrison until 1921, when, following the Treaty of Versailles in 1919, a civilian administration took over (Mackenzie [1927] 1987; see Figures 2.1–2.3). German residents could remain in their former colony during the occupation provided they signed agreements of neutrality. Plantations established by the Germans were expropriated and arrangements made for non-Germans to manage or purchase them, including returned Australian soldiers. Meanwhile, Rabaul itself became a 'backwater'. Despite some tension between the Australian occupiers and German residents in Rabaul, the garrison was largely untroubled by serious military hostilities, leaving the invaders bored and restless.

The Australian invaders took over a well laid-out town but, in doing so, they also inherited the volcanic-risk problem. No seismographs were installed, and systematic records of local and distant earthquakes and signs of local volcanic unrest—such as ground uplift or subsidence—do not appear to have been kept during the period 1914–21. Nevertheless, the garrison would experience the effects of a distant earthquake soon after taking control. On Wednesday 11 November 1914, ground-shaking was felt in Rabaul:

> It occurred at about 6.00 p.m. when most people were settled down to their evening meal. The force of the shake was sufficient almost to empty cups of tea and to shift other articles in buildings from their positions. The tremor lasted about thirty seconds, the first portion of which was of a mild nature—the like of which has been experienced a couple of times since our arrival—but it was the last few seconds that caused excitement. This is a somewhat novel experience for most of the troops and they are not desirous for a repetition thereof. (Anonymous 1914a, 2)

Furthermore:

> [A]bout 600 tons of rock has fallen from the old crater opposite Matupi Island. Several new fissures have also opened up, and from these sulphur fumes are being emitted. The members of the Expedition that has just returned from New Ireland report that they experienced a shock of earthquake on the same date as the one above. (Anonymous 1914b, 2)

These valuable descriptions correspond to an earthquake distant from Rabaul that was strong enough to cause landslips in the crater of Tavurvur but did not trigger any volcanic eruption. Signaller L.C. Reeves reported that the largest of several earthquakes felt in Rabaul soon after the invasion by the Australian troops occurred at 7.23 pm on 19 October 1914. It too is said to have displaced 'several hundreds of tons of rock and soil' in the crater of Tavurvur (Reeves 1915, 50). Reeves could here be referring to the same earthquake reported for 11 November 1914.

Figure 2.1. Australian troops after the 1914 invasion of Rabaul.
Australian soldiers pose for a photograph with a machine gun at Rabaul after the invasion of September 1914. Source: From the photographic collection of Albert Richards and provided courtesy of Bruce Young.

Major Howard Newport, director of the botanical gardens in Rabaul, published a report about an earthquake felt in Rabaul at 11.30 pm on New Year's Day 1916. It was

> variously spoken of by old residents as the most severe for five to seventeen years. Nevertheless, there is no record of any furniture, shelves, &c., having been shaken down, or crockery or glass having been broken worth mentioning; concrete basements to houses … were undamaged on this occasion … Owing to the buildings being of wood, and not rigid, none suffered in Rabaul itself, though one or two water tanks were displaced. Some 20 miles further west, however, several houses were damaged by being thrown off their piles and twisted, but some of this at least may be ascribed to the bad state of repair in which the buildings were at the time … These shocks are admittedly more severely felt at Namanula … Indeed it would appear that many mild shocks are observed there that pass unnoticed in the township below. (Newport 1916, 2–3)

Major Newport counted and recorded additional shocks that were felt through the night until morning, adding that 'most tall furniture is fastened by the walls by iron hooks by our predecessors [the Germans] evidently in readiness for them' (Newport 1916, 3).

Figure 2.2. Viewing the 1878 crater of Tavurvur.
Australian soldiers and a Tolai guide look into the 1878 crater of Tavurvur volcano after invading forces had taken control of Rabaul from the Germans in September 1914. The troops were evidently not well equipped for a tropical invasion judging by the puttees and thick shirts and trousers. GA negative references M2444-1-4 and M2447-31A.

Another military report, from the Australian Navy on this occasion, provides extra information about the effects of the 'somewhat severe' earthquake of 1 January 1916, including a related tsunami:

> The steamers alongside Rabaul wharf surged violently at their hawsers. The master of the S.S. *Masina* states that the water level fell 15 feet in a few moments, and rose again with tremendous rapidity; his vessel meanwhile being in grave danger. He noted that the top of a long corrugated iron shed on the wharf was actually moving 'in waves', and the entire wharf was on the move. The small steamer *Siar* was lifted bodily from Rabaul Harbour into Matupi Harbour, across the place where a strongly-built causeway had existed a few moments before. This causeway, which connected Matupi Island to the mainland of New Britain, entirely disappeared, leaving a depth of about 15 feet (H.M.A.S. *Una*, 1916). (Südsee-Handbuch 1920, fnt p. 20; see also Massey 1918; Fisher 1939a)

Another reporter, quoting from the *Rabaul Record*, a monthly newspaper of the Rabaul garrison, added that the tidal wave or tsunami caused by the 1916 earthquake did some damage to the foreshores and to vessels in port (Cilento 1937a, 1937b). Further, the water at the causeway, which carried a telegraph line, was deep enough to cover the telegraph line as well. Director

E.F. Pigot of the Jesuit-run Riverview Observatory in New South Wales concluded that, of 295 earthquake reports he had received for 1916 to June 1922, only six New Guinea earthquakes had notably high intensities and the highest value was for the 1 January 1916 event (Pigot 1923). However, Pigot did not rank the strong May 1919 earthquake that would also affect Rabaul. The epicentre for the 1916 earthquake (magnitude 7 ¾) was given in a later compilation of principal regional earthquakes from 1906 to 1962 as 154°E, 4°S—a point directly east of Rabaul and east of New Ireland (Brooks 1965, Table 1).

Another severe earthquake was felt in and around Rabaul on 7 January 1917:

> It lasted for several minutes, and is stated to have made lately arrived members of the force wonder what sort of place they had come to. One of them fainted, but otherwise no damage has been reported. The older hands of course knew all about earthquakes, and as a matter of fact would not feel at home in Rabaul if during the wet season they did not get a shaking up at least once a week. (Anonymous 1917a, 3)

Furthermore:

> According to statements made by German residents, Rabaul will be blown up in 1917 by volcanic forces. Should such a thing happen those [presumably Germans] who have always wished for a 'scrap' may be satisfied, if indeed they do not take exception to the latest inventions in heavy artillery. (Anonymous 1917a, 4)

This reaction by a writer in the *Rabaul Record* is an example of the tensions that existed between the Australians and Germans in Rabaul.

Two more earthquakes shook Rabaul, on 7 and 25 February 1917. The first one was the larger according to another article in the *Rabaul Record* (Anonymous 1917b, 3). It caused a landslip and interfered with telephone lines on the road to Namanula Hill, as well as throwing the supporting piles of a bungalow out of alignment. Advice was provided on the best way to evacuate from Rabaul. Major J.J. Cummins in the July issue of the *Rabaul Record* provided a brief history of the Rabaul area, including a summary of information on the 1878 eruption provided by Rev. George Brown (Cummins 1917). He repeated that Brown had been informed by local people that Tavurvur 'had been in eruption some 40 years previous [sic], but had been quiet up till the present ... On this fact is based the prophecy of another eruption which is due at present' (Cummins 1917, 10). Brown, however, had actually said that the previous eruption had occurred 30–40 years before 1878 (i.e. 1838–48) and he was apparently unaware that the informants were probably talking about the eruption at Sulphur Creek. As concluded above, the year of that eruption is unclear.

The rumour that an eruption would take place at Rabaul in 1917 or 1918 was criticised by Captain C.H. Massey, an officer in the Australian garrison who appears to have had previous geological training. He expressed disbelief that volcanic eruptions could be predicted accurately using a precise 40-year periodicity, writing 'that the periodic theory rests on a very slender basis' (Massey 1918, 8). Massey also noted that local earthquakes within the harbour seemed to originate from the area of Tavurvur volcano. Some local residents believed that earthquakes felt at Rabaul happened during heavy rain, but Massey did not agree. He did suggest, however, that there may be a relationship to high ocean tides and, therefore, to the north-west monsoon or 'wet' season, and so, in turn, to 'earth warping' or earth tides caused by the changing seasonal positions of the moon and sun relative to the earth.

Massey concluded that the large flooded depression of Simpson Harbour was the original 'crater'—actually a caldera—of Rabaul volcano and that it was the source of the thick deposits of volcanic dust, ash and pumice making up the surrounding countryside to the west and south, notably at Bitapaka and on the Namanula Road. These early geological observations by Massey relate to the origin of a *series* of calderas that form Blanche Bay, and to the especially large volcanic eruptions that accompanied caldera formation at Rabaul. They are matters that would occupy the attention of many subsequent volcanic geologists.

Another severe earthquake took place at about 5.40 am on 7 May 1919 after a fairly long period of seismic quiescence (Stanley 1923). It had a magnitude of 7.9 and its epicentre was at 154°E, 5°S—a point south-east of Rabaul and south-east of New Ireland (Brooks 1965, Table 1). The effects of the earthquake in Rabaul were reported in some detail by the administrator, Brigadier-General G.J. Johnston, in a memorandum to the Department of Defence in Melbourne, and later published by the Geological Society in London:

> It took the form of a preliminary shake, and was followed by a long shock, or succession of shocks. The ground rocked in a most alarming manner, trees swayed backwards and forwards, cracks appeared in the surface of the earth, and much damage was caused to property … Reveillé had blown at 5.30 A.M., and the men of the garrison

were preparing to go on early morning parade at 6 o'clock when the shock came. The wooden bungalows rocked to and fro, rifles were forcibly thrown from the racks, and a few who were snatching some minutes' extra sleep were pitched off their stretchers, which were overturned on top of them …

The shock was felt most severely on Namanula Hill, outside Rabaul, upon the summit of which Government House is built. The two portions of the house, separated by a wide gangway, rocked in opposite directions … When the earthquake subsided, most of the houses built on the hill presented an extraordinary appearance, the supports being tilted at all angles. Heavy 1000-gallon tanks were rolled over like toys, and our Government Printing Office was completely wrecked … In many places the hillside collapsed, completely filling [road cuttings].

These earthquakes appear to be closely associated with the volcanic belt in this region, and the earthquake of the 7th was followed by great activity in the sulphur-springs at the foot of Mount Mother [Greet Harbour], the green fumes spreading over the sea to a height of almost 100 feet … A tidal wave of some magnitude was experienced at Kokopo [and] Even at Rabaul, the wave was of such proportions as to leave thousands of fishes stranded above the high tide-mark. (extracts from Johnson [sic] 1919, v)

Further information on the 1919 earthquake was provided in another Australian military report:

At Rabaul wharf the water fell 6 to 8 feet and rose again rapidly; the shock made it impossible to stand on the wharf. The causeway to Matupi is now (September 1919) only knee-deep, and can be waded across (N.I.O., *1919*). (Südsee-Handbuch 1920, fnt p. 20; see also Stanley 1923; Official Handbook 1937; Fisher 1939a)

Geologist Evan R. Stanley added that many 'minor shocks' had been experienced at Rabaul and surrounding districts after the May 1919 earthquake, and he presented a list of 22 dated earthquakes—together with time of day, duration and intensity—that were felt in July–December 1920 (Stanley 1923, 44). Stanley also gave the following additional information in his description of the 1919–20 earthquakes:

Again on the 7th May, 1919, a severe shock was experienced at 5.40 a.m. after a fairly long period of quiescence. Tanks and houses were displaced from their piles, and the submerged Matupi causeway was elevated 2 or 3 feet. The shock was followed by great activity in the sulphur areas at Matupi and Tavurvur, and a tidal wave of some magnitude was experienced at Kokopo, whilst at Rabaul many fish were stranded on the beach. Many minor shocks have been experienced since at Rabaul and the surrounding districts …

General Johnston [reported on 29 September 1920] that the centre of disturbances is near the volcano of Ghaie (Tavurvur) in a probable line of weakness from the Father [Ulawun volcano] on the North coast of New Britain. He points out that the earthquakes are most severe about Rabaul when the Father is quiet. Most of the shocks recorded travelled approximately from North to South and a few from South to North. These latter appear to have originated in the Tavurvur. (Stanley 1923, 44–5)

Evan Stanley, an Australian, was the government geologist for the Territory of Papua. He was also a member of the Australian Commonwealth Scientific Expedition to the former German New Guinea, which arrived in Rabaul early in November 1920—its aim, an assessment of the natural resources of the newly acquired territory (Davies 1987). A 12-metre wooden ketch, the *Wattle*, was to be used as a mobile base, but fitting out the vessel took five months. The expedition finally set off from Rabaul in April 1921 for the north coasts of New Britain and New Guinea. During the delay Stanley was able to study the volcanic geology of the Rabaul area, as he had in both 1915 and 1920 (Stanley 1923). In the end, the expedition was a poor replica of those that had taken place in German times, and it turned out to be an embarrassment to the Australian Government and Prime Minister W.M. 'Billy' Hughes because of its cost and non-achievement of objectives (Davies 1987).

Stanley's report to the Australian Government was one of the expedition's few redeeming features. He summarised the volcanic geology of the Blanche Bay area where Simpson Harbour 'appears to have been the seat of a large volcano, remnants of which are left in the Dawapia Rocks' (Stanley 1923, 45). Stanley remarked that, together with volcanoes on the north coast of New Britain, Blanche Bay had had 'eruptions of the explosive type and extraordinarily stupendous judging from the number of parasitic cones and the huge quantity of fragmentary material deposited by them' (34). Notably, he also drew attention to the north-west–south-east alignment of the volcanoes Watom, Tovanumbatir (North Daughter), Kabiu (the Mother) and Turagunan (South Daughter) plus Tavurvur. This pointed

to the existence of a South-eastern rift, which has been responsible for the huge deposition of pyroclastic material, building up a great area of the land surface within recent geological times. (56)

Figure 2.3. Fort Raluana and background volcanoes.
Kabiu (left) and Turagunan volcanoes are shown in the background of this view
northwards across the entrance to Blanche Bay from Raluana Point (Mackenzie [1927]
1987, photograph facing p. 344). Tavurvur volcano is just visible down to the left of
Kabiu. Australian troops are seen in the foreground at a six-inch gun battery at Fort
Raluana. The battery was established in 1918, taking the place of a 4.7-inch battery on
Matupit Island. Australian War Memorial photograph H01987.

Stanley believed, however, that 'the whole series about Rabaul … appears to
be in the dying stages' (34).

Stanley's report contained a recommendation that a 'Geophysical
Laboratory be established in conjunction with a Volcano Observatory' for
public safety purposes in the New Guinea region (Stanley 1923, 92). This
recommendation had stemmed from an international Pan-Pacific Scientific
Conference in Honolulu on volcanology and seismology. Australian military
authorities did not take any specific measures to monitor the volcanoes
and earthquakes at Rabaul, although General G.J. Johnson, head of the
Australian garrison, in a letter written in September 1920, did recommend
the installation of a seismograph at Rabaul, evidently without success
(Stanley 1923, 44).

2.2. Civil Administration up to 1937

Rabaul and the former German colony came under Australian civil rule in 1921, which lasted until 1942. The captured territory—following the Treaty of Versailles in 1919—became a Class C Mandated Territory under the League of Nations, and this allowed the Australians to rule it almost as if it was part of Australia (Official Handbook 1937). The new civil administration's scientific interests were mainly restricted to health, agriculture, weather and anthropology, and, as in previous German and Australian military administrations, there was no interest in volcano monitoring—at least until after the disastrous 1937 eruption at Rabaul. There were, however, still signs of geological unrest at Matupit Island:

> After 1919, the causeway apparently rose, for by 1924 it was once again possible to drive a motor car to Matupi Island and this condition persisted until the 1937 eruption. Some observers stated that a slight amount of elevation continued to take place right up to May 1937. (Fisher 1939a, 18; see also Figure 2.4)

Earthquakes continued to be felt. Dr S.M. Lambert MD, for example, in his entertaining book *A Doctor in Paradise*, wrote the following about one in 1921:

> Rabaul was an extremely shaky Garden of Eden, geologically and politically. Jolly earthquakes came and went with seismic whimsicality, and were so frequent that every hotel, home, and office had its heavy furniture lashed to the walls. Otherwise, one might have waked up any morning and found a large German wardrobe in one's lap ... One morning in 1921 I saw some lumber that had been piled on Vulcan go scattering into the sea like a box of matches, and I saw the huge sheet-iron D.H. & P.G. store curl like a withered leaf. After that Eloisa and I agreed that at the next tremor we'd pick up little Harriette and make for the hills. (Lambert 1941, 79)

Vulcan Island by 1925–26 was a low casuarina-covered area used as a government quarantine station and occupied by local people under the charge of a resident medical assistant (Cilento 1937a, 1937b). Submarine disturbances were noted there in 1926:

> [S]ubmarine rumblings and explosions had been heard on the island, accompanied by earth tremors, and the following day it was observed that a reef of rocks had arisen at the north end of the island, in close relation to the area which [became] the site of the eruption of May [1937]. (Cilento 1937a, 8; 1937b, 45)

Rabaul Harbour from 3000 ft.

Figure 2.4. Aerial view of Tavurvur, Matupit Island and Vulcan Island.

This south-westward view of Tavurvur volcano (foreground) is undated but is thought to have been taken from a float plane probably sometime between 1920 and 1924. The height of the in-flight aircraft must have been lower than the stated '3000 ft'. The two flat islands are Vulcan Island in the left background and Matupit Island right of centre. The causeway between Matupit and the mainland appears to be absent, implying that the photograph may have been taken before 1924 (Fisher 1939a; see also unpublished discussions in R.W.J. Collection 25, Folios 49–50 and 55–57). Digital copy provided courtesy of the National Library of Australia (nla.pic-an20237763-55-v).

This brief note, if accurate, is significant, as it is the only known record that the Vulcan area had become volcanically restless at least 10–11 years before the volcano broke out in eruption in May 1937. The need for the installation of some sort of volcano monitoring of both Tavurvur and Vulcan seemed appropriate and further calls were made for instrumental earthquake monitoring at Rabaul—for example, in 1932 (Anonymous 1932)—but, like Stanley's and Johnston's earlier recommendations, they elicited no practical support. Earthquakes and tremors continued to be felt in Rabaul, as reported briefly in the local newspaper, the *Rabaul Times*, edited by Gordon Thomas: seismic events such as in 1931 (22 March, 14 June), 1933 (29 May, 13 December), 1934 (early January) and 1935 (15–16 and 20 September, and 8 and 15 November). Occasional reports of earthquakes felt at Rabaul were also published in the *Pacific Islands Monthly*, a regional magazine edited by R.W. Robson, and in other newspapers, including Australian ones.

The record of earthquakes or 'tremors' felt at Rabaul up to 1937 is incomplete given the absence of any systematic data recorded by a seismologist using a seismograph or, better still, a *network* of seismographs. Thus, the magnitudes and places of origin—the epicentres and depths—of earthquakes felt at Rabaul, and any sequences of related seismic events whether local or distant, cannot be identified readily from the limited records. Nevertheless, one difference between some of the reported events is that a few of them seem to have produced two shocks just seconds apart. This means that these particular earthquakes, at least, must have been distant 'regional' events—a 'primary' wave arriving before a 'secondary' one. A further problem with the existing earthquake record up to 1937 is the number of differences in recorded dates for particular events in the same year: whether these are errors between different records of the same event or signal multiple events in the same year is unknown.

A well-known Australian seismologist, Ian Everingham, in 1974 published a scientific paper in which he listed large earthquakes for the 1873–1972 period recorded from the New Guinea and Solomon Islands region (Everingham 1974). 'Large' meant earthquake magnitudes of greater than 6.9, which of necessity excluded many lower-magnitude events that may have been, or were, felt in the Blanche Bay area. Seventy-two earthquakes were listed for the period 1900–37, but none of these was in the Blanche Bay area—if the generalised coordinates used for a box-like area encompassing the bay are set at 152.0–152.5°E longitude and 4.0–4.5°S. A few earthquakes plot close to the box, including one of magnitude 7.0 (depth unknown) and dated 23 January 1937 whose coordinates are given as 153.0°E and 4.5°S—that is, in southern New Ireland.

Another notable feature of the large-magnitude earthquakes considered by Everingham is their relative abundance in the northern Solomon Sea area, including beneath north-eastern New Britain, southern New Ireland and Bougainville Island (Everingham 1974, Figure 2). This area also coincides with the deep-sea submarine trench and 'Planet Deep' discovered by the SMS *Planet*. The available earthquake information as a whole can be taken as evidence for ongoing seismic restlessness felt in the Blanche Bay area even though many of the European inhabitants in both German and Australian times seem to have taken the earthquakes for granted and not always with any accompanying concern that the volcanoes may again break out into eruption.

The geodetic record of ground uplift, subsidence and horizontal ground motion is also restricted to incomplete observations of the causeway area at Matupit and not, for example, to other parts of the shorelines of the island and of Greet Harbour. Finally, there are no systematic, time series records of changes to gas or vapour emissions at either Tavurvur or Vulcan Island or the development of new geothermal areas. There was, therefore, by early 1937, no strong reason for residents to be concerned about the possible outbreak of any new eruptions at the volcanoes of Blanche Bay.

Rabaul town had grown significantly during the 16 years following establishment of an Australian civil administration in 1921, as described in detail elsewhere (Threlfall 2012). There had been an increase in the size of the European population as people looked for new opportunities following the end of the war. New businesses were established and Burns Philp and W.R. Carpenter became the main trading companies (Buckley and Klugman 1983). The first administrator was Brigadier-General Evan Wisdom. The town expanded westwards along the head of Simpson Harbour, along both sides of Malaguna Road, to include new buildings and facilities established on land acquired from the Malaguna people; a new wharf was also built. Cars and other vehicles became more common on the roads and tracks. An electricity supply was developed for the town as well.

The year 1922 was significant due to the discovery by 'Sharkeye' Park and Jack Nettleton of commercially viable deposits of alluvial gold in Koranga Creek, inland from Morobe on the New Guinea mainland (e.g. Nelson 1976). This triggered a great deal of mining interest in the area and the development of the Morobe Goldfield. Gold extraction in the 1930s at Wau and Bulolo involved enhanced dredging technologies, aircraft for haulage and greater financial investment. New Guinea Goldfields and Bulolo Gold Dredging were among the most highly capitalised companies on the Sydney Stock Exchange (Garnaut 2010; Waterhouse 2010). Rabaul became a port of entry for overseas mineral explorers and miners before they moved on to Morobe. Increased shipping in the region meant that Rabaul also became the main coal refuelling depot in the Mandated Territory. The old hulk *Loch Katrine* was brought to Rabaul and anchored in the harbour for the purposes of coal storage.

Another significant date was January 1929, as this was when New Guineans from different parts of the Mandated Territory, and in a range of paid jobs in Rabaul, went on strike for better conditions (Threlfall 2012). Racism was still rife in the socially stratified Australian town, and the infamous White

Australia Policy still prevailed at home. The 'Rabaul Strike' caused great uncertainty and concern, if not real fear, among the white population, some of whom imagined the prospect of a non-peaceful uprising. Eventually, the potentially threatening, but actually quite peaceful, situation was brought under control. The strike leaders were confined to the *Loch Katrine* hulk where they received poor treatment under appalling conditions. Court cases were held on board and severe prison sentences were handed down.

The Wall Street Crash also occurred in 1929, triggering the Great Depression. Copra prices declined but government income was offset by the imposition of gold royalties from the Morobe Goldfields. The administration recorded as a significant historical event that the value of annual gold exports reached £1,897,244 in 1935 (Official Handbook 1937, 7). The Seventh-day Adventist Church, also in the eventful year of 1929, accepted the offer of a site at the south-eastern end of the island just across from Tavurvur volcano at the entrance to Greet Harbour (Oliver 2020; see also Figure 4.4). This part of Matupit Island would become a key instrumental measurement area for geodetic and earthquake monitoring in volcano-monitoring programs at Rabaul after WWII (see, e.g. Figure 7.4).

Figure 2.5. Aerial view of Rabaul town, Dawapia Rocks and Vulcan Island.

The town of Rabaul, including its tree-lined streets, wharves and shipping, is shown before the eruption of May 1937 in this undated aerial view of Blanche Bay taken from the north-east. Dawapia Rocks (the Beehives) and Vulcan Island (in the distance) can be seen in the upper-left quadrant. Sulphur Creek is adjacent to the centre of the left-hand border. GA negative reference GB3290.

The year 1934 was when the young Commonwealth geologist Norman H. Fisher began geological field work at Wau on the Morobe Goldfields, developing a strong interest in the volcanic rocks that had originally hosted the gold, and in volcanology in general (Fisher 1945; Wilkinson 1996). The science of volcanology had been advancing steadily after WWI. There was one significant advance, for example, when a New Zealand geologist working in the central part of New Zealand's North Island discovered, mapped and named a volcanic rock: *ignimbrite* (Marshall 1935). There was, however, no significant recognition of the international volcanological importance of this discovery—nor of its importance in understanding Rabaul's volcanic geology—until well after the end of WWII. The year 1934 was also when Walter McNicoll was appointed administrator of the Mandated Territory. Both McNicoll and Fisher would become involved in significant but different roles in managing the aftermath of the 1937 volcanic eruptions.

There had also been notable growth in the use of aircraft in general in the Rabaul area by this time, starting with float planes in the 1920s (Figures 2.4 and 2.5). Lakunai, south of Sulphur Creek and adjacent to Greet Harbour, was tried as a location for an airstrip but was abandoned in favour of one near Taliligap, later called Vunakanau, on a flat area with good approaches west of Simpson Harbour. Work on the Vunakanau Airfield was finally completed in April 1937. By then, radio communication technology had been improved too: the transmission-and-receiving wireless station at Bitapaka had been abandoned and Rabaul itself had two wireless radio stations—a commercial one for messaging run by Amalgamated Wireless (A/Asia) Limited (call sign VJZ), and one operated by the administration (call sign VHR2) for radio communication with district officers in remote areas and for receiving messages from Australia (Official Handbook 1937; Robson 1937; Threlfall 2012). Citizens commonly used shortwave radio for overseas communications and radio broadcast reception.

Rabaul town in 1937 was comfortable in its established routine as an enclave of the long-lived British Empire, but changes were ahead. Countries making up the empire had been surprised, if not shocked, by the abdication of Edward, Duke of Windsor, who had refused the throne of England so as to marry an American divorcee, Wallis Simpson. Further, there were press reports of Adolf Hitler spreading Nazism in Europe, leading to speculation that Germany, should it win any larger war with Britain, might want to reclaim its former colony in New Guinea. Also, Japan had invaded Manchuria in 1931 and would soon be expanding further, militarily, in east Asia and the Western Pacific. Nevertheless, life as a whole in Rabaul seemed reasonably stable and secure, at least until the last week of May.

3

Coping with the Unexpected:
The 1937 Eruptions

3.1. Rabaul in the Last Week of May 1937

Rabaul was an attractive town by 1937, surrounded by the towering volcanic cones and steep escarpments of majestic Blanche Bay and by the usually calm, blue waters of the sheltered Simpson Harbour. The town had matured since the Germans had laid out its network of streets. Trees had grown fully, providing much needed shade along major thoroughfares, such as Malaguna Road (Figure 3.1), Mango Avenue and Yara (or Casuarina) Avenue (Figure 3.2). Rabaul was a garden town where multicoloured croton hedges fringed carefully tended tropical gardens. Frangipani trees were common, accompanied by poinciana and other colourful plants. The Australians had developed the western part of the town, linking it with the Malaguna villages of the north-western shoreline (Figure 3.3). They had also developed the Rapindik, or Rapidik, area on the northern shore of Greet Harbour near Rabalanakaia volcano (which was also known as Rapindik Crater). Here, separated from Rabaul town, the laboratory of the Department of Health, the Rapindik native hospital and native labour quarters, as well as several official residences, had been established.

Figure 3.1. Malaguna Road, Rabaul, before the 1937 eruption.

Branches from the raintrees down the centre of Malaguna Road in pre–1937-eruption Rabaul provide extensive shade for both sides of the road. GA negative reference GB3299.

Figure 3.2. Chinatown and Yara Avenue, Rabaul.

Part of the Chinatown area of Rabaul at the northern end of Yara Avenue is seen as it was before the 1937 volcanic eruption. GA negative reference GB3298.

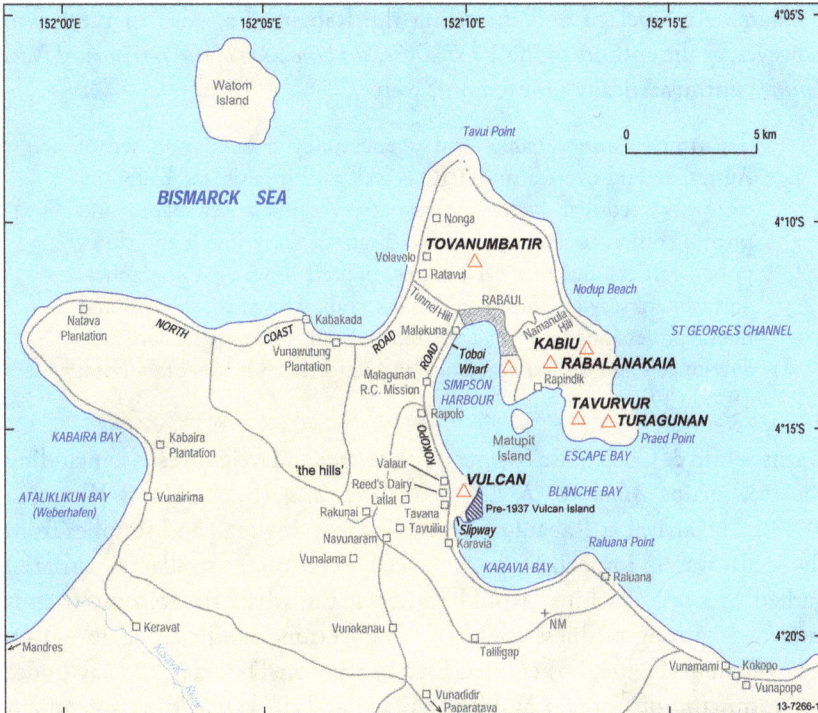

Figure 3.3. Map of north-eastern Gazelle Peninsula including Rabaul town.

Features of the Rabaul area in 1937 are shown here (adapted from Johnson 2013, Figure 42). Triangles represent volcanic peaks, and the main part of Rabaul town is shown stippled. Three villages called Malaguna, west and south-west of Rabaul, are not shown to avoid congestion on the map; one is close to the Malagunan Roman Catholic mission, and the other two are along the Kokopo Road between the Tunnel Hill turn-off and the Methodist District Headquarters at Malakuna (note the three different spellings of the name 'Malaguna' used in 1937).

A well-known place that attracted visitors and residents alike was the Rabaul market or 'bung' near the botanic gardens, where local women from the villages sold garden produce and souvenirs such as baskets and shells to curious tourists and others (Weetman 1937). Commerce in the town as a whole, however, was dominated by the giant Australia-based island trading companies of W.R. Carpenter & Co. Ltd, and Burns, Philp & Co. Ltd (Buckley and Klugman 1983). Australian Rabaul, as in German times, was still geared to the control and monitoring of commerce and to government administration of the native peoples of the territory. Commercial activity was still mainly in the production of copra and cocoa from the numerous plantations, mostly established by the Germans, along the shores of the

Bismarck Archipelago and including the Rabaul area. Also in the town, however, as the authors of the 1937 *Official Handbook of the Territory of New Guinea* enthusiastically pointed out, were

> department stores, banks and agencies for a large number of Australian and oversea suppliers, as well as shipyards, radio station, ice works, cordial factory, pharmacy, plumber, newspaper and printing office, restaurants, picture theatre, bakery, hotels and clubs, public carriers and hire car garages, billiard rooms, photographers and camera service, library and book club. The medical, legal and dental professions are sufficiently represented. There are accountants, auctioneers and insurance agents [and so forth]. (Official Handbook 1937, 139)

Burns, Philp & Co. Ltd was owner of passenger and freight vessels, including the steamships *Montoro*, *Macdhui* and *Malaita* (Buckley and Klugman 1983). The arrival at Rabaul of large ships carrying mail and supplies from the south was an important, even a social, occasion: Australian expatriates, garbed in tropical whites, would gather at the wharf to welcome friends and meet visiting business contacts and government officials (Figure 3.4). The *Montoro* (Figure 3.5) entered Blanche Bay on Thursday 27 May under the command of Captain William Michie and docked at the government wharf. Among the passengers disembarking was Miss Melville Chaseling from Sydney, who was met by her husband-to-be, missionary Jack Trevitt, and others from the European Methodist community in Rabaul (Trevitt 1937; H.L. Jones 1937). Melville and Jack were to be married in Rabaul on the following Saturday, 29 May. The usual cluster of coastal ships and inter-island schooners occupied the smaller wharves of the waterfront or lay at anchor in the harbour. The vessel *Durour*, 820 tons, belonging to W.R. Carpenter & Co., was in Blanche Bay too, but, more correctly, on the Carpenter's slipway on the north-western shore of Karavia Bay a few hundred metres from the western end of Vulcan Island.

Figure 3.4. Meeting the arrival of a vessel at Rabaul's main wharf.

Europeans gather to welcome disembarking passengers at the main wharf some time before the 1937 eruption (McCarthy 1963, copy of photograph opposite p. 164). New Guineans are standing to one side.

Figure 3.5. Painting of the SS *Montoro* at sea.

This photograph of SS *Montoro* in stormy seas was kindly supplied by Mr G.A. Clarke in 1980. The original painting is by A.J. Grant (1932). GA negative number GB2426.

Still in charge of the Mandated Territory, and of the Rabaul administration, was Brigadier-General Walter Ramsay McNicoll, CB, CMG, DSO, VD, who had been appointed administrator on 13 September 1934. He and his wife lived in Government House—previously the German Residency—on Namanula Hill and therefore part of the north-eastern caldera rim between Tovanumbatir and Kabiu volcanoes. Here they enjoyed spectacular views of St Georges Channel and New Ireland to the east and of Simpson Harbour to the south-west (Figure 3.6). The sea breezes were a relief from the, at times, heavy heat in the town, which was hemmed in below by the caldera walls. Administrator McNicoll was not in Rabaul during this last week of May 1937. He had travelled to the Morobe Goldfield on the mainland of New Guinea for an inspection, and his position back in Rabaul was being filled temporarily by Chief Justice D.S. Wanliss. Judge F. Beaumont ('Monty') Phillips was also based in Rabaul.

The Territory of New Guinea as a whole in 1937 was administered from Rabaul by a Legislative Council consisting of the administrator, government officers of an Executive Council and seven non-official members appointed by the Governor-General of Australia (Official Handbook 1937). The Honourable John C. Mullaly, for example, was one such 'non-official' member in 1937 (Hopper 1986). He was president of the New Guinea Planters' Association and owned Natava Plantation out west on the north-coast road (Figure 3.3). The director of agriculture, G.H. Murray, and the director of district services and native affairs, E.W.P. Chinnery, were among the five most highly paid first division officers on the administration staff; the other three were the director of public health, E.T. Brennan, the government secretary, H.H. Page, and the treasurer, H.O. Townsend. The administration hierarchy also included the secretary of the Department of Land, Surveys, Mines and Forests, an anthropologist, an entomologist and an economic botanist; missing from the list of science-related officers was a volcanologist. This omission reflected the general attitude of the administration to volcanic hazards at Rabaul: namely, that volcanoes had not been active since Australia took control after WWI; and the previous volcanic eruption had been in the previous century, even before the Germans had claimed the protectorate. There were earthquakes, certainly, but these *gurias*—the Tolai word for earth-shakes—added to Rabaul's character and, since volcanic eruptions did not seem to follow them, were, to an extent, taken for granted.

Figure 3.6. Entry gate to Government House in 1927.
Government House on Namanula Hill in prewar Rabaul commanded superb views of Simpson Harbour and St Georges Channel. This photograph was taken by Mr C.D. Meares in 1927. GA negative reference GB3300.

On Empire Day, 24 May, Judge Phillips visited Rabaul Public School, reminding the children that the empire stood 'for justice, a square deal and playing the game'. The junior choir sang 'Golden Wattle', the seniors sang 'Australia' and all joined in for 'Rule Britannia' (*Rabaul Times* 1937). No Asians or New Guineans attended the public school. Rabaul and its surrounding area were still a racially stratified society in which the colonial *mastas* were the 700 or so Caucasians of the town, mainly Australian and German. Next in line came about 1,000 Asians, mainly Chinese and a few Ambonese, and last about 8,000 New Guineans, including many indentured labourers from as far away as the Sepik area and Manus Island (Official Handbook 1937, supplement, p. 1). The celebration of the coronation of King George VI and Queen Elizabeth that had been held on 12 May in Rabaul was evidently enjoyed by all, irrespective of social status—a parade of ships, fireworks, torchlight procession and a dragon dance from the Chinese. There had been a splendid procession through the town in which the float decorated by Father Murche of the Sacred Heart Mission and his Eurasian pupils had won first prize (MSC 1937; Arculus and Johnson 1981, 21).

3.2. Volcanic Unrest and Early Warning Signs

Volcanologist N.H. Fisher, using the benefit of hindsight, wrote in 1939 that:

> The 1937 eruption was undoubtedly preceded by many warning phenomena whose significance would have been appreciated by a trained volcanologist. Many of the phenomena which may have occurred, such as changes in the temperature, the quantity or the composition of the gases being given off by the fumaroles in and outside of the craters, and possible increases in the gas ebullitions and water temperatures around Matupi Bay and in the wells at Rapindik, have not been recorded and consequently there is no means of telling whether they took place or not. (Fisher 1939a, 18)

Long-term Rabaul resident and editor of the *Rabaul Times*, Gordon Thomas, wrote in the newspaper in the first week of June 1937 that:

> The Karavia district has always been noted for the severity of earth tremours [sic] ostensibly due to a geological fault which runs across Blanche Bay to the Matupit crater [Tavurvur]. It is in the area that Vulcan Island is situated—that island which 'came up in a night' 59 years ago, and was most appropriately named. (Thomas 1937a, 2)

However, two years later, volcanologist N.H. Fisher portrayed the situation at Karavia and Vulcan Island before the weekend of 29–30 May in a somewhat different light:

> [P]remonitory symptoms were observed whose importance would never again be underestimated by the people of Rabaul. Very small earth tremors were experienced before the eruption and Father George Boegershauser [sic] reports that three days before the eruption, at Tavailiu [Tavuiliu], on the top of the hill opposite Vulcan, tremors apparently vibrating in a vertical direction were noticed. (Fisher 1939a, 18)

The absence of volcanological assistance was indeed a major limitation, and Fisher's comments are relevant in asking the question of when these early warning signs—whether earthquake activity, thermal changes or changes in coastal elevation—might have first been identified: were they days, months or years beforehand? For example, how significant was the submarine disturbance at Vulcan in 1926 noted by Cilento (1937a) or the large earthquake of 23 January 1937 listed much later by Everingham (1974)? Further, how significant were the elevation changes to the causeway at Matupit after the 1919 earthquake and up to 1937, mentioned briefly by Fisher (1939a), when no systematic geodetic surveying of the island is known to have been undertaken in the intervening 18 years?

Friday 28 May dawned on what promised to be a normal working morning in Rabaul. The night had passed uneventfully. People geared themselves for their routines during this last full working day of the week, but they also looked ahead to the coming weekend when there would be leisure activities, social engagements and religious obligations to fulfil. Catholics would be attending confession on Saturday afternoon in preparation for Mass on Sunday during the Corpus Christi celebrations. Methodists, too, would have church services on Sunday, and their mission staff and others would be in the congregation at the Rabaul Methodist Church for the Chaseling–Trevitt wedding on Saturday afternoon. Europeans and Tolai would be playing and watching the baseball (the Commonwealth Bank team was anxious to defend an unbeaten record), and others would be engaged in the wide range of other sporting activities so much enjoyed by Australian expatriates. The Boy Scouts had an outing planned. There was also to be a traditional *tubuan* ceremony at the coastal villages of Tavana and Valaur—a gathering of Tolai that was not greatly approved of by the missions, as such dances and feasts were the activities of secret societies established in pre-Christian times. The social clubs and hotels would be busy, assisting many Europeans in enjoying another favourite Australian pastime: consumption of liquor and beer. Others, however, had work commitments. Captain Michie would require the cargo of the *Montoro* to be off-loaded so that he could steam on to Kavieng in northern New Ireland that evening. Captain Eugene M. Olsen would be preparing to load copra on board his American ship, SS *Golden Bear*, at Carpenter's Toboi Wharf in the north-western part of Simpson Harbour (Olsen n.d.; see also accounts by Third Officer A.C. Willson [1937] as well as a cadet in *Mill Valley Record* [1937]).

The Friday lunchbreak passed, and then, at about 1.20 pm, as office workers returned to their duties, the sharp and sudden shock of a *guria* was felt. Miss Carol Coleman, who worked upstairs at Carpenter's offices on Mango Avenue, said the earthquake was

> [a] terrific bang, as though two lorries had collided. We soon realised what was happening when everything swayed and rumbled. Carpenter's building was a two-storied wooden building, and so we rocked very heavily, and we were all very upset. (Mason 1937, 1)

One moment Amy Anthony was squatting on her heels, cleaning paint brushes on her lawn, and the next she was flat on her back (Anthony 1981). Coconut palms across the road shook, the roof of her house rattled, and water slopped back and forth in the rain tanks. Brett Hilder and other officers on board the *Montoro* had been discharging and tallying cargo down in the holds during the morning, but were up on deck for a late lunch,

> when all of a sudden the ship started to tremble, and the tremble got more and more violent until the plates and cutlery on the table started dancing along ... in front of our eyes, which was rather disturbing—made a great clattering noise. (Hilder 1980, 28)

The earthquake lasted only about 30 seconds and there was no reported damage to buildings and roads in Rabaul. However, damage was reported from out on the Kokopo Road near Karavia where Mr and Mrs Furter's house was partly thrown off its piles; a wardrobe fell, trapping the elderly Mrs Furter, who was bedridden with a broken leg, and she was brought for treatment to the European hospital on Namanula Hill. Landslides blocked parts of the Kokopo Road, and, on Vulcan Island, a cliff on the lagoon disappeared and cracks opened in the south-western part of the island. A village school building collapsed at Raluana. The exact location of the earthquake's focus remains unknown because of the absence of seismographs in the Rabaul area, but the above descriptions are sufficient to indicate that the shock must have taken place under the general vicinity of the Karavia–Vulcan area. The earthquake was a premonitory expression of volcanic disturbance.

Fluctuations in the level of the water in the harbour, presumably caused by the earthquake, were noted at about 2 pm. The water receded gently to 100 metres out from the shore before returning to the normal waterline. The sea went out slowly at Vulcan Island, came in again to about 150 metres past the high-water mark, and then receded back to normal level. Fish were left stranded. The water at Karavia Bay at 1.30 pm rose over the road to about 2 metres higher than its usual level.

The remainder of Friday afternoon in Rabaul was more or less free of felt earthquakes. Normal activities in the town resumed, although the earthquake was a topic of conversation into the evening, and tremors were felt continuously out at Karavia and Taliligap, as well as at Rapindik south of Rabalanakaia. Rabaul inhabitants went to bed as usual and had an earthquake-free night, more or less, until about 5.05 am on Saturday 29 May when another sharp shock occurred, waking sleepers and causing pictures and bottles to fall to the floor. Saturday mornings were part of the working week for administration staff and banks, as well as for shops and businesses, but this Saturday morning was disturbed by a continuous series of felt earthquakes of different intensities. Most of the earthquakes were short, lasting no longer than about 10 seconds, but some went on for more than a minute. Intervals between shocks were about three minutes or less, but even during the breaks there was a gentle ground vibration. Tremors were especially severe around 10 am.

During each shock, storekeepers saw their stock fall from shelves and drivers felt their cars wobble; filing cabinets rattled and pedestrians staggered in the shaking streets. Medicine bottles fell off shelves at the Rapindik native hospital and bank tellers found it difficult to count notes and coins. The Commonwealth Bank building swayed and windows rattled; Virgil King, branch manager, decided to shift all the records into the strongroom and to evacuate his staff from the banking chambers as soon as possible (King 1937). Lawyer James Cromie took a legal paper, just signed by Chief Judge Wanliss, to the Bank of New South Wales only to find 'the staff running out and the manager leaping over the veranda rail' after a severe shock (Threlfall 2012, 212).

Judge Wanliss, who was still acting in the position of administrator because Ramsay McNicoll was visiting the Morobe Goldfields on mainland New Guinea, sent a telegram at 11 am on Saturday 29 May to the Department of the Prime Minister in Canberra informing the Australian Government of the disturbances the previous day (Wanliss 1937). The message contained the misleading, but somewhat prophetic, statement that there had been a 'small eruption' on the 'shores [of] Rabaul harbour'. The term 'eruption' may have referred to small ejections of water along the shoreline or in areas of shallow sea floor affected by the earthquakes. Wanliss's intended meaning has not been resolved:

> EARTH TREMORS VARYING DEGREES EXPERIENCED CONTINUOUSLY FROM FIVE A.M. TODAY. ONE OF UNUSUAL SEVERITY EXPERIENCED SHORTLY AFTER NOON TWENTY EIGHTH RESULTED SMALL LANDSLIDES KOKOPO ROAD SHIFTING ONE RESIDENCE SLIGHTLY FROM PILE FOUNDATION AND CAUSING SMALL ERUPTION SHORES RABAUL HARBOUR. NO OTHER DAMAGE REPORTED AND NO CAUSE ALARM. CAUSED PROBABLY BY LONG ABSENCE RAIN. WANLISS. (Wanliss 1937)

Damage was becoming more widespread. A section of rocks fell from the larger of the two Beehives at 12.45 pm and further landslides took place along the Kokopo Road. Rain tanks began leaking. Cracks appeared in the floors of the native labour quarters at Rapindik and also on Matupit Island, where, before lunch, Mr Chinnery and two Crown law officers heard deep rumbling noises that appeared to come from the bottom of the harbour in the direction of Vulcan Island. More especially, there had been some startling happenings at Vulcan Island during the course of the morning: land was rising out of the sea.

Elevation of the sea floor around Vulcan is thought to have begun between 8 and 9 o'clock in the morning. Fringing reef became visible around the island, and several small islets appeared. The effects were most noticeable between the mainland and the island. Brother Averbeck was at the Sacred Heart Mission shipyard at Karavia but abandoned his work at about 10 am because of the constant shaking (MSC 1937; Arculus and Johnson 1981, 22). An overseer of a New Guinean road gang noted a seething of the sea on the foreshore near the W.R. Carpenter slipway west of Vulcan Island occupied by Carpenter's 820-ton steamer *Durour*. Australian Medical Assistant C.W. Lambert and his New Guinean orderlies left Vulcan Island, abandoning the quarantine station. News of these happenings reached Rabaul and, although there were no officials able to evaluate their significance, there was concern, as the disturbances of the previous few hours had created tension in the town.

Abortive efforts were made to release the *Durour* from Carpenter's slipway on the afternoon of 29 May. Harry Hugo, a labour contractor in Rabaul at the time, was asked as a matter of urgency to provide a gang to assist in refloating the vessel. He recalled the event in an interview recorded for an Australian radio program broadcast in 1980–81:

> The Captain of the boat was aboard, an old chap called Tom Proctor, and the old engineer, Fred Northey. And, anyhow, one of the boys was working away and he said to me, 'Masta, have a look at the end of the passage', and I looked at the end of the passage and there was the bottom of the ocean. I thought, 'That's rather strange'. So I sing out to the Captain, 'Tom, come and have a look at this', and while we were looking there was another earthquake, and you could see this rise again—about another foot. And old Tom, being who he was, an old New Zealander, and happening to know a little bit about volcanoes and things, called: 'Jesus Christ! Let's get the bloody hell out of this!', and with that he and old Fred Northey—

both elderly men in their sixties—made up to the Kokopo road where I put the lorry, and we—me, myself, and all the boys followed them, and, young as we were, couldn't catch those pair of old fellows. (Hugo 1980, 8; see also Nelson 1982, 85)

Mr Murray, the director of agriculture, called at the Chinnerys' Rabaul home after lunch on Saturday and invited Mr Chinnery and his wife, Sarah, a keen photographer, to drive with him to Taliligap so as to inspect the new land being created around Vulcan (Chinnery and Chinnery n.d.; Ripley 1947; Chinnery 1998). They stopped at Rakunai Mission station and admired the decorations being erected for the Corpus Christi celebrations, reaching Taliligap, where Mrs Chinnery took photographs of the rising island below them to the north (Figure 3.7), by mid-afternoon. The party decided to return to Rabaul by the coast road and take a closer look. They drove on, passing the Furters' dislodged house at Karavia, and reaching the *Durour* on the slipway opposite Vulcan Island. Two young engineers of the *Durour* were going out in a launch for a closer view and invited the Chinnerys and Mr Murray to join them, together with some New Guinean helpers.

The new raised land was yellow and brown, up to 2 metres above water level, and had the familiar fetid stench of exposed live coral. Mrs Chinnery took another photograph. Some people from nearby villages, including Tavana and Valaur, were on the newly elevated reef and islets collecting stranded fish, apparently unaware of the significance of the raised land (Neumann 1996). Others were still on shore preparing for the *tubuan* ceremony. Mrs Chinnery noticed a peculiar rippling of the water and suggested turning back, to which the men eventually agreed. But she was startled on the return to see that one of the larger islets they had passed a few moments earlier had risen higher. Further, her husband noticed that another small islet had appeared. They were still about 200 metres from the *Durour* slipway at about 4.10 pm when a causeway of rocks suddenly rose out of the water completely connecting Vulcan Island to the mainland and cutting them off from the jetty (Figure 3.8). The party then

> turned for the nearest point on the shore, about 70 yards away, and as they did so heard a dull explosion, and there, about 100 yards behind them and just a few yards north of the islets they had just passed a jagged-edged column of black liquid-like ejecta about 30 feet in diameter was shooting 40 feet or so into the air. Eruption followed eruption, the height of the column increased rapidly, and the base crept towards them with continuous sharp explosions.
> (Chinnery and Chinnery n.d., 2)

Figure 3.7. View from Taliligap of the rising Vulcan Island.
Mrs Sarah Chinnery took this photograph from Taliligap northwards to Vulcan Island about 45 minutes before the beginning of the Vulcan eruption on the afternoon of Saturday 29 May 1937. Patches of newly raised reef are seen clearly around most of the north-western and eastern shores of the island. Tovanumbatir and Dawapia Rocks are in the distance. Published courtesy of the Chinnery family and the National Library of Australia. GA negative reference GB3292.

The boat party managed to reach the mainland, splashing on foot through shallow water to the shore. Mr Chinnery grabbed his wife's camera in one hand, his wife in the other and, with one of the engineers holding Mrs Chinnery's free hand, they all three raced with some of the others for the Rabaul–Kokopo Road and eventually reached the car. Flight back to Rabaul was now impossible by the coast road, for the eruption cloud had, within a few minutes, already risen thousands of metres and was dumping boulders, pumice and ash to the north-west over the road and beyond. Instead, they drove southwards to Vunapope. Mrs Chinnery stayed with the sisters while her husband attempted to reach Rabaul by tracks west of the caldera, but the attempt was futile because of blocked roads ahead.

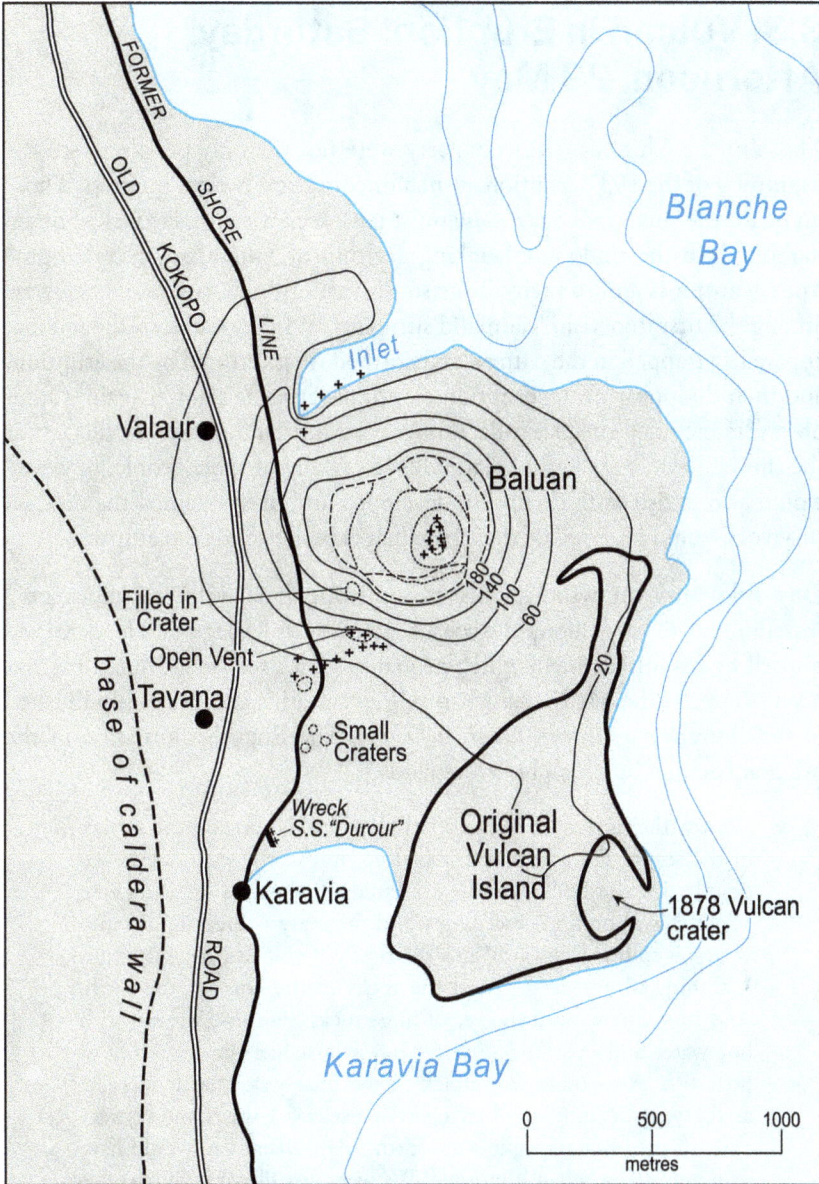

Figure 3.8. Map of the 1878 Vulcan Island and the 1937 Vulcan cone.

The old shorelines of the mainland and of the former Vulcan Island are shown in this map by the thicker lines. They are shown in relation to the new volcanic cone — called here 'Baluan' — that grew out of the sea to the north-west of the island on 29 May 1937. The contours for the cone are in 40-feet intervals. The map is a detail adapted from a large chart published by Fisher (1939a, Plate A1). The pre-eruption positions of the villages of Valaur, Tavana and Karavia have been added and are only approximate. Death tolls of 186 and 184, respectively, for Valaur and Tavana were the highest of all the settlements impacted by the 1937 eruption. North is to the top.

3.3. Vulcan in Eruption: Saturday Afternoon, 29 May

Those in the Chinnery–Murray party were not the only people to see the beginning of the 1937 eruption at such uncomfortably close quarters. Those on board the small sailboat *Kavivi* must have seen it too as she tacked down harbour into the trade wind, taking a group of Vunamami people home from Matupit Island to their village south-east of Raluana Point. They were among the first to perish. Mainland survivors said they had seen the sailboat apparently trapped in the pattern of disturbed waves created by the eruption, and then disappear in the eruption cloud, never to be sighted again. Other observers included the fish collectors on Vulcan Island, as well as villagers at the shoreline at Tavana and Valaur villages. Many of these people, however, would also perish as the eruption developed and overwhelmed them. Tolai survivors would later tell of their personal experiences of the eruption.

One Tolai survivor wrote of his escape from the developing disaster by travelling northwards along the coast road towards Malaguna. He identified himself by his initials I.M., which probably stands for Isikel Mulas. He was head printer at the Methodist Mission Press and his account was published in the church magazine *A Nilai ra Dowot*. The English translation of the original Kuanua in which he wrote reads:

> We saw the signs of it first as a bubbling and boiling on the surface of the sea, at the part where it is deep, not in the shallows. It was exactly like the boiling of a great cauldron which is on a hot fire. When it was boiling exceedingly and the sea was breaking up into waves, it spouted upwards into the sky. When it had risen up thus, it thundered and it shattered the rocks in the crust of the earth. And so it threw out a shower of huge rocks high into the air, with hot water and dust and pumice. It was just like the scattering of little fish when other fish attack them. That was what it was like at the very beginning, when I saw it because I was standing very close to the place where it burst forth. Man! What you would have seen among us, when it erupted! We were just like fish which have been benumbed with poison. We were astonished and we were in confusion. (I.M. 1937, 2)

Lesley ToGe from Karavia No. 1 village was a young boy in 1937. In 1984 he recalled that, on 29 May 1937, he had paddled out with his brother and others to the newly exposed reef

> and filled up their canoe with fish. They were just about to turn back, when the volcano began to erupt from Ralebe, a place on the Valaur side of Rakaia … The big waves created by the initially submarine eruption capsized the brothers' canoe and they had to swim. (Neumann 1996, 22)

They tried to stay together and swim to the shore, eventually taking hold of a drifting canoe that had to be abandoned when the volcano began emitting ash and pumice. 'Terrifying whirlwinds [*kalivuvur* in Kuanua] lifted objects out of the water' and one of their group was 'virtually sucked out of the water … and did not survive' (Neumann 1996, 22). Such disturbances evidently were also experienced by other small vessels, including the doomed *Kavivi* sailboat.

William To Kavivi, who was at the beach at Vunamami near Vunapope Mission when the eruption occurred, provided an accurate description of the initial development of the eruption as seen from the south. He saw a column of white vapour wreathing up from the direction of Vulcan:

> First it shot up straight, like the smoke from the funnel of an old, coal-burning steamer. Then it spread out at the top like an umbrella. Then the top of it leaned over towards Latlat, and the cloud became black and spread all over the area … I could see rocks coming up through the smoke and shooting in all directions, like sparks flying everywhere. (Threlfall 2012, 215)

This is a good description of what volcanologists call a 'plinian' eruption (Figure 3.9). Plinian eruptions can be particularly powerful, reaching heights of tens of kilometres, well into the stratosphere where volcanic materials are spread by high-level global winds. The Vulcan eruption in 1937 was smaller, perhaps 'reaching a height of 25,000 feet [7,620 metres] or more' (Fisher 1939a, 20), or '26,000 feet' (almost 8,000 metres), the height measured from the Baining Mountains by a surveyor using triangulation methods (Joycey c. 1937). The term 'sub-plinian', then, can be used for the small Vulcan eruption cloud.

Figure 3.9. Plinian eruption at Vesuvius in 1822.

The term 'plinian' was first used in volcanological references for the great volcanic eruption at Vesuvius, Italy, in 79 CE. This somewhat idealistic illustration is of a similar plinian eruption at Vesuvius as sketched from Naples (in the foreground) during October 1822. The 'umbrella' shape consists of a central, rising stalk that is spread out horizontally by higher-altitude winds, together with lightning discharges and pumice ash fall to the leeward side of the column. Note how some ash falls out of the sides of the ascending column too, contributing to the growth of the volcano itself. The illustration was used as a frontispiece in a volcanology book by G.P. Scrope (1862). The plinian column at Vesuvius in 79 CE eventually collapsed, producing the famous Roman volcanic disasters at Pompeii and Herculaneum.

Father K. Schlüter had been standing on the verandah of the Fathers' House high up in Vunapope when, at about 4.30 pm, a colleague drew his attention

> to a pillar of cloud to the northwest, which was becoming enormous and resembled a vast cauliflower … We immediately alerted the Fr Procurator and drove with the Bishop and the Fr Superior in the direction of the eruption. A short way behind Raluana there was a clear view of the new cone. We had to stop there, as a hail of pumice in the form of minute balls was beginning. A terrible and beautiful sight lay before us. Pitch-black masses of ash were shooting up in all directions like great arrows to a height of over 1000 m, to fall slowly back again and build up a considerable mountain. Massive hot stones shot down like comets, leaving a white plume of smoke behind them and sending up great splashes in the sea. Lightning flashed continuously and thunder growled and cracked …

> We saw how the clouds of smoke and light ash, spiralling up slowly to a height of several thousand metres, were carried by the prevailing southeast wind mainly over the area between Karavia and Malagunan. Nothing could be seen of Rabaul as the eruption cloud blocked the view of the whole harbour. In spite of the heavy rain of ash it was possible to observe clearly constant new outbreaks of new masses of debris [floating pumice] in the direction of Matupit, so that it seemed as if the whole harbour of Rabaul would be cut off. Near to the newly-formed island we saw a white sail, which soon vanished again [presumably the *Kavivi*]. (MSC 1937; Arculus and Johnson 1981, 8)

Captain Eugene Olsen was on the starboard side of the lower-bridge deck on the *Golden Bear*, which was tied up at Toboi Wharf about 6 kilometres north of Vulcan Island, when he saw, at about 4.10 pm, a small white speck on the water near the island (Figure 3.10). The speck grew larger in convolutions, assuming the appearance of a large cotton ball, and when it began to rise Captain Olsen realised that a submarine volcano was in activity. Three or four minutes lapsed, and then

> it suddenly burst wide open, and with a tremendous roar sent a column of white steam and black lava thousands of feet into the air. Spreading and becoming larger it enveloped Volcano Island and, with the light southeast breeze blowing at the time, drifted in a northwesterly direction over the main land [i.e. to the south-west of the *Golden Bear*]. (Olsen n.d., 1)

JULY 24, 1937   THE ILLUSTRATED LONDON NEWS   151

A LOW ISLAND TRANSFORMED INTO A HIGH CRATER IN A FEW MINUTES.

THE ERUPTION WHICH DEVASTATED RABAUL VIEWED AT CLOSE QUARTERS.

THE FIRST OF A SERIES OF PHOTOGRAPHS TAKEN IN FIVE MINUTES—SHOWING THE SUDDENNESS OF THE CATASTROPHE AT RABAUL: VULCAN ISLAND BEGINNING TO ERUPT:— PHOTOGRAPHED FROM A STEAMER IN THE HARBOUR.

PUFFS OF STEAM AND SMOKE BEGINNING TO RISE FROM VULCAN ISLAND ACCOMPANIED BY A LOUD EXPLOSION.

STEAM AND SHOWERS OF PUMICE DUST RISING FROM VULCAN ISLAND, A LOW DARK BORDERING RABAUL HARBOUR.

fifty bodies have been recovered. Two Europeans and one Chinese were also killed. Recently it was learned that Dr. Stehn, head of the Netherlands Indies Volcanological Department, had been invited to investigate the question of whether the authorities would be justified in maintaining Rabaul as the capital of New Britain in view of the danger of volcanic eruptions and earthquakes. We give below the account of an eye-witness, who has supplied the photographs (reproduced on this page) of the beginning of the eruption. "Although much has been said, and many descriptions printed of the volcanic eruption in New Guinea, few people realise the rapidity with which this disaster occurred. The illustrations here will give some idea of the suddenness with which Vulcan Island in Rabaul harbour came into eruption. They were taken from [Continued on right.

VULCAN ISLAND, FORMERLY LEVEL AND INNOCUOUS, TURNING INTO A VOLCANO VOMITING PUMICE DUST AND ASH.

THE DISTURBANCE ON VULCAN ISLAND ASSUMES HUGE PROPORTIONS.

EMITTING A THICK COLUMN OF SMOKE AND DUST: VULCAN ISLAND BECOMES STILL MORE MENACING.

THE series of volcanic eruptions which broke out, near Rabaul, the capital and seat of Government of New Britain, at the end of May led to the evacuation of the town, as recorded with illustrations in our issue of July 3. The 5000 inhabitants, of whom 700 are whites, were moved quickly to Kokopo, a village some twenty miles down the coast. According to a recent statement by the New Guinea Administration, an exhaustive check of the population of Rabaul and the neighbouring villages has shown that 424 natives—adults and children of both sexes—are missing. Only [Continued above.

the deck of the 'Golden Bear,' which I was on at the time. We were anchored close to Vulcan Island and admiring the beauty of the scene when a terrific explosion occurred and Vulcan Island and the sea near by began to smoke and send out showers of pumice and ash. Our little ship was in a very precarious position. Anchors were quickly weighed, and we steamed out of the harbour, pumice and ash falling thickly. The pictures reproduced here were taken within a total space of less than five minutes—just time to turn the film from one number to another. On [Continued below.

THE EFFECTS OF THE ERUPTION AT VULCAN ISLAND BECOME APPARENT: A HIGH CRATER WHERE BEFORE WAS ONLY A LOW FLAT, SURROUNDED BY STEAMING WATER.

FIVE MINUTES AFTER THE FIRST PHOTOGRAPH WAS TAKEN: THE CURTAIN OF STEAM AND SMOKE FROM VULCAN ISLAND COMPLETELY BLOTTING OUT THE SKY.

May 29 Vulcan Island was little more than a mud bank. On June 2 it was a huge crater fully 1000 ft. high. Two mountains just behind Rabaul, known as the Mother and the Daughter, are also in eruption, sending up showers of pumice and ash, so the town is right in the middle of the disturbance. The pumice dust is the finest dust that one can imagine. It finds its way everywhere—it is even said, it will get under an apple-skin. All cars and machinery have been disabled owing to the dust penetrating everything, while a vile smell of sulphur fumes from the craters has made the place unbearable. The weight of the

pumice dust has broken all the leaves of the palms and other beautiful trees about Rabaul, and in many cases buildings have collapsed under it. Many coconut and rubber plantations have been completely ruined. One man who had a dairy farm close to the Mother crater was forced to leave when the crater began to erupt. On returning to the place a few days later he found it impossible to find even where his farm had been. The eruption in the sea had caused a tidal wave, while lava from the crater had covered parts of his farm to a depth of from twelve to fifteen feet."

Figure 3.10. Magazine photograph compilation of Vulcan in eruption in 1937.

This page from the *Illustrated London News* of 24 July 1937 (p. 151) includes the collection of photographs taken by a crewman on board the *Golden Bear* at Toboi Wharf minutes after the initial outburst from Vulcan at about 4.10 pm on 29 May 1937 (see also the collection of photographs in the 3 July 1937 issue of the same magazine, pp. 8, 9). Note the dark, spire-like ejection of ash in the upper-left photograph, which is the first of the series of seven shots taken over a five-minute period. The bottom left-hand photograph was apparently taken later and seems not to be part of the series. Courtesy of the *Illustrated London News* Picture Library.

Captain Olsen, as a precautionary measure—in case of a shift of the wind—gave word for the engines to be made ready so that he could move the vessel away from the wharf if necessary. The wind direction next changed to the southward, and, by 4.46 pm, a light rain of ash was falling and the engines were ready. Suddenly the ship was

> enveloped in complete darkness and a solid downpour of ashes, accompanied by a strong sulfurous odor. This, shutting out all visibility, forced me to give up the idea of moving the ship or [getting] underway, and not knowing what may happen, I phoned the engine room to kill the fires under the boilers, shut down the auxiliaries, and for all hands to come on deck. (Olsen n.d., 2)

The majority of the *Golden Bear* crew were in Rabaul for the baseball, but those still on board:

> left the ship under conditions hard to describe [and], crawling along the wharf on hands and knees, feeling our way along the car tracks, we finally reached the shore. There, some of us nearly overcome, we found it impossible to proceed any further, or to remain outside in the suffocating air. Fortunately for us the wharf office door was unlocked, and we all took shelter there against the heavy downpour of ashes. (Olsen n.d., 3)

This 'first phase' of the eruption did not last long—perhaps less than one hour—but it was distinctive in having the characteristics expected for explosive volcanic eruptions that begin from new vents on relatively shallow sea floors, as shown in several of the photographs taken at the time. The technical name for such eruptions is 'phreatomagmatic' because they involve the interaction of hot pumice and volcanic ash with cold seawater that vaporises explosively. The interactions are so violent that the eruption vent can enlarge significantly, expelling large rocks and allowing more water to enter the sea floor vent, thus causing yet more water-rich pyroclastic material to be emitted. The eruption column itself then widens, momentarily losing upwards momentum. This causes erupted pumice and ash to collapse onto the sea surface, its lateral movement radially outwards from the vent in some cases creating significant tsunamis. Eventually, however, and especially for shallow-water vents, the erupted pumice and ash build up around the vent and a new volcanic cone grows above sea level. The number of large tsunamis would decrease in these circumstances.

The precise time when the new cone first appeared at Vulcan is unknown, although by dusk—when observations became difficult anyway—the eruption seems to have started to become a 'normal' subaerial plinian one. The technical term 'phreatoplinian' has been used for the early stages of such water-affected eruptions (Walker 1981). The original depth of the young active Vulcan vent in 1937 is unknown, but it was probably only a few hundred metres from the old shoreline (Figure 3.8), in which case water depths may have been only 50–100 metres (see the bathymetric contours given by Fisher 1939a, Plate A1).

The mechanism of 'lateral movement radially outwards from the vent' referred to above corresponds to what volcanologists today would call 'pyroclastic flows'. This term was not used in any of the observer reports from 1937; nor was it used by Fisher (1939a) in his pioneering volcanological account of the 1937 eruption at Vulcan. 'Pyroclastic flows', however, are seen clearly in several of the photographs taken at the time (Figures 3.10–3.13). Further, an eyewitness of the eruption, Burns Philp employee Bernard Ryan, described clearly in a telephone interview in 1981 seeing the process of collapse of the Vulcan ash column and 'avalanches rolling down and outwards from the eruption centre' (Ryan 1980–84, 2). He also saw: 'Now and then "very large blocks" being flung out of the column.' Another eyewitness, Brett Hilder, wrote:

> The strangest thing we saw was a white mass coming over the edge of Vulcan's crater and tearing down the side like a Roman chariot in a cloud of dust. It met the sea in a cloud of hissing steam. (Hilder 1961, 57)

Figure 3.11. *Golden Bear* photograph of Vulcan in eruption.

This is an enlargement of one of the photographs taken by the crewman of the *Golden Bear* at Toboi Wharf to the north of Vulcan (see the middle image in the left-hand column of Figure 3.10). It was taken a few minutes after the beginning of the Vulcan eruption on 29 May 1937, and already the eruption column has grown to a considerable height. Note, however, that some of the cloud extends laterally out over the sea to the left (east). These are pyroclastic flows caused by the early collapse of parts of the column. GA negative references M2444-3-2 and M2447-1A.

Figure 3.12. Vulcan eruption as photographed from Rapindik.

This was the view south-westwards from Rapindik about 10 minutes after the start of the Vulcan eruption on 29 May 1937, and before the short-lived fallout of ash on Rabaul and Rapindik, in the early evening of the same day. This timing is given by medical assistant Douglas C. Joycey (1981). The photographer, however, was fellow medical assistant Roger Davies (Joycey c. 1937), who took a series of excellent photographs from Rapindik showing the development of the Vulcan eruption (see also Figure 3.19). The eruption column is still developing in this shot, and its top has not yet filled the camera's field of view. The eruption is still at the 'phreatoplinian' stage, its base having expanded laterally to produce pyroclastic flows (see also Figure 3.11). A print of this image is in the photograph album donated to the National Library of Australia by Dr H. Champion Hosking (see also GA negative reference G2007).

Figure 3.13. Pyroclastic flows from Vulcan as seen from Rabaul.
Two kinds of volcanic 'flows' are seen moving laterally outwards from the base of the Vulcan eruption column in this tilted photograph that was taken, almost certainly, on 29 May 1937 from the wharves at the north-eastern corner of Simpson Harbour. The 'old burnt wharf' is behind the one with the man on it. A light-coloured cloud on the right moving to the north appears to be less dense than the one moving to the south on the extreme left. GA negative references M2447-3-4 and M2447-41A.

Meanwhile, Miss Chaseling had married Mr Trevitt and was beginning her new life in unusual style (Figure 3.14). Since disembarking the *Montoro* with her friends Hazel and Albert Jones, she had been staying at Kabakada over Tunnel Hill Road, on the north coast. Rev. Jones and his wife 'went to no end with preparations', and Albert made a bouquet of frangipani for the grateful bride-to-be (Trevitt 1937, 1; H.L. Jones 1937). The car arrived just about on time at 4 pm at Rabaul Methodist Church, which was 'beautifully decorated with frangipani and maiden-hair fern—it was a lovely picture of good taste'. Two earthquakes shook the church, one just as Mr Trevitt was saying 'I will'. Miss Chaseling, now Mrs Trevitt, recalled the ceremony and its aftermath in a letter to a friend in Australia:

> Hazel was scared bowls of flowers at the back of the minister, Rev. Lewis, would fall down on top of him—the whole place shook and trembled. I thoroughly enjoyed it all & wasn't in the least bit nervous & just managed to keep back the giggles. While we were in the vestry signing up there was a most terrific explosion which shook everything rather drastically. We just thought it was another—

worse—'quake but after we got away from the front steps of the church we saw a most marvellous sight—huge, thick dense clouds of smoke coming from the sea or an island in the harbour. Gosh it was a sight! Our car started off first for Kabakada where the breakfast was to be—but ran a little way along the coast first to see the sight. I can't describe the hugeness and denseness of the clouds which were circling up to the sky and over towards the land. (Trevitt 1937, 2)

Hazel Jones herself was no less impressed by the eruption at the time of the Trevitts' wedding:

The sight was—well I can't find a word, but out from the sea was a gigantic column of white, grey, thick cloudy smoke, which seemed half as wide as the harbour, and hundreds of feet high. The heavy white clouds rose and twisted and curled higher and higher straight from the sea. Everyone was interested and thrilled and yet awed and already the road was dotted with natives hurrying away, and cars leaving for our road as quickly as possible.

The wind began to carry the smoke over our way, so we left and turned up Tunnel Hill and made for home [at Kabakada]. It was beginning to darken and already the pumice and stones were beginning to fall on our garden and yet we still intended to carry on [with the wedding reception]. Lots of our Kabakada natives came hurrying across with their children and babies and came to our verandah. Mrs. Pearce spoke to them and I sent them to the garage. (H.L. Jones 1937, 2)

People elsewhere in Rabaul, and at other viewpoints not yet affected by the fallout, were similarly awe-struck by the great convoluting column of vapour and black ash that was rising from the sea north of Vulcan Island. Bernard Ryan of the Burns Philp office in Rabaul was at the baseball ground where the games had been interrupted by earthquakes during the afternoon (Ryan 1980–84, Sleeve 49). He had seen 'waves' running roughly north-westwards through the kunai grass outside the ground, and towards 4.30 pm was surprised to see excitement among the Tolai spectators, who began streaming down towards the harbour before the game had ended. Office worker H.E. Burgess was also watching, and noted that 'about three minutes after, the man on second base had the whole field to himself. A car came full speed past the sportsground, with the horn working overtime' (Burgess 1937, 1). People hurried to the harbour foreshore and found, says Mr Burgess, 'a spectacle so colossal that it held us all spellbound'. Mr Ryan likewise recalled vividly, more than 40 years later, the sight of a brilliantly white vapour column of ash that towered upwards against a cloudless sky.

Figure 3.14. Trevitt wedding photographed at Rabaul.
Jack and Melville Trevitt pose for photographs on the steps of the Rabaul Methodist Church after their wedding on the afternoon of Saturday 29 May 1937, just after Vulcan was beginning its nearby activity. On the steps behind them are, from left to right, Mrs Lewis (wife of Rev. Lewis), Rev. Howard Pearson (best man), and Mrs Hazel Jones (matron of honour). The cameras of some wedding guests that were used to take shots like this one subsequently became inoperable because of the entry of volcanic dust and damage to shutter mechanisms. Published courtesy of Mrs Melville Walker. GA negative reference GB3294.

Father Franz Utsch Jnr had just come out of the confessional at about 4 pm at Kabaira on Ataliklikun Bay well to the west of Rabaul. He looked up at the sky for a sign of good weather for the Corpus Christi procession on the next day, but saw instead

> a peculiar cloud … It stood there like a mushroom, or rather it grew from second to second into something gigantic. It billowed and rolled forward. The lightning flashed and thunder rolled from it in a terrifying way. At the top it was shining white, the stem of the mushroom was dirty yellow. The thing advanced at tremendous speed. The thought flashed through my mind: if that is a waterspout and it falls on us then we can say goodbye to the procession for everything will be washed away. (MSC 1937; Arculus and Johnson 1981, 33)

Figure 3.15. Fully developed eruption column from Vulcan as seen from Kokopo.

Vulcan is seen in full plinian-like eruption in this impressive view from Kokopo on 29 or 30 May 1937. A densely laden eruption column rises in convolutions, darkening the sky and dumping its pumice load over the land mainly to the north-west. The ship in the foreground is the old coal hulk *Loch Katrine*, which had been towed out to Kokopo from Rabaul for use as a breakwater for a swimming pool, as shown here. This image is a copy of a print in the photograph album donated to the National Library of Australia by Dr H. Champion Hosking. GA negative references M244-5-3 and M2447-38A.

Another Sacred Heart priest, Father Mayrhofer, who was at Lamingi in the southern Baining Mountains more than 50 kilometres from Rabaul, had a spectacular view of the whole northern part of the Gazelle Peninsula, and of the

> gigantic pillar of cloud which was rising up in the direction of Rabaul. It grew so rapidly to an enormous height—certainly 4 to 5000 m—and showed such a whirling movement on the lower side that I immediately thought of a volcanic eruption ... The edges of the cloud shone white in the sun, but the interior was pitch-black. It consisted of a massive column of steam, mud and stones, amongst them rocks as big as a house. There was at the time a fairly strong southeast wind and so the rain of stones, ash and mud was driven northwest ... At the beginning of the eruption a dreadful thunderstorm developed in the cloud, that was more intense than any I have ever experienced ... Even here in Laminqi we were dazzled by the weird light of the lightning. At the same time whirlwinds developed in the affected area, and these, together with lightning, stonefalls and mud rain, caused the most damage. (MSC 1937; Arculus and Johnson 1981, 36–7)

Few of these observers who saw the rapid ascent of the Vulcan eruption column from different vantage points at the beginning of the eruptive activity had opportunity to reflect fully on the implications of the spectacle before them (Figure 3.15). The sight to many of them was so unbelievable that, from their grandstand viewpoints, they must have felt detached from the happenings that were developing before their eyes. Yet few of them would escape the consequences of the eruption, for in the hours ahead Rabaul township would be evacuated, the MSC missionaries at Vunapope would be called on to look after thousands of refugees and hundreds of Tolai would die in the west.

3.4. Fallout from the Vulcan Cloud and Evacuations to the North Coast

'Dear Lord', cried Father Nollen at the Malagunan Catholic Church, 'if you go on roaring and groaning like this I can't hear confessions any more!' (MSC 1937; Arculus and Johnson 1981, 38–9). The priest's somewhat lighthearted reaction to the beginning of the Vulcan eruption would soon give way to fear for his life, as volcanic fallout from the eruption carried to Malagunan:

The people rushed into the church to pray. For a moment I debated whether to summon them to flee to Rabaul or Volavolo. But the black wall came on so fast, that it was no use to think of flight. The cloud banks reached Malagunan in ten minutes … A few seconds more and there was Egyptian darkness. Then the rain of ash began. Our church has no glass windows. The openings are just filled in with criss-cross laths. Thus the ash came into the church from all sides unhindered. The lamps and candles, which were lit, could hardly be seen. We were all covered and had to shake the ashes all the time from our heads and shoulders. Meanwhile the people prayed very loudly and urgently for help.

The earthquakes had ceased when the eruption occurred, but the ground trembled continuously and moved all the time like flowing lava. Powerful electric discharges followed one after the other, and the sharp claps of thunder made the ground and the church vibrate. Then came a rain of hot ash, so thick that we could not see a gleam of lamplight. I thought 'This is the end', felt my way through the darkness to the front, and asked the people, as loudly as I could, to beg for God's forgiveness and to vow to fulfil His commands better, if we should escape with our lives. I led an Act of Contrition and gave General Absolution …

The people never ceased to pray aloud for help. But at times the ash was so thick that they could do no more. Fortunately the rain of ash ceased and the weather became calmer; but after a few minutes there came fresh masses of ash with thunder and lightning: however, the ash was no longer hot. During one of the pauses, I opened the door a little and saw a faint glimmer in the air where the moon ought to be rising. It may have been half past eight or nine. Then some refugees arrived and wanted to get into the church. First of all they had to clear away ash with their hands in order to be able to open the door. Meantime there was enough faint light for me to find the way to my house, where I got a shovel with which we cleared the thick layer of ash from the door. More and more refugees came from the direction of Karavia and reported that several people had been killed by collapsing huts and copra stores. Then ash and rain fell together and we had to go into the church again.

Trees bent under the heavy layer of ash, coconut palms snapped off and shattered what lay beneath them. The broad branches of the mango tree crashed down. There was a hellish noise, for the thunder was also still rolling. In addition there was now a great flood of water over the land, which was presumably due to the sea being driven far inland by the eruption. About 11 o'clock there was a slight pause, and the rain of ash ceased as the wind had gone round. The Sisters were now able to take the children home to rest for a time. The natives did not want to leave the church. About half past two the weather became dangerous again. There was further rain of ash and mud. The Sisters had to leave their house hurriedly as it was cracking and swaying on all sides. The verandahs collapsed. We all went again into the church, which offered the best refuge. (MSC 1937; Arculus and Johnson 1981, 29–31)

Methodist district headquarters was at Malakuna. After officiating at the Trevitt–Chaseling wedding, Rev. F.G. Lewis, district chairman, was anxious to return there, despite the ominous north-westerly growth of the Vulcan eruption cloud and 'though we lost much valuable time in doing so' wrote Rev. Albert Jones later (A.S. Jones 1937a, 2). Lewis was driven to Malakuna, but Jones was eager to take the women passengers who were also on board his utility vehicle, to greater safety at Kabakada where the Trevitt–Chaseling reception was to be held. So he left Rev. Lewis and three others at the mission residency and drove back up Tunnel Hill Road. An exodus was taking place from Rabaul to the north coast, and Tunnel Hill Road was already

blocked with natives making their way out, and a huge line of cars. It was a terrible experience to see natives we knew, but could not take with us as the car was full. (A.S. Jones 1937a, 2)

Meanwhile, Rev. Lewis at Malakuna saw the eruption column become denser:

[F]rom the black rising columns there were shooting out in all directions stones and cinders, like multitudinous rockets or comets, and then this great moving, towering mass was seen to be making fast towards us as if the Malakuna Mission Station was in the direct line of its objective. (Lewis 1937a, 6)

Rev. Lewis decided to abandon the residency and began racing on foot with the others along the road northwards to Tunnel Hill. But the cloud soon reached them, and they were enveloped in complete darkness as pumice ash showered down, irritating their eyes and penetrating into ears, hair and pockets of clothing. Visibility improved at the top of Tunnel Hill Road. Several cars sped dangerously down the hill. Lewis and his colleagues were then relieved to see ahead the district car coming in from Kabakada to pick them up. But progress back out to the north coast was impeded, again by the cloud of darkness, as well as by fallen palm trees across the road and, finally, by rain, lightning and thunder (A.S. Jones 1937a, 1937b). The Lewis party eventually joined a group of stranded refugees at the nearby home of planter J.O. Smith whose house at Vunawutung was refuge for several groups of motorists that night (Lewis 1937a). Another planter who took in refugees further along the north-coast road, at Natava Plantation, was the Honourable John C. Mullaly, who organised the feeding of large numbers of homeless people from the areas devastated by the Vulcan eruption (McNicoll 1937e).

Newlyweds Jack and Melville Trevitt had driven up over Tunnel Hill Road soon after leaving the wedding and seeing Vulcan, for they were expected back at Kabakada for their reception (Trevitt 1937; H.L. Jones 1937). Hazel Jones had prepared three tables on the back verandah; set with crystal, silver and china, and edged with ferns, the tables were enhanced with bowls of salmon-pink lilies. There were salads and meats, fruits in jelly and a fruit salad that nearly filled a wash dish. Fresh cream had been whipped, all ready to be put straight onto plates. The weather had been glorious and the view out across the Bismarck Sea was delightful. Jack and Melville were on the front lawn having their photographs taken and waiting for the guests to arrive when the ash fall from Vulcan began. Showers of pumice drove them inside, and soon after a frantic Rev. Laurie Linggood arrived in his truck, urging them to board immediately and head inland to the hills: 'Rabaul is blotted out with smoke and utter blackness and it's coming this way— have already tried two roads to the Hills ... but ran into walls of blackness' (Trevitt 1937, 2). Linggood's pregnant wife had been unwell and had not attended the wedding, and now he was cut off from her and their small son at Raluana. The bride and Hazel changed their clothes at once. Albert Jones had arrived with his load of ladies and, quickly assessing the situation, began to load up his utility again, this time with two cases into which he piled the wedding cake and other food from the tables (Trevitt 1937; A.S Jones 1937a, 1937b). His wife's work and planning for the reception had almost been for nothing!

Melville Trevitt scrambled into the front of Linggood's truck, alongside Linggood and Mrs Pearce, who had been helping with the reception, and Jack Trevitt climbed into the back with the suitcases of clothes packed for the honeymoon in the hills. The other ladies urged them on, but Linggood needed no encouragement—he 'drove like mad fury in his anxiety to get to his wife'. However, the road inland was impassable, and they decided to attempt the drive around the north-coast road to Vunairima, which was to be Melville Trevitt's new home. Mrs Trevitt would later write:

> The smoke and sulphur fumes and falling mud and stones were blinding … the thick cloud of smoke impenetrable … Native people, panic-stricken, were fleeing in hundreds, down from the hills, not knowing where to go, and cars and lorries impeded our progress.
>
> We were nearly blind for we had to open the windscreen in an effort to follow the road. Mud was falling thickly and lay in inches over the truck. It was pitch black and the man just drove furiously while we shut our smarting mud-filled eyes. I kept my hand on the horn, sounding it at short intervals for the safety of those fleeing, stricken people. We grazed past lorries and cars and ran into trees and gutters and finally were ahead—Jack continually getting out and scraping away mud from the headlights so we could go a little further. No words could describe the thunderous deafening roars like terrific explosions and the lightning—great streaks which seemed to reach from the sky to the ground. We were travelling through a coconut plantation and could see, in the flashes of light, coconut palms bending under their weight of mud. Soon they began falling with great crashing thuds and behind us the trees on either side of the road were on fire.
>
> We suddenly came to an abrupt halt. The men, by feeling, found a tree across the road and were able to move it but very soon we were stopped by another palm, sixty feet or so, lying across the road. There was nothing we could do but sit in that truck and wait— possibly until morning or until the end! … We … knew (but didn't voice it) that, at any moment, coconut palms could crash over us, and we knew too there was a danger of a tidal wave. In the lightning flashes we could see the water only a few feet from where we were stuck.
>
> We sat in that truck for approximately one and a half hours—each with his own private thoughts … I, feeling so thankful that my husband was beside me, experienced no feelings of fear whatsoever— Mrs Pearce not knowing where her husband was—and Laurie Linggood with anxiety for his wife and child and the one shortly to be born. (Trevitt 1937, 3–4)

Figure 3.16. Vulcan ash on car and coconut-tree damage.

This car was one of many abandoned along the roadside during the fallout of Vulcan pumice and rain on Saturday 29 May. Plantation coconut trees in the background have been damaged by the same fallout. GA negative references M2444-2-5 and M2447-3SA.

A truck driven by a New Guinean eventually stopped a short distance behind the Linggood truck. Some of the Kabakada wedding guests were on board, including the Jones's as well as Ron and Helen Wayne, Methodist lay workers, who had come over from the Duke of York Islands for the Trevitt–Chaseling wedding and reception. Those on the truck had left the abandoned reception at different times and, fortunately, had been picked up by Vincent To Papa in his truck along the road at various points where their own cars could not negotiate obstacles. The party also included Mr and Mrs Atherton and their 10-month-old baby. Trevitt and Linggood struck out to nearby Kabaira, returned with borrowed axes and cleared the fallen palm trees, but a mile further on yet another tree blocked the road. All then began to negotiate the last few kilometres to Vunairima on foot, scrambling, Mrs Trevitt recalled, 'over and under logs and stumbling along often wondering if we were still on the road or somewhere in a plantation' (Trevitt 1937, 5). Ron Wayne's diary of these events provides a valuable source of information about the movements and experiences of European Methodists along the north coast (Wayne 1937), particularly where supplemented with information from other sources (Figure 3.16).

Ron Wayne called the walk

> a nightmare, especially for Mrs. Atherton and her baby. Mr. Atherton
> being in like plight as all the men, his wife was frightened to let
> him carry their girlie lest she catch cold. Thanks to [an] umbrella
> the child was still more or less dry. Trees, palms, and fronds were
> tripping us every few yards. Often we stopped while some tree that
> cracked ahead of us decided what to do. Once or twice we thought
> palms were coming down on top of us. At last we reached Vunairima
> [at 8.35 pm] and went to the Sister's [sic] House, whence Linggood
> had gone ahead to warn Miss Mills of our imminent arrival ...

> When we entered the Sisters' big living room what a sight we proved
> to be! Nice new marocain and georgette frocks had shrunk upwards
> and inwards! Hats were beyond description. Trevitt's helmet was
> three times its normal weight with pumice that had adhered to
> it, but his buttonhole was still in place. A source of considerable
> comment and amusement, this being the only sign that he was a
> brand new bride-groom ... Those of us who lacked hats were in
> the worst plight because our hair was plastered as with cement, and
> much mud had stuck to our necks. (Wayne 1937, 5)

Their appearances were so comical that Hazel Jones said they 'just roared'
with laughter (H.L. Jones 1937, 4). Tensions and anxiety were temporarily
relieved. The party dispersed for the night after hot soup and quinine, the
bride sharing the guest room in the crowded Trevitt house with Hazel.
Yet, recalled Mrs Trevitt:

> Sleep was out of the question for us although we were dog tired—for
> the crashing thunderous explosions were deafening and lightning
> circled around all night and the house shook and trembled. I lay on
> the edge of my bed on the alert, ready to escape should the house
> begin to tumble. (Trevitt 1937, 5)

The Methodist mission station reception back along the north coast at
Kabakada was deserted, and Mrs Lulu Miller, of Samoan heritage, had been
having a worrying time (Miller 1980). She had hurried from the plantation
at Kabakada at the beginning of the Vulcan fallout to her home to rescue
her dogs, cockatoo and goura pigeon, but was enveloped in darkness, and
spent the terrifying evening in her small bath hut huddled together with her
pets and a young New Guinean boy. Despite this company, she felt lonely
and isolated in the deserted bush among ash-laden fallen trees and bamboo.
Late in the evening, she left the hut and went to the shed where a village
headman kept a hot-air copra drier, and 'during the night we had rain—was

salty the water. Came down here like a stream' (Miller 1980, 35). Seawater was mixed with the pumice dust and rain had evidently been thrown up from the harbour during the eruption.

April and May had been particularly dry months. The rain that fell on Saturday over and north-westwards of the new volcano might have been regarded as a welcome boost for water tanks had it not been for the pumice ash and salt water that was mixed with it. The rain fell mainly from beneath the eruption canopy, affecting neither Rabaul nor other places outside the ash fallout zone. Most of the rain probably formed from condensation of volcanic water vapour in the upper parts of the Vulcan eruption cloud in the higher and cooler parts of the atmosphere, and from seawater—at least some of it vaporised—that had been caught up in the rising column of ash and carried over the hills to the north-west. It all fell as a deluge, including over the volcano itself (see, e.g. Figure 3.35). Torrential rain was common during Rabaul's wet season, but its force was generally lessened by the dense vegetation that allowed the water to percolate through foliage and grasses into porous humus and soil. But this protective natural screen had been destroyed by the pumice fallout, and the deluges of rain and seawater formed floods that raced across exposed surfaces into dry gullies, gouging out deep ravines. Raging torrents quickly built up and rushed down to the coast, washing away the north-coast road in places. These floods added considerably to the difficulties of using the north-coast road as an evacuation route out of Rabaul.

Father Utsch's afternoon fears of a 'waterspout' ruining his Corpus Christi procession at Kabaira on Ataliklikun Bay were, therefore, somewhat justified during Saturday night, as the torrential rain contributed to the volcanic devastation (MSC 1937; Arculus and Johnson 1981, 33). The 'storm' quickly reached Kabaira during the late afternoon, and stones that sounded like hail, rather than rain, clattered on the roof of the church occupied by Father Utsch:

> People put their cloths over their heads and staggered in under my house. Amongst the stones was dust and flakes like cement. In an instant all was dark; only towards Mandres [to the west] was the sky clear. We were all terrified. We could not flee; as there were only a few canoes and they would be sunk by the weight of this rain. So we stayed where we had a solid roof over our heads. Trees were already crashing down. Deadly fear seized us all. It got even darker. I thought of Pompeii. Lord God, save us from such a fate! In the church all started praying as never before. I sat in the confessional.

Wesleyans were praying in one corner. I lit a candle for them too. Then cars began to arrive. We now learnt that the volcano near Karavia had erupted.

The whites were fleeing from Rabaul to the north coast. My house was soon full. They could go no further. One could hardly see one's hand before one's face. Constantly we could hear the noise of trees breaking off. Then a terrible cry, which echoed through the church in spite of the frightful thunder. The roof had fallen in. I could not see what was happening but it was impossible to keep the people in the church any longer. I directed them into the school and under my house. Then I discovered that only the verandah round the church had collapsed. In the house I had to prepare some food for the children. I looked like a miner, muddy from head to foot. I brought my horse into shelter. The poor animal was trembling violently. But I could not let it loose; it had only been two days on the station. The night was everything but quiet. I threw myself onto my bed exhausted. But I started up involuntarily at the next clap of thunder and lightning. It might have struck us, the whole house rattled so much. And all night long this noise, now softer, now louder. The cement still kept on falling. (MSC 1937; Arculus and Johnson 1981, 33–4)

This same Catholic Church along the north-coast road may have been the refuge for staff from the Bank of New South Wales and Commonwealth Bank. The bankers had been to visit a 'sing-sing' in the hills, had heard the 'violent roar [of] the new volcano', had seen the eruption column—'a fascinating yet eerie experience'—and had been unable to return to Rabaul by the north-coast road (King 1937, 5). Kerevat was chosen as a more appropriate destination but, like others on this evacuation route along the north coast, they were caught in the fallout zone. Hundreds of confused New Guineans sheltered in the church—'seventeen miles out'—with the bankers. To raise their spirits, Virgil King

thought the best thing to do was to hurry and get them singing. What an accompaniment they had—the electric grand organ of hell, together with the fire and brimstone—His Satanic Majesty was surely reigning. (King 1937, 5)

Other Rabaul evacuees who had earlier taken the north-coast road were enjoying the safety of Keravat, which lay well to the south-west of the fallout area. One of these was Miss Carol Coleman (Mason 1937). She had left the baseball game with a group who had piled into Alf Dowsett's open tourer car and headed out over Tunnel Hill Road for the north coast. They encountered the same unpleasant fall of ash and mud as had the others on the road but managed to escape from beneath the volcanic canopy before the road became impassable. Lightning and thunder nevertheless disturbed their night at the plantation at Keravat run by Mr and Mrs Green.

The frightening experiences of all these people on the north-coast road must have been insignificant compared with the terror of those who died within the maximum zone of devastation, where pumice was finally to accumulate to depths of more than 2 metres. Many Tolai saved themselves by flight away from the area of maximum fallout, but hundreds were caught there. Among those who escaped was Father Laufer of Rakunai Mission, only about 4 kilometres upslope and west of Vulcan Island. He had, like his colleagues fathers Nollen and Utsch, been taking confessions in church, but was soon 'in the midst of the "Hell"':

> With the women and children who were standing nearest I tried to run through to Navunaram. But stones started to fall after only ten minutes. First we crept into some holes in the road embankment, but nearly squeezed ourselves to death there. So with the men and boys onwards to To Puia's wooden house [To Puia was a catechist]. Violent drumming on the roof. Still bearable. About 50 people crowded into the tambolo (the space under the house, which stands on piles). Then suddenly a wall as black as shoe-polish. Dreadful fear. Everyone clung to me. Repentance and general absolution. Then the flood of mud rained down. Continuous lightning with whirlwinds that drove the mud under the tambolo. We lay flat on the ground, beside and on each other, held sheets of tin and corrugated iron over our heads for shelter, tying our shirts round eyes, nose and mouth, damping them with mud against the sulphur.

> About ten minutes, then the sky became a bit red. By the lightning flashes we could see that all the trees and coconut palms were snapped off and lying on the ground. Without our noticing, two coconut palms had fallen right across our house and broken through the roof, but had held the structure firm in the wind. We ate first a little coconut, to get the sand and mud out of our mouths. Until ten or half past we still sat together and prayed as well as we could. In the east the moon and stars gradually appeared. By the gleam of lightning we set out and felt our way through the uprooted bush as far as Navunaram corner. Then clear road and steady march to the school at Navunaram, where I found Sister Carola with some girls whom I sent to Vunakanao. I tore on with my people to Vunalama.

We arrived there at midnight. It was not possible to sleep on account of the frightful thunder and lightning. (MSC 1937; Arculus and Johnson 1981, 26–7)

An attempt was made 40 years ago to illustrate in simplified form the possible nature of the eruption cloud that had been experienced by the missionaries up on the plateau west of both Vulcan and the intervening caldera wall (Figure 3.17). The missionaries' descriptions given above contain some features relating to passage of a pyroclastic flow, but this would require the flow to climb the caldera wall before levelling out and continuing westwards. In Figure 3.17, the advancing cloud is shown trapped between the low eruption cloud and the plateau west of Vulcan and escaping laterally to the west.

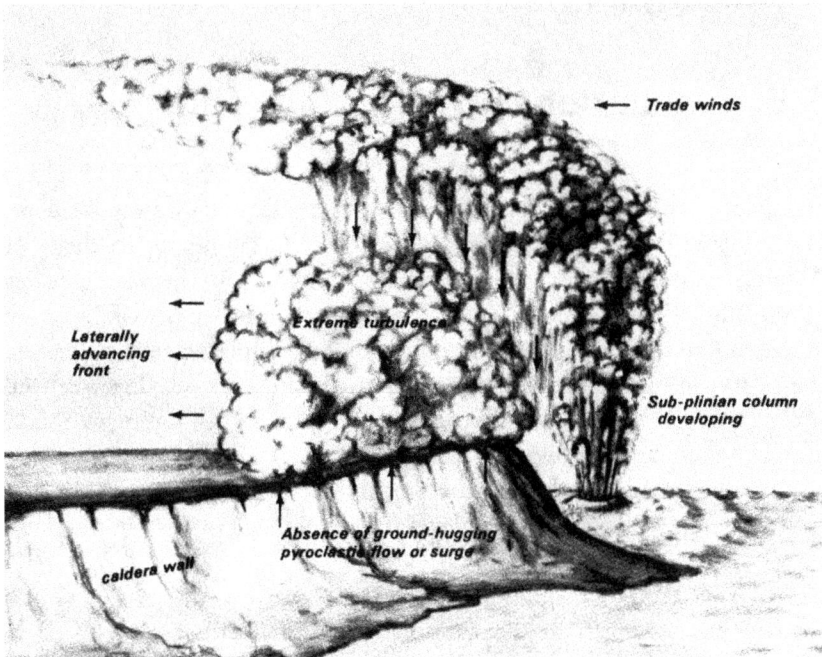

Figure 3.17. Sketch of components of Vulcan eruption cloud.

Schematic representation of the Vulcan eruption cloud from the south-east at about 4.30–5 pm on 29 May 1937 showing the formation of turbulent, laterally advancing ash clouds (McKee et al. 1985, Figure 11). The main elements of the eruption have been artificially exaggerated and separated so that they can be identified more clearly. This interpretation does not mean that the lighter parts of otherwise dense pyroclastic flows (which are not shown here) could not have drifted westwards over the caldera wall and contributed to the laterally advancing cloud. The sketch was drafted in the Bureau of Mineral Resources, Geology and Geophysics Drawing Office (GA reference number 24/B56-2/23).

The hundreds of deaths in the villages closest to Vulcan were probably caused mainly by suffocation as pumice falls rarefied and heated the air, by crushing and suffocation as roofs fell on the occupants of lightly built shelters, and possibly by burning of lungs and suffocation as hot avalanches from the base of the eruption column overran helpless victims. Some people were probably struck down by flying blocks and boulders, and some of those who sought to escape Vulcan Island and the new islets by swimming to shore may have drowned. There may also have been fatal heart attacks. The body count would amount to only about a tenth of those missing. Most corpses would remain buried under the ashes.

It took many days for news of the devastation and casualties north-west of Vulcan to become known in Rabaul, for the townspeople had difficulties of their own to deal with—including a major evacuation of the town.

3.5. Evacuations up Namanula Hill

Clive Meares was chief clerk of the Department of District Services and Native Affairs; E.W.P. Chinnery was its director. Meares lived on Namanula Hill with his wife, close to Government House, the residence of Administrator McNicoll, who was still away on his inspection tour (Meares 1968, 1980, 1980–84, n.d.). Meares had inspected the stumps of his house at various times during the afternoon for any displacements caused by the earthquakes that had taken place that day, but was relaxing when he became aware of people from Rabaul arriving in their cars hoping to obtain a view of the rapidly developing scene at Vulcan Island. Meares grabbed his government-owned camera but, by the time he had focused it, the top of the billowing clouds over Vulcan were 'already too high to show in the view-finder' (Meares n.d., 3):

> It was an awesome spectacle. Belches of dense smoke were thrusting straight up into the sky while, simultaneously, what looked like curling, black waves of still thicker smoke were shooting out from the base of the 'pillar' at all angles like rockets. We could see what appeared to be huge rocks, varying in size from five-ton trucks to small houses, hurtling through the air and, apparently as they met the cooler atmosphere, disintegrating like bombs and falling back into the sea.

> Higher and higher wreathed the main column of smoke until, approaching an altitude of about 25,000 feet, it slowly began to spread, like a mushroom, over the harbour, the town and the

neighbouring country-side … the main body of it was steadily carried away to the north-west. In a matter of minutes some fifty to a hundred residents had joined us on Government House lawn, but as the 'mushroom' began to spread across the sky [towards Rabaul and Namanula] we moved back to our home, and the crowd broke up into groups as those of us living on 'the hill' invited various ones to stay with us. What the future had in store for us no one knew. (Meares n.d., 3–4)

W.B. Ball was acting superintendent in charge of police and, like Meares, was spending a restful afternoon at his home on Namanula Hill when an agitated servant from Government House telephoned to say that 'something was very wrong in the harbour' (Ball 1937, 1). Ball hurried to Government House and from the verandah there he saw the 'huge and sinister looking column of smoke coming out of the sea from the neighbourhood of Vulcan' (1). He telephoned for a car and set off for the police station, passing numerous carloads of people making their way up to Namanula Hill. Ball was met at the station by Acting Inspector W.B. Prior, and they made off with a driver towards Kokopo Road to investigate the column of 'smoke'. Progress was slowed by traffic making its way towards and up Tunnel Hill Road, but just past the turn-off they saw

coming towards us from the sea and along the Kokopo road, something that is very hard to describe; it was a dark brown almost black cloud, so dense that it looked more like a moving mass than a cloud. It has been called the 'black-out' and that is actually what it did, for immediately one was enveloped in it daylight suddenly became blackest night in which it was impossible to see anything at all, even the powerful headlights of a car failed to pierce it for more than a few inches as we found to our cost a few minutes later. Realizing that it would be futile and dangerous to go on, I ordered the driver to turn and go back to the station. He succeeded in turning the car and managed to drive a few yards along the road before we were enveloped in the cloud which by this time had overtaken us. A few yards more and the car landed in the ditch. It was impossible to see even a few inches. I said to Prior that we should get out and try to walk back. We did, and groping our way on the tarmac commenced to stumble along. It is hard to describe my feelings. The cloud seemed to be composed mostly of grit that filled my eyes and nose causing great discomfort and adding to the horror, but I remembered to keep my mouth shut and to refrain from rubbing my eyes. I observed that the blackness was as yet almost entirely free from poisonous fumes, but at the same time I feared it could only be a matter of moments before we would detect them and be overcome. It seemed a horrible way to die and I wondered if one could keep from panicking. All this time we were stumbling and groping our way, trying to keep on the road, but not always being successful. On one occasion I collided with the back of a lorry that was quite invisible. All of a sudden we emerged into daylight; it was just as though the blackness had been cut with a knife. (Ball 1937, 2–3)

Ball and Prior ran back along Malaguna Road but 'again ran into darkness and our gropings recommenced'. They finally boarded a Department of Public Health vehicle and managed to return to the police station, but 'there was very little that could be done at that time except to take shelter from the rain of dust and ashes that kept coming over the town' (Ball 1937, 3). The 'almost black cloud' noted by Ball that was moving northwards along the Kokopo road—'so dense that it looked more like a moving mass than a cloud'—was caused by large amounts of pumice and ash being dumped from the sides of the rising column that was now changing its shape to something more cauliflower-like. The falling masses then hit the surface of the land or sea and were diverted laterally along the ground as 'flows' of pyroclastic material having different densities (Figure 3.13).

Ash began falling on Rabaul a little after 5 pm, caused by a change in direction of the normally south-east winds. Rabaul people then began to experience some of the fear, uncertainty and disorientation that was being encountered to a much greater degree by those in the main fallout zone to the west. There was a 'slight smell of sulphur'—probably hydrogen sulphide, the volcanic gas known for its bad-egg smell (Fisher 1939a, 30). Amy and Mercer Anthony were at their home when the Rabaul fallout started, and they decided to stay in the town because there was no available transport (Anthony 1981). They joined their next-door neighbours. The weight of fallen pumice outside was breaking limbs from large trees, but the wind direction eventually changed, and the Anthonys were able to return to their own home, clambering over fallen branches in their garden that 'had been completely covered with pumice, and looked like the sea shore, with humps here and there where there had been shrubs or groups of flowers' (Anthony 1981, 3). Dry pumice had blown into bedrooms and they had to gently tip a layer of it from the top quilt of their bed. The Anthonys stayed the night in Rabaul, wakeful on account of the Vulcan thunder and lightning over the harbour, and by morning found that not only was their garden a 'wreck', but also they 'could see from our back verandah right through to the far end of the town, [that] all the Casuarina and Mango trees were just trunks'

(Anthony 1981, 3). They also discovered that many people had left Rabaul the previous evening and were now on Namanula Hill or down the other side on St Georges Channel at Nodup.

Some of the people who drove up to Namanula shortly after the beginning of the eruption obtained a spectacular view of the eruption before the fallout began on Rabaul. Gladys Forsyth and her husband, for example, 'gathered up a few things for the child … gathered up the cat and the dog, got into the car and set off to find [Gladys's] brother Adrian Field' (Forsyth n.d., 9), but, failing to do so, returned, left a message and set off for Namanula. Field was caught in some of the fallout, but he received his sister's message and, packing a haversack with a revolver and bottle of water, began walking up Namanula Hill (Field n.d.). Several cars passed him at high speed, and eventually he was offered a lift.

Bernard Ryan and two friends had also decided to drive up Namanula, but they were caught soon after the start of the climb by the fallout on Rabaul (Ryan 1980–84, Sleeve 49) and were enveloped by pitch blackness like night. Wet gobs of ash had at first spattered the car windscreen like bird droppings, but the cloud soon became so dense that headlights could not penetrate it, and fine ash entered the car, getting into eyes and lodging behind ears and in hair. They all got out of the car, but were still unable to see ahead. There was a hot, clammy, stifling sensation, as well as a distinctive, repetitive, swishing sound in the ash cloud. Handkerchiefs in front of their mouths helped prevent inhalation of ash. Ryan remembers that this was the most frightening part of the eruption for him, but fortunately the wind swung away to the north-west and they were able to drive again up onto Namanula Hill. Mr J. Hoogerwerff, manager of the Rabaul Printing Works, had a very similar experience to that of Ryan and his friends while climbing up Namanula Hill (Hoogerwerff 1937; Langdon 1973).

Hal Evans, a junior administration officer in the Rabaul District Office, became involved—with others—in ferrying evacuees out of Rabaul to Namanula Hill 'where nearly the whole [white] population had gathered' (Evans 1937, 2). Rabaul town, as seen down below the hill, became smothered in pumice and ash while he was there and

> about 15 minutes later it reached Namanula and completely enveloped the Hospital and surrounding country. Everyone shut themselves in the wards and everything became dark. The marvellous thing about it was that all the women and children were wonderfully calm, and though we all thought that it was just a matter of time before we were

suffocated, not one even screamed … However, the Pumice stopped
falling, and in time we were able to go outside with handkerchiefs
as masks and get water for the sick. (Evans 1937, 2–3)

A layer of Vulcan dust on a dining room table in Rabaul was found useful
at about 8 pm when Gordon Thomas, editor of the *Rabaul Times*, wrote
a message in it for his wife whom he had been unable to find (Thomas
1937b). Thomas had spent a stressful few hours searching for his wife,
driving, stumbling and losing himself in the volcanic 'blackout' near the
bottom of Tunnel Hill Road. Having climbed up over the hill towards the
north-coast road, he heard, at several spots, the eerie 'murmur of praying
natives'—in Kuanu, *'A kulou U na belaure aret, ma U na tulue mule pire aret
ne a kapa* ('O God, protect us and send us light again')—and 'the crooning
music of hymns' (Thomas 1937b, 37). Thomas, on one occasion, when the
air became charged with sickly warm 'sulphur fumes', believed it 'was the
beginning of the end' for him. However, he eventually reached Volavolo
Mission, and then met an oncoming car driven by the new proprietor of
Roberts' Garage; he accepted a lift back to Rabaul over Tunnel Hill in
conditions much easier than those of his outward trip, but he was still
anxious to know the whereabouts of his missing wife.

Jean and Keith McCarthy also had an eventful time. Mrs McCarthy had
been in Rabaul for only a week, and was newly wed to Keith, who was
waiting for a government posting as a patrol officer elsewhere in New
Guinea (McCarthy 1963, 1971a, 1971b, 1980). The McCarthys' wedding
gifts were still crated under the house and Mrs McCarthy's wedding dress
was hanging on the wardrobe door when the Vulcan eruption began. They
ran down to the wharf, joining the crowd there, but fled back to the house
when they saw the eruption cloud rolling towards Rabaul. Keith McCarthy
'grabbed a couple of towels with some idea that they might help us to
breathe while we made for safety', and joined the throng of people climbing
Namanula Hill,

> many wild eyed and very near panic, going at a jog trot as they
> hurried up the hill; there were whites and Chinese, the men carrying
> bundles and the women carrying babies; and there was a long line of
> cars and trucks. (McCarthy 1963, 173–4)

The McCarthys abandoned their climb because of the crowds and set out
to try a little used track through kunai grass across the caldera wall. Then:
'Here comes the first of the ash, Jean! Use your towel. It might be hot!'
(McCarthy 1963, 174). Fortunately, the ash did not affect them, and the

McCarthys changed direction again, resolving to weather out the pumice fall back in town. They reached Rabaul and joined a party of perhaps 30 Europeans at the Cosmopolitan Hotel.

Acting Superintendent Ball would subsequently have problems with some of the drinkers in the Cosmopolitan bar (Ball 1937). The general opinion there, recorded J.K. McCarthy, seemed to be: 'If this is our last night let's make it a good one!' (McCarthy 1963). Proprietor Morton Wilmot was kept busy serving drinks to customers, especially to many of the crew of the *Golden Bear*. H.E. Burgess was at the hotel but left at 8.45 pm with a flashlight to check on any damage to his office. This was normally a three-minute walk, but it took him 25 minutes because of ash and broken branches in the street, and because he had to be watchful for fallen electricity wires (Burgess 1937). Waiau Ahnon in 1953 recorded that he went to the Cosmopolitan Hotel and saw some enthusiasts

> still playing billiards on top of a dusty billiard table—and I saw somebody, too, being appointed as one of the volunteer police and all that, with the red band now, and he is carrying a revolver with him, and he is trying to tell people to get out of the pub, but no-one seems to pay any heed to him. He came inside then, and is still trying to tell people to get out. He even fired a shot through the window to tell them to get out, but nobody wants to move. They said, 'We'll move as soon as we feel like moving'. (Ahnon 1953, 39)

Acting Superintendent Ball did not detail this theatrical incident in his later report to the administrator, but he did decide to temporarily close the Cosmopolitan Hotel that night because of the 'very unsavoury crew of the Golden Bear' who were 'in an excited state' (Ball 1937, 4). The intoxicated sailors were ejected and driven to Toboi Wharf. One of the sailors, Victor M. Costner, radio operator, did not board the ship. He was never found and his disappearance was never fully accounted for, although Captain Olsen reported dispassionately that 'he must have walked overboard from the wharf and drowned' (Olsen n.d.). Costner was evidently not one of the drinkers at the Cosmopolitan, for Captain Olsen stated that Costner was on board the *Golden Bear* just before the remaining crew had had to leave the ship a little before 5 pm and crawl to the wharf office on hands and knees. Olsen's clinical account counteracts the belief expressed in some reports that Costner had been drinking too heavily at the Cosmopolitan and had foolishly attempted to reach the ship by either leaping across the gap between wharf and deck or walking across the pumice-covered water.

Captain Olsen and his skeleton crew had been able to leave the wharf office at about 5.15 pm and had secured the gangplank to the *Golden Bear*, but they were showered by light to heavy downpours of pumice for several hours (Olsen n.d.). Then:

> At 7 o'clock it cleared some and everyone on board were [sic] called out to assist in getting lines to the wharf. We were at the time experiencing a series of small tidal waves, causing a tremendous surge. The water would raise to the level of the wharf, then run out to about 8 or 10 feet below. The vessel would surge fore and aft about 50 feet, until she fetched up in the anchors and the stern breastline. It looked several times as if the propeller would get foul of the corner of the wharf. We had great difficulty holding the vessel alongside and parallel with the wharf and were steadily at it until midnight, at which time we managed, with the assistance of the men who had been absent but had then returned and were assembled on the wharf, to get the vessel secured. (Olsen n.d., 4)

Other vessels were also buffeted by the surging waters in the harbour during the evening. Waters were reported to be up to almost 3 metres higher than normal high-water at the Rabaul waterfront, and down to well below the usual low-water mark by roughly the same amount (Figure 3.18). Waterfront store sheds were flooded and some were swept away. Waiau Ahnon was surprised in the evening to find boats and boxes knocking around his house alongside the customs boatshed and to discover seawater almost up to his bed. Small inter-island boats were dashed against wharves and several of them sank (Ahnon 1953). The schooner *Meto* was washed up into the front garden of Treasurer H.O. Townsend. *Induna Star*, drawing about 2 metres of water, was dumped on top of and crushed a small jetty by one surge, but was then lifted clear again by another. Carpenter's *Duris* was sunk at her moorings, and another fleet member, the wooden motorship *Desikoko*, was battered but survived. The *Durour* was now under a thick layer of pumice on the slipway near Vulcan. *Golden Bear* left Toboi Wharf at 1.36 am and anchored amid empty oil drums, drifting vessels, pumice and debris of all kinds floating in the harbour (Olsen n.d.). The crew salvaged two schooners and anchored them. Cadet Pederson had taken down the Australian ensign on one of them—the two-masted *Nereus*—and had nailed up the American ensign in its place, but as the vessel was the one used by Administrator McNicoll, Captain Olsen very quickly had the ensigns changed round again (Olsen n.d.).

Figure 3.18. Tsunami debris and grounded boats at Rabaul.
Small boats and debris were left stranded on the Rabaul foreshore, apparently swept by the tsunamis that took place on the evening of 29 May 1937. Rabaul town and Tovanumbatir volcano are in the background. The original print used for this image is slightly damaged. GA negative reference GB2591.

Captain Roy Kendall of the *Induna Star* saw that the rises and falls of the water level in the harbour corresponded to periods of activity of the eruption at Vulcan. The eruption was beginning to build a volcano out of the sea. Kendall noted:

> During the first three hours active eruption was practically constant and no changes were observed in water level (after the initial fluctuation at the beginning of the eruption). At about 7 p.m., however, definite periods of active eruption were followed each time by corresponding periods of dormant steaming, and it was during this period, and prior to the rising of the rim of the crater above sea level that the tidal waves were observed. During the dormant periods enormous quantities of water must have been pouring into the crater itself … This was followed each time by a violent and sustained eruption which stopped the flow of water into the crater … When the rim of the crater appeared above the sea the waves ceased. (Fisher 1939a, 22)

Kendall was able to proceed out of the harbour that night but had to stop frequently so that injection pipes could be cleared of pumice. He passed within about 1 kilometre of Vulcan and noted that the new volcano was about 100 feet above sea level.

Vessels at Kokopo were well clear of the disturbances in Simpson Harbour, and two of them—pinnaces *Theresa* and *Paulus* belonging to the Vunapope Mission—had already been commissioned by the Kokopo police to sail around to the far side of Crater Peninsula and begin ferrying evacuees from Nodup back to Kokopo and Vunapope. Father Schlüter accompanied the pinnaces 'and as a precaution I took with me the holy oils and baptismal water' (MSC 1937; Arculus and Johnson, 1981, 9). Chinnery had consulted with the district officer at Kokopo, D. Waugh, about the possibility of a Rabaul evacuation, and Waugh began making preparations with mission staff for receiving evacuees. Chinnery returned to Rabaul by mission boat that evening, landing at Nodup after midnight. *Theresa* and *Paulus* would together make several return trips during the night.

Rabaul was now under the charge of Judge Beaumont ('Monty') Phillips (Phillips 1937a, 1937b, 1937c; Thomas 1937a). Ramsay McNicoll was still on mainland New Guinea and Chief Judge Wanliss would have normally deputised for him. However, Wanliss had been in poor health and on Saturday, before the eruption had begun, he had been allowed by doctors at the European hospital to drive to Kokopo accompanied by Sister Latta. He attempted to return to Rabaul using one of the back roads to the north coast, but was blocked by the pumice fallout, and even though he gained passage on the Japanese trading schooner *Asakaze*, his return was further thwarted when the vessel ran aground near Natava. Judge Phillips, therefore, assumed the role of administrator on Saturday evening, and he was soon faced with the problem of what recommendations should be issued. Many people who had evacuated from the pumice-covered town were now congregated on Namanula Hill and down at Nodup, and Phillips had little information on the whereabouts and condition of those on the eruption-affected west side of the harbour, or of those who had joined the exodus up Tunnel Hill Road to the north coast. Kokopo and Vunapope, however, were clear of the pumice fallout, and were together an obvious choice as a refuge for the townspeople.

Acting Superintendent Ball and other officers at police headquarters would have little rest during the 48 hours after the initial Vulcan outburst. Ball 'placed a guard over the Telephone Exchange to see that the operators remained on duty' (Ball 1937, 3), but later commended them for their 'quiet courage ... sitting in that tiny room with only a couple of hurricane lamps as light through the horror of those two nights' (4). Mr 'Pug' Noble from the Department of Public Works was placed in charge of the exchange early in the evening, but was later replaced with Mr Walsh from the post office (telephone branch), and between them they supervised the operators who kept open the lines, particularly between the town and Namanula Hill. Maintaining power in the town was a difficult task because of falling wires, but Jack Barrie, manager of the Rabaul Electricity Supply Co., kept up the supply by mending fuses and restoring many circuits. Broken wires were, as Burgess (1937) noted, a hazard, especially where they touched water tanks and roofs, but maintaining lights in the streets, in the hospital and in Namanula residences, and keeping telephone lines open, was considered necessary. The risk was an acceptable one. There is a report of one New Guinean being killed by electric shock, but this is unconfirmed.

Administration medical staff at Rapindik native hospital did not know of the massive exodus from Rabaul. Medical assistants Doug Joycey and Roger Davies had seen the initial rise of the Vulcan eruption column, and orderlies and patients had run out from the hospital to view the eruption, 'including our pet Osteomyelitis case, who was supposed to never to [be] able to walk again' (Joycey c. 1937, 1). Thunder and lightning kept Joycey and Davies awake during the night. They attempted to play chess but found that this was impossible too. Davies, an enthusiastic and careful photographer, took spectacular photographs of the eruption, including long-exposure shots of Vulcan lightning by standing his camera on a chair (Figure 3.19). His friend Joycey noted that the lightning 'streaks stood in the air and quivered for minutes at a time, and were practically continuous' (Joycey c. 1937, 1). Both of them had retreated hastily from the hospital cookhouse, to which they had been drawn by swishing noises, on discovering harbour water suddenly rushing in over their ankles and back out again just as quickly. They spent 'terrifying' periods during the Rabaul pumice fallout with towels around their mouths and noses. Not until Sunday morning did they receive a telephone call: instructions were to proceed to Namanula to assist in evacuating the European hospital.

Figure 3.19. Lightning in night-time eruption cloud at Vulcan.
This dramatic shot of forked lightning in the Vulcan eruption cloud was taken by medical orderly Roger Davies at Rapindik on the night of Saturday 29 May 1937 during the severe 'electrical disturbance' phase of the eruption. The area west of Vulcan is being devastated by the fall of pumice out of the electrically charged eruption cloud. GA negative reference G2008.

Clive Meares on Namanula Hill heard a car drive up outside the house at 6 pm but, on going outside, 'walked into the blackest "blackness" one could imagine' (Meares n.d., 4). He could hardly see the car's lights on full beam from just 2 metres away. The car's driver reported that Rabaul town appeared to be practically deserted, but he thought many residents, in their desire to get as far from the volcano as possible, 'had driven to the north coast, not realising that they were travelling right into the path of the prevailing winds blowing from it' (Meares n.d., 4). Meares next noted that:

> The explosions from the volcano seemed to diminish after a while [and the atmosphere at Namanula to clear a little] but they were replaced by the still louder noise of a storm—perhaps an 'electrical disturbance' would be a better description—which raged above and around the centre of the eruption until the morning hours [Figure 3.19, this volume]. I had often experienced severe thunderstorms and had seen the sky lit up by a sequence of flashes of chain

lightning, but never have I seen the sky illuminated by a dozen or more 'chains' at the one time as we witnessed, time and time again, that night. It seemed as though they were playing 'chasings' in and out of their hiding places behind the huge banks of whirling and wreathing clouds. Each flash set off a tremendous crash of thunder and, when a dozen flashes appeared at once, so was the sound of the ensuing thunder multiplied. In the occasional short lulls that did occur between the thunderclaps, still another sound thrust itself upon us: the continual crashing of falling trees and branches all over the hills on either side of the ridge [Namanula] as they were no longer able to bear up under the increasing weight of the debris that settled on them from the volcano.

By this time the atmosphere at Namanula had begun to clear, either because the wind from the south-east had freshened and was taking more of the ejecta away from us, or because the rain nearer the centre of the disturbance was forcing it down into a more restricted area. Although there was no rain in Rabaul that night, it was impossible even to guess how much fell on the hills across the harbour and near the volcano itself ... the stormwaters swept down the hillsides either into the harbour or onto the beaches and then into the sea on the north coast. In the latter area many Rabaul residents, who thought they would be safer there, found themselves held up by wide rivers where not even a small stream had existed before. They were forced to abandon their cars and trucks and managed to wade through the swift flowing torrents by holding hands and forming human 'chains'. (Meares n.d., 4–5)

The extreme volcanic 'weather' and land-surface flooding described by Meares and others can perhaps be regarded as an evening 'second phase' of the eruption. The growing volcanic cone may have appeared above sea level by this time, but the volcanic cloud above it was drenched with seawater derived from the harbour—'was salty the water', Lulu Miller had said—and with water precipitating from volcanic vapour dissolved previously as part of the magma before its eruption. Water then rained out of the cloud copiously, causing the flooding. Vivid lightning is known in many other explosive eruptions where friction between solid volcanic particles generates static electricity, but distinguishing this kind of electrical discharge from that formed in what might be called 'normal', non-volcanic thunderstorms can be difficult. The combined volcanic/weather cloud may also have created different wind directions and strength in the surrounding atmosphere, thus perhaps accounting for the Vulcan pumice and ash that fell for a short while on Rabaul while north-westerly winds were blowing.

Keeping the two wireless radio stations in operation at Rabaul had been impossible. Operator E.B. Alexander at the administration radio station on the foreshores of Rabaul evidently made every effort to man the equipment, but the pumice fallout overwhelmed him. No longer able to see, and with the power plant broken down, he abandoned the task at about 6 pm (Threlfall 2012). The operator on duty at the AWA radio station at Malaguna, L.E. Coleman, was advised to leave by the officer in charge of radio operations in Rabaul, J.K. Twycross. Coleman experienced the same intense fallout as had Father Nollen and others at Malaguna, but he stayed at the station, and a fellow operator took the women of the station over Tunnel Hill to the north coast.

Judge Phillips urgently required communication links with centres outside Rabaul. The radio on board the *Golden Bear* was still working, but Radio Operator Costner was missing. However, the *Golden Bear* was still at Toboi Wharf, so Phillips sent round to Captain Olsen the experienced Rabaul radio operator S.W. Faulkner whose ship was the Carpenter's unfortunate *Durour*. Over the next few hours, Faulkner sent a backlog of messages using the *Golden Bear* radio, which Olsen placed at the disposal of the administration. Phillips was thus able to send out two important messages that evening: one at 9 pm to the Department of the Prime Minister in Canberra informing the Australian Government of the Vulcan eruption (Phillips 1937a), and the other, at 11.30 pm, to Captain Michie on board the *Montoro* informing him of the eruption and requesting that he 'hasten Rabaul and stand by' (Phillips 1937b):

> AFTER CONTINUOUS EARTH TREMORS SINCE 4 A.M. THIS DAY VULCAN ISLAND ERUPTED ABOUT 4 P.M. EMITTING DENSE VOLUMES SMOKE AND COVERING RABAUL WITH VOLCANIC DUST MAKING DAY INTO NIGHT. SHIP GOLDEN BEAR STANDING BY AND MONTORO RECALLED FROM KAVIENG EXPECTED HERE TOMORROW. AT NINE P.M. ACTIVITY MUCH MODERATED AND ERUPTION STILL SMOKING BUT COMPARATIVELY QUIET. SEEMS NO IMMINENT DANGER THOUGH CONDITIONS EXTREMELY UNPLEASANT AND FUTURE POSITION OF COURSE OBSCURE UNPREDICTABLE. NO CASUALTIES REPORTED SO FAR MOST OF RESIDENTS OF TOWN COMING TO NAMANULA WHERE CONDITIONS LESS UNPLEASANT AND SECURITY GREATER. ESSENTIAL SERVICES

FUNCTIONING REASONABLY. ADMINISTRATOR ON TOUR CHIEF JUDGE HELD UP KOKOPO SIDE BY LAND SLIDES. PHILLIPS. (Phillips 1937a)

VULCAN ISLAND ERUPTED TODAY COVERING RABAUL WITH VOLCANIC DUST. SLIGHT TIDAL WAVE ALSO OCCURRED BUT CAUSED NO LOSS OF LIFE. THOUGH NOW QUIETER POSITION STILL DOUBTFUL. REQUEST YOU HASTEN RABAUL AND STAND BY. IF YOU CANNOT RAISE RABAUL RADIO OFFICE COMMUNICATE THROUGH SHIP GOLDEN BEAR RABAUL. PHILLIPS. (Phillips 1937b)

People on Namanula Hill and those still in Rabaul would be advised to evacuate to Kokopo and Vunapope via Nodup using a flotilla of boats that would be made available.

3.6. Evacuations from Nodup

Mr and Mrs Meares on Namanula Hill had taken in one crowd of people, and when ash began falling on the hill at about 5 pm on Saturday they

> collected buckets and basins of water and passed round towels, flannels, pieces of sheeting … to about 25 persons congregated in the dining room where we had shut every door and window. (Meares n.d., 4)

The party included

> four or five young children whom we perched on top of the piano so as to keep them in sight, but I remember how vainly we tried to get them to submit to having wet flannels held to their faces. (Meares n.d., 4)

However, the European hospital on Namanula Hill was the centre of greatest activity that night for most who had come up from Rabaul. Gordon Thomas later wrote:

> [T]he work done by the hospital staff and their able and willing volunteers will never be forgotten by the two hundred-odd people who were cared for during Saturday night and Sunday morning. Tea and refreshments were always available throughout the night; enquiries for missing relatives given the promptest attention and nothing was left undone which could have been done by Matron

McKinnon and her helpers. Fortunately no serious accidents occurred to residents; many sufferers from dirt-infested eyes were treated and sedatives administered to those suffering from shock and nervous prostration. (Thomas 1937a, 2)

Thomas himself had reason to be grateful for he was there reunited with his wife (Thomas 1937b). Others were equally appreciative. Gladys Forsyth, for example, was rejoined by her brother Adrian. Her friend, a sister at the hospital, gave Mrs Forsyth her room in which the grateful mother installed her baby, together with the cat and dog. Mrs Forsyth retreated to the hospital room when the ash fell, and 'waited to be "buried alive" (as described in the "Last Days of Pompeii") as did everyone else who had read the book' (Forsyth n.d., 9). Hospital doors were closed to keep out the dust, and the children were gathered in rooms more protected from the fumes and ash, but the heat indoors became unbearable. After the worst of the ash fall passed, Mrs Anthony, like others near and on Namanula Hill, was kept awake by the lightning and thunder (Anthony 1981). By morning, the demand of the crowd of refugees on Namanula was so great that the hospital could no longer cope. Acting Superintendent Ball received a telephone call from Dr H. Champion Hosking advising that

he couldn't carry on ... and demanding that they be sent down to the beach at Nodup to await the transport that was said to be coming from Kokopo to take them there. (Ball 1937, 4)

The extent of the fallout on Namanula Hill and the effects on Rabaul town could be fully appreciated in the morning light when, shortly after daybreak, Clive Meares and several others drove down into town. Vulcan was still in eruption, but noiselessly, and driving was eerily quiet as the pumice cover on the road deadened the sound of the car wheels. Meares wrote:

Looking across the town, not a vestige of green was visible in the forests on the hillsides and, in fact, everything in sight was covered by a mantle of death-like grey ... In the town itself, motor traffic was at a standstill as the roads were almost completely blocked by fallen trees. (Meares n.d., 5–6)

Meares made his way to the police station where Judge Phillips had hastily convened a small meeting of leading government officials and principal businessmen (Ball 1937; Meares n.d.), including, among others, R. Melrose, the district officer at Rabaul; Acting Superintendent Ball; and Gordon Thomas, editor of the *Rabaul Times*. Phillips informed the meeting of his decision that Rabaul should be evacuated and that people should assemble at Nodup to be taken by boat to Kokopo and Vunapope.

From then on, Acting Superintendent Ball reported, the

> whole of Sunday morning was spent in a terrific effort to clear the
> roads which were in a frightful state from pumice, rain and fallen
> trees. Every lorry that could be found was pressed into service, every
> prisoner was used, every Officer was at work. Our idea was to clear
> the main routes to expedite the work of the evacuation ... other
> members of the Force were directing the movement of refugees to
> the beach at Nodup from Namanula and their work by all accounts
> was highly creditable and very necessary, particularly in the handling
> of the Chinese who showed every inclination to panic. (Ball 1937, 5)

Ball had been concerned the previous evening to hear that Senior
Gaoler H.C. McFarlane had evacuated the native prison. McFarlane and
J.H. Theckston had mustered the prisoners when they were on the point of
becoming unmanageable and had marched them up to Namanula Hill. Ball
was impressed with their work, but nevertheless instructed McFarlane on
Sunday morning to return their charges to prison. Doug Joycey and Roger
Davies at Rapindik Hospital had also, on Saturday night, released the nine
or so inmates of the native asylum so that they could have the same chance
of escape as the other patients; Davies is said to have remarked that 'there
were many worse cases outside, anyway' (Joycey c. 1937, 3).

The McCarthys returned to their home from the Cosmopolitan Hotel to
collect some possessions, and then began walking again, up over Namanula
Hill to Nodup. Mrs McCarthy collected her fur coat and marriage certificate,
but later could not remember any of the other more important items they
took (McCarthy 1980). She recalled losing one of the heels of her white
high-heeled shoes so that she had to hobble along. The Anthonys emerged
from their home and saw the changed view of Rabaul from their verandah,
their bleary-eyed servants drifting out from their *boi* house after a sleepless
night and reporting for duty (Anthony 1981). The cook boiled some water
on a primus and an attempt was made to brush away the ash from the
verandah. A car picked its way along the street and the driver, Ted Cook,
surprised to see the Anthonys still in town, told them of the evacuation
and offered them a lift to Namanula Hill. Mrs Anthony packed a small
case, threw a loaf of bread and other food into a shopping basket, and the
party, which included their fox terrier (which later sat on the bread), its two
puppies, a servant and two children from next door, drove up to Namanula
and joined the crowds milling round the hospital before going on to Nodup.

Some people who had overnighted on Namanula Hill took the opportunity of going back into town to collect a few clothes and essential items. Mr Hoogerwerff, for example, rushed back down the hill for extra clothing (Hoogerwerff 1937). Few had time to consider emptying toilet pans, and ice-chests and refrigerators that would be without power. Perishable foods were left to spoil in most homes. Mr Burgess had time to think of packing only a 'toothbrush, comb, and a clean shirt' (Burgess 1937, 1). Bernard Ryan also returned to town and was there appointed, with others, to assist in the evacuation. Ryan and John Cox were assigned the task of removing from the Cosmopolitan a residue of *Golden Bear* drinkers who were 'recovering slowly' (Ryan 1980–84, Sleeve 55). Ryan and Cox drove them over to Nodup ready to board their ship. Chinnery had also been busy since his arrival at midnight from Kokopo. However, at about 9 am on Sunday, his car crashed and he received a severe head injury, resulting in his evacuation to Kokopo as a hospital patient.

Mrs Kathleen M. Bignell, proprietress of the Rabaul Hotel, came down from Namanula and resumed catering at the hotel, as well as taking in a range of household pets that had been deserted by their owners (Bignell and Clarence 1981–84; Clarence 1982). She would remain in Rabaul and experience further difficulties related to the eruption over the next few days. Waiau Ahnon was reluctant to leave his house at the foreshore, but Messrs Ball and Prior made their intentions clear, and he soon began walking to Nodup (Ahnon 1953). Joycey and Davies at Rapindik, after receiving instructions via telephone to help evacuate the European hospital, also set out on foot but became lost after taking a short cut; they finally arrived at Namanula at about 10 am, just as the last of the patients were being driven down to Nodup, and were admonished for their lateness (Joycey c. 1937). Davies was ordered to Kokopo with the patients. Joycey was told to stay in Rabaul with Dr R.W. Cooper, who had been put in charge of medical and sanitary arrangements.

People were now congregating in their thousands at Nodup, streaming down the road from Namanula on foot, in private cars and in trucks, waiting for the boats to pick them up (Figure 3.20). Eric Hopkins placed his fleet of hire cars at the disposal of the administration, and personally drove the Rabaul–Namanula–Nodup Road many times conveying people without means of transport out of Rabaul to the Nodup evacuation point (Hopkins 1937). L.W. (Bill) Heinicke of Burns, Philp & Co. also, among others, provided valuable assistance in transporting refugees between Rabaul and

Nodup. Hopkins stayed in the Rabaul area after the Nodup evacuation and, with some others, experienced the next volcanic eruption that would affect Rabaul township.

Captain Olsen on board the *Golden Bear* in Rabaul Harbour received a request at 7 am on Sunday to proceed to Nodup to help pick up refugees; about an hour later, he steamed into the channel between Crater Peninsula and Vulcan, which was still in full eruption (Olsen n.d.). Later, upon attempting to leave the harbour, they 'narrowly escaped being enveloped in the flow from the volcano' when a 'column of steam and lava' was sent across the channel directly ahead of them (Olsen n.d., 8). Olsen 'asked for all speed possible, trusting to luck', and ordered everybody off the decks. Government officials on shore lost sight of the ship in the flow and feared she was lost. But the cloud dissipated and the *Golden Bear* managed to exit the harbour, covered in several inches of pumice and looking a 'sorry sight'. Her holds had been left open at Toboi Wharf to dry out in readiness for copra loading; the pumice later found in holds two and four had to be hoisted up through the hatches in almost 50 loads of 500-pound (225-kilogram) rattan baskets. *Golden Bear* rounded Praed Point and arrived off Nodup at about 9.30 am.

Figure 3.20. Car parking at Nodup after Rabaul evacuation.
Cars are parked in an orderly fashion at Nodup on the morning of Sunday 30 May 1937 while Europeans wait for their evacuation by boat. Many such vehicles were requisitioned by the administration for use in and around Rabaul. Some of those left at Nodup were used by people who returned to Nodup from Kokopo and who required transport to Rabaul so as to retrieve personal possessions, business documents and other essential items. GA negative reference GB2603.

Captain Olsen attributed the 'flow' of ash that enveloped the *Golden Bear* to a sudden change in the direction of the wind affecting the Vulcan eruption column. Local changes in wind direction around the volcano may well have been caused by the hot column of ash and pumice rising from Vulcan itself. However, a more probable explanation for the origin of the dust-laden flow is that it came directly from the base of the collapsing eruption column. Brett Hilder noted the *Golden Bear* had been 'blasted all over the starboard side as she passed Vulcan, making her a rusty grey colour all over that side while her port side was clean and brightly painted' (Hilder 1980–81, see letter to R.W.J., 6 April 1981, 1). It is possible that a large amount of pumice was dumped out of the Vulcan eruption column and hit the ground, sending gusts of dust-laden air from the base of the column and plastering the ship on one side, perhaps like the 'light-coloured cloud' shown in Figure 3.13. Another possibility is that the lateral flow was a fast-moving, low-density, turbulent cloud that had detached from the top of a denser pyroclastic flow. Volcanologists call such clouds 'surges' (among other terms) and they can cause great destruction on account of their searing heat. The flow that plastered the side of the *Golden Bear*, however, appears to have been cold—more like a lateral gust of dust.

Judge Phillips's radio message sent the previous night from the *Golden Bear* to the *Montoro* reached Captain Michie at 11.30 pm while at sea between Kavieng and Salamaua. Captain Michie was described by Purser George Clarke as a 'slow-spoken gentleman of invariable calm and even, pleasant temperament' (Clarke 1960, 1; 2001, 42). Not surprisingly, Michie responded well: within 12 minutes he had set course for Rabaul; advised the administration, via the *Golden Bear*, of his estimated time of arrival of 2 pm Sunday; and informed Burns Philp's head office in Sydney of his new schedule (Michie 1937). According to Clarke, the chief engineer was instructed to 'get every possible revolution out of the engines. The "Montoro" never throbbed as heartily in her life' (Clarke 2001, 42). Second Mate Brett Hilder, who came on watch at midnight in bright, full moonlight, noted that the ship was curiously whiter than usual, and soon discovered that the cause was fine white volcanic ash that had carried north-westwards over 240 kilometres from the eruption at Rabaul (Hilder 1961). Hilder later suggested that there may have been a relationship between the earth tides of the full moon and the volcanic eruption at Rabaul, a general hypothesis that Rabaul-based volcanologists would consider in later years (Hilder 1980).

Meanwhile, Judge Phillips—who was much less concerned about the science of earth tides—sent further instructions during the morning to Captain Michie to proceed to Nodup to take on evacuees (Michie 1937). Stock was quickly taken of water and provisions for the expected thousand or more Europeans and Asians (Clarke 2001). All New Guinean hands on board were moved to 'tweendecks' after breakfast so that the decks were clear, and every cabin and sleeping place was prepared for passengers. All six lifeboats were stripped of gear, except the steering oar, and lowered to the waterline ready for immediate release when the ship stopped. The cargo boats and two pinnaces were also readied for heaving out.

The *Golden Bear*, a cargo vessel, experienced difficulties at Nodup in taking on refugees. The beach at Nodup had only a small jetty so the loading of boats had to be undertaken in the shallows. There was, fortunately, none of the surf that can pound that part of the coast during the south-east season, but the freighter had no launches for towing her two lifeboats and work boat, and there was some initial difficulty in heaving out the boats, as the davits and pulleys were seized up with pumice. One boat davit gave way, injuring Ordinary Seaman Samuel O'Neal. Further, the lifeboats had tall sides and were difficult to load in the shallow water. Amy Anthony recalled seeing her husband and another man grappling with

> a very heavy lady, and as her weight spreadeagled their legs in the soft sand, they finally heaved and tossed her over the side of the boat, whilst willing hands on the lifeboat caught her and broke her headlong tumble. (Anthony 1937, 5)

The morning sun was strong, and those without hats felt it badly. Evacuees had to clamber up rope ladders and cargo nets that had been slung from the freighter's sides. The *Golden Bear* had, by the time of its departure at about 2 pm, taken on board about 750 people, mainly women, babies in arms and children, both European and Chinese. There were also the patients from the hospital, the doctor, several nurses and an assorted group of family pets.

Other smaller craft also took on board evacuees for Kokopo. These included Vunapope Mission vessels *Paulus* and *Theresa* that had begun ferrying people to Kokopo during the night, the Chinese-owned *Kwonchow* (or *Kwong Chow*), the plantation boat *Muruk* and the Japanese *Asakaze*, which had been refloated and had transported Judge Wanliss to Nodup. Wanliss went on to Kokopo, leaving the evacuation in charge of Judge Phillips. Captain Kendall was there too with Oscar Rondahl's *Induna Star*, which he had brought round from Kabakaul Plantation. The administrator's *Nereus* had made her way out of Rabaul Harbour during the morning—Captain Jackson, said Gordon Thomas, 'making a very fine dash past the dense smoke clouds of the volcano for the open sea' and Nodup (Thomas 1937a, 2).

These vessels could carry only small numbers of people, so when the 360-foot, 4,088-ton *Montoro* hove into view off Nodup at about 1.15 pm, shortly before the fully loaded *Golden Bear* was set to leave, there was increased optimism among the thousands still on and near the beach. As the coastline was poorly charted, Captain Michie was forced to move slowly; the anchor was dropped to 30 fathoms and the *Montoro* edged in carefully until, at 1.40 pm, the anchor touched bottom and the ship was halted (Michie 1937; Hilder 1961, 1980). The people on shore, already anxious to board the old ship, grew even more anxious when they became aware of a dark eruption cloud curling upwards behind Kabiu and towering up from the shore close to Nodup. Tavurvur had joined Vulcan in eruption at about 1 pm!

3.7. Tavurvur in Eruption: Sunday Afternoon, 30 May

Many women on board the *Golden Bear* and small boats on their way to Kokopo were badly affected by the sight of the Tavurvur eruption: '[I]t was a terrorfying [sic] sight', recalled Amy Anthony who was then aboard one of the small boats, 'and very frightening and most of the women passengers disolved [sic] into tears, as our husbands were all back at Nodup' (Anthony 1981, 6). Some thought that Kabiu herself—the Mother—might also break out into activity. George Clarke on board the *Montoro*, 'right under the lee of the Mother', was anxious about that possibility: '[I]f the old lady blew her top whilst we were there, the *Montoro* would cop the lot' (Clarke 2001, 44).

Evacuation down to, and from, Nodup had been orderly throughout the morning, but excitement mounted as the *Montoro* anchored and lowered her boats and as Tavurvur continued to discharge a considerable eruption cloud (Figures 3.21–3.23). Yet the crowds on shore stayed controlled, and their embarkation proceeded steadily and without significant interruption. The presence and directions of the police were influential in maintaining order. Acting Superintendent Ball (1937, 5) wrote that the Chinese showed 'every inclination to panic', but he and others said that the general behaviour of the evacuees was exemplary. George Clarke, who was appointed beachmaster, said that the Burns Philp

> labour line caused some trouble through claiming priority because the *Montoro* was a company vessel, and at one stage put on a demonstration and rushed the boats. However, the vigorous but unconventional methods of the police restored order. (Clarke 2001, 44)

Young European male civilians such as Adrian Field, who armed himself with a stick, were also involved in maintaining order (Field n.d.).

The lifeboats, pinnaces and cargo boats were lowered into the water as soon as the *Montoro* came to a halt. Boats were towed two at a time by the launches, and each was filled in the shallow water by the beach with more than 100 people at a time, although the boats were certified to carry only fifty-four. Rope ladders, cargo nets, lifelines and the two gangways enabled boats to be quickly discharged when they reached the *Montoro*. Captain Michie was requested to take on board all the New Guineans assembled on the beach, as well as the Europeans and Asians, and only after five hours of running the lines of boats between beach and ship was the task completed. Brett Hilder was impressed with the quiet, controlled manner of the evacuees during the ferrying and embarkation: 'There was no excitement of any sort. Everybody was overawed; they were speechless … they were all pretty frightened, but all very quiet—amazingly quiet' (Hilder 1980, 31–2). Passenger D. Stewart, sales manager for Holbrooks Ltd, Sydney, also remarked that the 'discipline and lack of panic was most marked'. He attributed this to European control, adding, patriotically, that 'one felt proud to be British' (Stewart 1937, 72).

Figure 3.21. Tavurvur eruption cloud as photographed from the SS *Montoro*.
This was the view seen from the decks of the *Montoro* as she approached Nodup in the early afternoon of Sunday 30 May 1937 (compare with Figure 3.22). Refugees from Rabaul are assembled on the beach as small boats take some of the evacuees out to two of the smaller vessels used in the Nodup evacuation. Eruption clouds drift north-westwards (to the right) from Tavurvur, which is hidden behind the slopes of Kabiu (extreme left). The hill in the centre of the photograph represents the north-eastern side of Palangiangia within which is Rabalanakaia. The photograph was supplied courtesy of B. Hilder. GA negative reference GB3301.

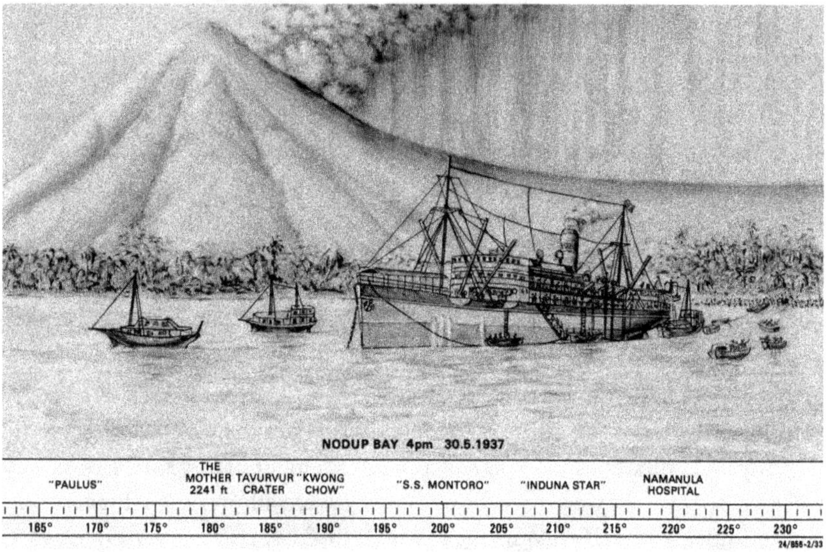

Figure 3.22. Brett Hilder sketch of the scene off Nodup.

This is a previously unpublished sketch (based on compass readings) made by the late Brett Hilder of the evacuation from Nodup at 4 pm on Sunday 30 May (compare with Figure 3.21). Small boats are relaying refugees from the shore to the SS *Montoro* shortly after Tavurvur volcano, on the other side of the Mother, had broken out in explosive eruption. Note that Palangiangia and Rabalanakaia volcano are not shown in the sketch.

Figure 3.23. Evacuees at Nodup being ferried out to the SS *Montoro*.

Ghostly figures in this underexposed photograph congregate at the Nodup evacuation point. Heavily laden small boats ferry evacuees out to the SS *Montoro* (top left) as the *Golden Bear* in the top right, carrying other evacuees, steams off to Kokopo in the early afternoon of Sunday 30 May. GA negative reference GB2506.

There was no opportunity to count all those on board the *Montoro* by the time she weighed anchor at 6.35 pm, but most estimates range between 4,000 and 6,000 people, about 200 of whom were Europeans. Brett Hilder wrote:

> And we had them standing on every deck in the ship—all the way up to the bridge. We had them in the 'tween decks, but we didn't put any in the lower holds; but all these people standing up, like cigarettes in a packet, or a bit like a Sydney bus in rush-hour ... We had the police boys who came aboard with them, and they kept a gangway clear so we could get up the companion ways up to the bridge and back ... There was absolutely no trouble—no noise. Everybody just stood rooted to the spot, like the good commuters coming down to the office from the North Shore line ... [Y]ou couldn't believe you could load that many people, at an anchorage—just with ships' boats—and get them all to stay on deck and stay in position all night. (Hilder 1980, 32)

The assembled refugees 'were a motley sight', said Mr Stewart, 'some maimed and crippled—carrying their baskets, parrots, dogs, primus stoves, hurricane lamps, camphor boxes, etc.' (Stewart 1937, 72). Many carried 'a bundle of clothes, or a little suitcase, or a kit-bag, or something with them—or a baby', listed Brett Hilder (1980, 32). An elderly Chinese man carried a very heavy bucket, apparently full of potatoes, but the potatoes were only a top layer covering a hoard of coins.

The refugees on the beach at Nodup on Sunday 30 May 1937 could not see Tavurvur volcano directly, but others in Blanche Bay were able to see the start of the eruption and even to photograph it. Mr and Mrs A.G. Vagg of Madang were in the Rabaul area at the time of the 1937 volcanic eruptions, awaiting the arrival of the *Neptuna*, which would take them back to Australia, and Mrs Vagg took an exceptional set of snapshots of Tavurvur breaking out into explosive eruption, using a small Brownie camera (Figure 3.24; Vagg 1981). Mr Vagg also took photographs. Another photograph of Tavurvur at this time was taken from Matupit Island by a missionary (Fisher 1939a, 57; see also Figure 3.25). Vulcan meanwhile was still producing its sub-plinian eruption (Figure 3.26).

A European man, William Elworthy, had gone missing out at Tavurvur on Saturday (Joycey c. 1937; Saxton 1937). Elworthy, an electrical engineer who had been assisting Chief Engineer Jack Barrie of the Rabaul Electricity Supply Co., was due to return soon to Australia. He had decided to hire a canoe at Matupit Island early on Saturday afternoon so that he could be paddled across Greet Harbour to Tavurvur volcano to take photographs. Elworthy had not returned. Barrie found it impossible to maintain the electricity supply and by Sunday had shut down the generators, freeing him to search for his young colleague. He drove out to the causeway leading to Matupit Island early in the afternoon, accompanied by, among others, Doug Joycey, who wanted to return to Rapindik to retrieve some medical books, but they became bogged when driving onto what they thought was a solid roadway (Joycey c. 1937). The causeway, in fact, no longer existed, and the water was covered in floating pumice from Vulcan, across the harbour, which was still in full eruption.

Figure 3.24. Six photographs of Tavurvur eruption taken by Mr and Mrs Vagg.

This set of six photographs of Tavurvur in eruption on 30 May 1937 was taken by Mr and Mrs A.G. Vagg using two cameras (Vagg 1981). The photographs have been arranged here top left to bottom right in what may be a time sequence. Note how the initial upward impulse of the Tavurvur eruption column in the upper-left photograph has been arrested by the south-easterly wind that blows the ash towards Rabaul. The ash fall is dense and the column completely bent towards the north-west in the bottom right-hand photograph. The peak in the distance is Turagunan (South Daughter). A similar photograph to these was taken by Pastor G. Peacock from the Seventh-day Adventist mission on Matupit Island (Fisher 1939a, lower photograph on p. 57; see also Oliver 2020). GA negative reference GB3293.

Figure 3.25. Tavurvur in eruption probably as seen from Rapindik.

Tavurvur volcano is seen in eruption in this photograph attributed to Dr R.W. Cooper (Fisher 1939a, 60) and shot apparently from Rapindik. The time that the photograph was taken is uncertain. Both Dr Cooper and photographer Roger Davies were probably on Namanula Hill at the time of the first outbreak of the eruption, so this photograph may have been taken later. Note the damage to vegetation in the foreground presumably caused by fallout of the mud-ash in the early afternoon of Sunday 30 May. The photograph, therefore, may be of the early, still-daylight part of the later phase of activity that was witnessed so vividly after dusk from the *Montoro* off Kokopo (Figure 3.27). GA negative reference GB3296.

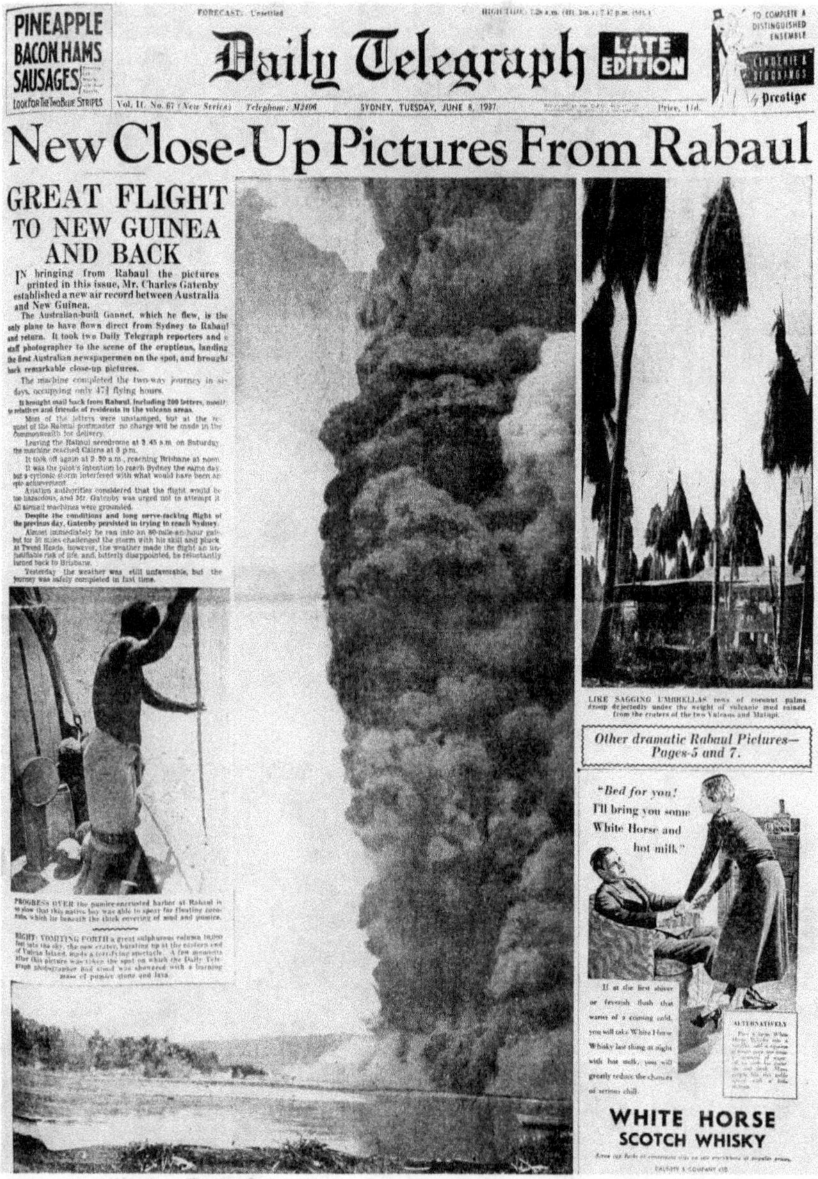

PINEAPPLE BACON HAMS SAUSAGES

LOOK FOR THE TWO BLUE STRIPES

FORECAST: Unsettled

Daily Telegraph LATE EDITION

Vol. II. No. 67 (New Series) Telephone: M2406 SYDNEY, TUESDAY, JUNE 8, 1937 Price, 1½d.

TO COMPLETE A DISTINGUISHED ENSEMBLE

by Prestige

New Close-Up Pictures From Rabaul

GREAT FLIGHT TO NEW GUINEA AND BACK

IN bringing from Rabaul the pictures printed in this issue, Mr. Charles Gatenby established a new air record between Australia and New Guinea.

The Australian-built Gannet, which he flew, is the only plane to have flown direct from Sydney to Rabaul and return. It took two Daily Telegraph reporters and a staff photographer to the scene of the eruptions, landing the first Australian newspapermen on the spot, and brought back remarkable close-up pictures.

The machine completed the two-way journey in six days, occupying only 47½ flying hours.

It brought mail back from Rabaul, including 200 letters, mostly to relatives and friends of residents in the volcano areas.

Most of the letters were unstamped, but at the request of the Rabaul postmaster no charge will be made in the Commonwealth for delivery.

Leaving the Rabaul aerodrome at 3.45 a.m. on Saturday the machine reached Cairns at 5 p.m.

It took off again at 2.30 a.m., reaching Brisbane at noon.

It was the pilot's intention to reach Sydney the same day, but a cyclonic storm interfered with what would have been an epic achievement.

Aviation authorities considered that the flight would be too hazardous, and Mr. Gatenby was urged not to attempt it. All airmail machines were grounded.

Despite the conditions and long nerve-racking flight of the previous day, Gatenby persisted in trying to reach Sydney.

Almost immediately he ran into an 80-mile-an-hour gale, but for 30 miles challenged the storm with his skill and pluck. At Tweed Heads, however, the weather made the flight an unjustifiable risk of life, and, bitterly disappointed, he reluctantly turned back to Brisbane.

Yesterday the weather was still unfavorable, but the journey was safely completed in fast time.

PROGRESS OVER the pumice-encrusted harbor at Rabaul is so slow that this native boy was able to spear for floating coconuts, which lie beneath the thick covering of mud and pumice.

RIGHT: VOMITING FORTH a great sulphureous column 10,000 feet into the sky, the new crater, bursting up at the eastern end of Vulcan Island, made a terrifying spectacle. A few moments after this picture was taken the spot on which the Daily Telegraph photographer had stood was showered with a burning mass of pumice stone and lava.

LIKE SAGGING UMBRELLAS rows of coconut palms droop dejectedly under the weight of volcanic mud rained from the craters of the two Vulcans and Matupi.

Other dramatic Rabaul Pictures—Pages 5 and 7.

Figure 3.26. *Daily Telegraph* front page featuring the high Vulcan eruption column.

Part of the towering eruption column from Vulcan dominates the front page of the Australian newspaper the *Daily Telegraph* for Tuesday 8 June 1937. The exact time that the photograph was taken is not certain, but most probably it was sometime on Sunday 30 May—that is, after both the initial 'phreatoplinian' phase and the night-time 'electrical disturbance' phase of the eruption. The eruption column in this photograph appears to be rising in typical 'plinian' fashion from the centre of the newly formed volcanic cone of Vulcan (compare with Figure 3.9). A small pyroclastic flow seems to be running down (or has run down) the eastern side of the cone as seen by the darker lobe on the light-coloured flanks of the new cone. Collapsed crowns of fronds on the tops of plantation coconut trees are shown in the photograph on the right, and a small boat moves through floating pumice on the left. The photograph may have been taken by Eric Hopkins, who is known to have given photographs to the *Daily Telegraph* (Hopkins 1937). Published courtesy of Australian Consolidated Press Limited, Sydney.

Joycey recalled that, just as they were dragging out their vehicle, Tavurvur, which was nearby,

> went up with a loud explosion. Now, Matupi was different altogether from Vulcan. Vulcan shot out light pumice which floated away in whatever direction the wind was blowing. Matupi produced thick, heavy, slatey-blue mud, which rolled down towards Rabaul, in a cloud. We turned the utility round as quickly as possible, filled it to overflowing with the boys who had been helping us to drag it out, and raced for Rabaul. I would say the clouds travelled slowly, because we beat it in the ute.

> Well over one hundred boys from the Compound and Matupi village were running along this road, imploring us to pick them up; but, as we had natives on the running board and hanging on the back, we could not stop.

> The cloud of mud was a terrifying thing to watch, billowing along the road a thousand or so feet high, and faster than a man could run. I thought I was witnessing a great tragedy … but Matupit sent up just one puff of mud which lasted about half an hour, and then only steam; and not one of the natives on the road was lost. (Joycey c. 1937, 2)

Other witnesses said the initial explosion was extraordinarily quiet and continued this way 'except for a sort of whirring sound' (Fisher 1939a, 23). The Tolai of Matupit Island nevertheless were in a turmoil as the Tavurvur cloud grew and developed. Many of them believed the eruption was the work of the *kaia*, To Lagulagu, who lived in the volcano and that the eruption was somehow a punishment directed at them (Mennis 1972). Matupits who had not already left the island on Saturday afternoon after Vulcan had begun its activity streamed across to the mainland carrying children and belongings (Mennis 1972). Those left behind—the old, sick and crippled who had no way of escaping—worried that the volcano would engulf them. However, the 'whirring' cloud missed the island, advancing instead on Rabaul, causing a blackout, and dumping on the town an unpleasant, wet, mud-like ash, accompanied by the stench of sulphur dioxide that made breathing difficult (Fisher 1939a). More than 10 centimetres of this sulphur-rich, moist ash would eventually accumulate in the southern part of town, covering the lighter-coloured and mainly drier ash of Vulcan that had fallen the previous evening. Gordon Thomas noted later that the message he had left his wife indoors on the table written in the dust-like ash from Vulcan was still there, whereas none of the heavier, wet mud-ash had entered the house to cover the message (Thomas 1937b).

The Tavurvur eruption cloud rose higher and could be seen clearly from the *Montoro* to the north-east as she rounded Tavui Point and approached the Nodup pick-up point between about 1 and 1.30 pm on Sunday (Hilder 1961). Father K. Schlüter at Vunapope to the south-west also saw the Tavurvur cloud:

> Suddenly—at exactly 13 hours—we noticed a small white cloud over the small crater of Kaia (to the right of Matupit). The eruption followed only a second later. A black mass of stones and ash shot 1000 m straight up into the air. Shortly afterwards we could see the splashes of the boulders into the sea. It recalled pictures of naval battles; except that here there were several hundred splashes. I estimate that the water shot up 30–40 m. The clouds rose up even higher above the crater and then slowly sank over Matupit and Rabaul. At that moment two pinnaces belonging to the Adventists, lying off Matupit, which were ready to sail, did set sail and fortunately escaped destruction. (MSC 1937; Arculus and Johnson 1981, 10)

Acting Superintendent Ball, in Rabaul, heard of the Tavurvur eruption soon after 1 pm and was notified that the prisoners, who had been taken back to the prison, were becoming restive (Ball 1937). He therefore instructed that all police, prisoners and rations should be brought to the Rabaul Police Station and, following hurried consultation with Acting Inspector Prior, decided to confine the prisoners to a convenient building close at hand: the government store. Trucks were sent to bring everyone in, but the Tavurvur blackout and showers of mud caused 'indescribable confusion' and only one truck reached the police station; another two reached Namanula, and thus many New Guinean police and prisoners were, contrary to Ball's intention, evacuated from Nodup to Kokopo. Ball reported:

> There followed in Rabaul another afternoon and night of horror far more dreadful than the previous afternoon and night. No words I have at my command can adequately describe the effects of the incessant and nerve-racking thunder; of the lightning of an intensity never seen by any of us before; of the almost incessant boiling over of the new volcano at Vulcan Island accompanied by the most sinister

rumblings. To have remained in Rabaul, beneath it as it were, during the hours of darkness, was a test of endurance. The horror was increased by having to listen throughout the night to the crash of falling trees or huge branches as they gave way under the weight of mud with which Rabaul had been deluged during the afternoon and evening. (Ball 1937, 6)

Hal Evans was in Rabaul when the Tavurvur eruption took place:

> We, the 18 remaining persons in Rabaul closed ourselves in the Police Station which had become the Administrative Headquarters, and the only habitable building in Rabaul. The second eruption was worse than the first, because it took place within about 2 miles of Rabaul Township. However, we survived it [and] after being in the darkness for about 3 hours, we were again able to carry on. (Evans 1937, 3–4)

The fallout of the volcanic material from Tavurvur on Rabaul was different to that of the mainly dry, dust-like pumice ash from Vulcan, which blew about and had fallen earlier that day for a short while. Rather, the Tavurvur mud, once it landed, could not be resuspended by any breezes. Further, most of the Tavurvur mud-ash was restricted to a fairly narrow band that extended north-westwards from the volcano, over nearby Rapindik to the central and south-eastern parts of Rabaul town. Namanula Hill was much less affected, as were the evacuation routes down to Nodup and over Tunnel Hill to the north coast. The mud had a sulphur-like or sulphuretted smell, possibly including the 'bad-egg' smell of volcanic hydrogen sulphide, but most probably was dominated by volcanic sulphur dioxide, much of which had been dissolved in the odoriferous mud. The Tavurvur mud was, therefore, acidic—sulphuric acid—which would have a corrosive effect on exposed metal surfaces such as on motor cars left outside (Fisher 1939a).

People remaining on Matupit Island close to Tavurvur had a frightening night:

> The Catholics locked the Church doors and went over to the Methodist church where they spent the night huddled in the darkness. They were very frightened of the large fiery rocks which continued to be hurled out of Matupit. They had no food and only a little water, and they thought they would soon die … Tavurvur continued to emit steam, mud and rocks all through Sunday and the following night. A terrible choking gas accompanied the explosions so that many thought they would suffocate. Earthquakes continued and lightning flashed as the rumbles of thunder and explosions continued. (Mennis 1972, 64)

Those who had left the island took any route they could over to Nodup. Hardly any ash from either Vulcan or Tavurvur fell on Matupit Island but the island was nevertheless damaged by tsunamis. Some Matupits came to the conclusion that the old *kaia*, To Lagulagu, 'was not so angry with them after all' (Mennis 1972, 67).

Following a visit two days later to nearby Rapindik, across the now submerged causeway to Matupit, Doug Joycey observed that:

> The mud and the pumice, on the palm leaves, had bent them down like pine-trees. Hopping about in the mud were thousands of birds, their wings so weighted with mud that they could not take off. The whole landscape was grey mud, right to the top of the hills—a most depressing sight. (Joycey c. 1937, 3–4)

The successful evacuation of Rabaul on Saturday night and Sunday morning meant that most Rabaul people were spared the effects of the Tavurvur fallout during Sunday afternoon and night, as many of them had already landed safely at Kokopo or were awaiting disembarkation from the *Montoro* on Monday morning. The picture of Rabaul presented at first light on Monday to the few who remained there was, according to Ball, one of 'desolation … the roads were jungles and ankle deep with a thick and treacherous mud' (Ball 1937, 6). Nevertheless, Ball organised his resources and responded to the following priorities:

1. *Sanitation*: to remove the contents of sanitary pans and ice-chests, and to bury them, or dump them in the sea.
2. *Road clearing*: to open trafficable routes so that supplies in Rabaul could be transferred to Kokopo via Nodup.
3. *Police work:* to establish order among the small community—the 'garrison'—who remained in Rabaul, to prevent unauthorised people from coming into Rabaul and to safeguard property.

A police mess was organised and run by 'Soldier' Williams and Charley Bye, both of Carpenter's. Mrs Bignell carried on the mess at the Rabaul Hotel, rationed by the administration and supported by the police.

The *Montoro* had headed out to sea late on the Sunday afternoon. Unable to discharge her passengers at Kokopo until daybreak the next day, she stood off Blanche Bay waiting for the night to pass. A severe wind and rain squall struck the *Montoro's* starboard side at about 2 am, and George Clarke, the ship's purser, woke to find the ship keeling over to port. He leapt out of his bunk and landed on a New Guinean who had sensibly crept into his cabin after lights out. Much later, he recalled:

On deck the scene was bedlam. The locals on the starboard side were pressing to port so as to avoid the driving rain and the more they pressed the greater the list became, and a few Europeans awake were trying to drive them back to trim ship. Fortunately the native police aboard rallied magnificently and once understanding what was wanted, the excess weight on deck was again evenly distributed in a short space of time. The ship's head was also brought into the wind promptly, but the few minutes over which this happened were as tense as any I have known. (Clarke 2001, 45)

A magnificent pyrotechnic and electrical display was performed all night by the volcanoes for those *Montoro* passengers who stayed awake to watch the rare spectacle of two volcanoes in full, simultaneous eruption. Brett Hilder drew a sketch of the view (Figure 3.27) and recorded that:

Most of us stayed up to watch the satanic celebrations; the two volcanoes on each side of the entrance, were throwing up a solid jet of red-hot dust and stones to a great height and the two columns appeared to meet somewhere over Rabaul. The lightning was fantastic, some flashes bursting like bombs, others running horizontally around the ascending columns, while forked lightning zig-zagged down to the surface of the sea. (Hilder 1961, 54–5)

Clive Meares also saw the same 'grandstand' view from the comparative safety of the *Montoro*, adding that a new cone must have been built up at Vulcan because there was a 'repeated appearance of a ball of fire which then broke in two and "poured" back into the sea as though running down the two sides of an isosceles triangle' (Meares n.d., 7). Meares was describing some form of subaerial pyroclastic flow down the new cone of Vulcan.

Viewing the eruptions from Kokopo, George Clarke felt that:

As a spectacle the volcanoes were their best at night. At two or three minute intervals each would erupt with a mighty blast, shooting a molten and glowing stream over a thousand feet into the air, and as the thrust diminished, red-hot boulders could be seen falling back into the crater. Each night whilst the volcanoes were in major eruption a terrific electric storm would develop over them between 8 and 10 o'clock, producing the most vivid lightning I have yet seen. At regular intervals a chain of lightning would strike downwards from high above them and at a distance of up to 10,000 ft divide into two forks each simultaneously penetrating the core of both craters. These storms when at their peak were an awesome sight. (Clarke 2001, 45)

Figure 3.27. Brett Hilder sketch of Vulcan and Tavurvur eruptions at night-time.

This sketch by navigator Brett Hilder, showing both Vulcan (left) and Tavurvur in activity, is of the northward view, as seen by Hilder and thousands of others on board the *Montoro* off Kokopo overnight on Sunday 30 May 1937 (Hilder 1961, 57). Note the difference in the eruption style between the two volcanoes represented in the sketch: violent, jet-like explosions at Vulcan together with an eruption cloud that extends above the top margin of the sketch; Tavurvur, whose eruption cloud is almost entirely within the field of view. Winds at different levels are blowing the eruption clouds off to the north-west or west-north-west. The distance between the two volcanoes is about 6 kilometres, so the height of the Tavurvur eruption cloud where it starts to be blown on a more westerly vector is about 4 kilometres. The volcanic cones of Tovanumbatir, Kabiu and Turagunan (the Mother and two Daughters) are clearly visible on the right. GA negative reference GB2189.

An important aspect of these reports of eruptions from both volcanoes after dusk relates to the likelihood that Tavurvur had changed its style of eruptive activity from its initial wet 'ash-mud' eruption—which lasted only half an hour according to Doug Joycey—early on the Sunday afternoon, to ongoing explosive activity of a different kind. The evening/night-time drawing by Brett Hilder (Figure 3.27) is significant with regard to the near-vertical eruption column from Tavurvur, which reached a height of about 4 kilometres before being driven north-west by high-level winds and merging with the Vulcan cloud. This height of 4 kilometres is far higher than either the Mother or South Daughter cones on the right-hand side of the sketch, and higher than the clouds that produced the mud-ash clouds early on the Sunday afternoon. Finally, the later, night-time Tavurvur eruption clouds were incandescent, consistent with the Matupit observation of 'large fiery rocks' being expelled during the night (Mennis 1972, 64). There is, however, some uncertainty about how much of the new ash from these later Tavurvur eruptions fell on Rabaul during the night when so few people had remained in the town, or indeed on Matupit Island itself where there was comparatively little damage. Much of it may have fallen into Simpson Harbour. Hal Evans recalled that the darkness in Rabaul, apparently caused by the eruption, lasted only about three hours (Evans 1937). In any case, judging by photographs, the ash-producing Tavurvur eruption had more or less ceased by daylight the next morning, although water vapour and volcanic gases continued to be expelled, whereas Vulcan continued to produce explosive eruptions.

3.8. North-Coast Rescues and Back to Herbertshöhe and Vunapope

Those Rabaul townspeople seriously affected by the eruptions of both Vulcan and Tavurvur were in effect abandoning—at least until the future became clearer—the site of what had been the new German capital in 1910 in favour of the old German capital, Herbertshöhe, now known as Kokopo, which appeared to be a decidedly safer place. It was, after all, situated outside Blanche Bay and the area of its active volcanoes and protected from any volcanic fallout by the south-east trade winds at that time of year. Other people seriously affected by the Vulcan eruption to the west also focused their attention on reaching the facilities of the Kokopo area. This included the MSC headquarters of the Catholic Church and hospital at Vunapope, the 'place of the Pope', which was run by German fathers, led by the Bishop Gerard Vesters, a Dutchman. Disembarkation from the rescue vessels in the anchorage required the transfer of refugees by small boats to the shoreline beaches (Figures 3.28 and 3.29).

The *Montoro*'s anchor was let go at the Kokopo anchorage at 6 am on Monday 31 May, and the first of the 5,000 or so New Guineans—this number is Captain Michie's estimate—began leaving on the ship's small boats (Michie 1937; Hilder 1937, 1961; Clarke 2001; MSC 1937; Arculus and Johnson 1981). About 200 Europeans were given breakfast before landing, but many European women and children stayed aboard—mostly wives and families of residents who could not be accommodated ashore. The *Montoro*, said crew member George Clarke, became for a day or two 'a floating boarding house and restaurant' (Clarke 2001, 45). Seventy people, apart from ticket-carrying passengers, were fed and housed on the boat that normally had room for only 50 passengers. The *Montoro* shared anchorage with the cargo vessels *Golden Bear* and *Polzella* and with other smaller vessels (Figure 3.30).

Figure 3.28. Rabaul refugees at Kokopo beach.

A multiracial group of Rabaul refugees has reached Kokopo beach after evacuation from Nodup on Sunday 30 May 1937. GA negative reference GB2618B.

Figure 3.29. Launch being unloaded at Kokopo beach.

Expatriate men help unload a launch at the beach at Kokopo. One European is still wearing his baseball shirt (V-shaped insignia) from the games played at Rabaul's sports field just before the outbreak of eruptive activity at Vulcan. A European woman, possibly a patient from Namanula Hospital, lies on the roof of the launch. This photograph is one of a set accompanying a 15-page article on the 1937 eruption published in the *Sydney Mail* of 9 June 1937. GA negative reference GB2590.

Figure 3.30. Ships anchored at Kokopo and hulk of the *Loch Katrine*.
The *Montoro*, the *Golden Bear* behind it, and other smaller vessels are seen anchored off Kokopo in this photograph taken from above the shoreline on 30 May 1937. The listing hulk of *Loch Katrine* in the left foreground acts as a break for the enclosed swimming area at the beach edge. The *Montoro* can be identified by the distinctive white band around its funnel (compare with Figures 3.5 and 3.23; see also Figure 3.41). The dated photograph was supplied courtesy of Mrs G. Forsyth, who was evacuated from Nodup on the *Golden Bear*.

The *Montoro* became the radio link between Kokopo and the outside world until Mr Twycross, the officer in charge of the wireless station at Rabaul, established operations on shore at Kokopo on Thursday 3 June. L.C. Coleman at the Rabaul receiving station and C.B. Alexander at the transmitting station had stayed at their posts. By 12.15 am on Sunday, Alexander had re-established power from an auxiliary generator and, with the assistance of C.H. Sturgeon and H.S. Burgess, messages were being transmitted and received—though under trying, if not heroic, circumstances. The *Montoro*'s wireless office 'was besieged by people trying to send wireless messages, mostly without money to pay for them', reported Captain Michie (1937, 2), and the wireless operator, Norm Odgers, worked heroically, non-stop, on board the *Montoro* from Sunday night until Tuesday morning coping with the deluge of radio traffic (Clarke 2001).

Another role taken on by the *Montoro* was that of provider of stores to the growing population at Kokopo. The holds contained stores destined for Salamaua and Lae, but Mr P. Coote, manager of Burns Philp in Rabaul, gave orders that they be off-loaded for use at Kokopo. Both bulk freezers at Rabaul had been put out of action after the electricity station became inoperable, so fresh meat, vegetables and fruit were supplied to Kokopo from the *Montoro*, together with other goods, which were all carefully tallied and sent ashore. There was a special plea from European mothers for milk powder for their babies. One small box of it was on the Lae manifest, but it was somewhere among more than 300 tons of total cargo. Purser Clarke was relieved when: 'We opened the first hold of the Lae cargo and there perched right on top of the stow was our case of Glaxo' (Clarke 2001, 45).

Planning by administration officers at Kokopo to receive the influx of evacuees had begun as early as Saturday evening when Mr Chinnery returned to Kokopo after his escape from the Vulcan eruption. District Officer Don Waugh and Assistant District Officer K.C. McMullen were in charge of arrangements for the evacuees, but the administration had little in the way of facilities to house and feed them. However, the Sacred Heart Mission at Vunapope kept large supplies of food in bulk store for distribution to outlying stations and had several large buildings in the mission grounds. Bishop Vesters and his mission staff agreed readily to a request that they assist with the expected evacuees. Vunapope staff were to play a major part in the relief effort, and they subsequently received high praise from many grateful individuals. Vunapope Mission was the lynchpin of the administration's plan for coping with the refugees of the Rabaul evacuation (Figures 3.31–3.33). Civilian participation in the relief efforts was also facilitated through Mr Waugh and others establishing a citizens' committee that included representatives from the administration, business community and townspeople.

Many Rabaul refugees had already reached Kokopo and Vunapope before the off-loading of the *Montoro* began on Monday morning. Those from the *Golden Bear* and the other small vessels had disembarked on Sunday. Some European families were initially accommodated in homes on plantations following spontaneous invitations from individuals, but subsequently a billeting system was organised. Amy Anthony, for example, landed at Kokopo on Sunday without her husband, but accompanied by her two puppies ('while sitting on the beach and feeling sorry for myself', she met fellow evacuee Jean McCarthy, who ended up having a puppy given to her). The Anthonys were eventually accommodated together at Carpenter's plantation at Ralabang where they had to sleep on the verandah—the first night on a sofa using store-issue blankets, and the second on mattresses made from copra sacks stuffed with banana leaves. They 'could hear the booming sounds of the volcanoes all night [Sunday], and feel the shakes of the tremors. It was a pretty grim night', and by morning they found the verandah crowded with others wrapped in blankets and in a similar plight. Newlywed Jean McCarthy was billeted at Ulaveo plantation for almost three weeks, while husband Keith stayed on at Rabaul to assist with relief work there.

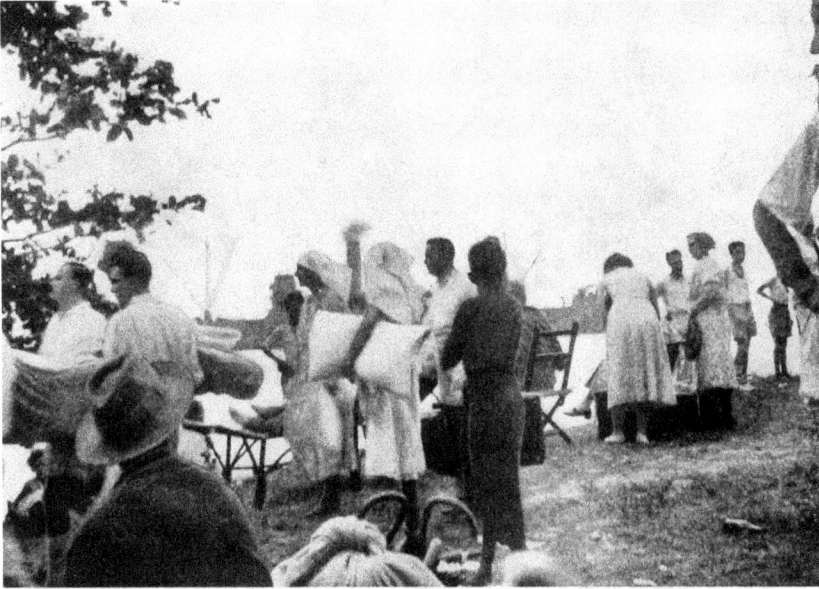

Figure 3.31. Nurses attending to refugees at Kokopo.

A first-aid station on the foreshore at Kokopo has nurses and others attending to needy refugees brought in by vessels from Nodup. One of the ships in the lower background (just visible behind the people) appears to be the *Golden Bear*. GA negative reference GB2592.

Figure 3.32. Camping out at Kokopo.

Tent accommodation such as this was provided for many people evacuated to Kokopo. This photograph was taken by Mr J. Hewett. GA negative reference GB2940.

Figure 3.33. Vunapope procession and Vulcan eruption cloud.

A Corpus Christi procession went ahead at Vunapope on Sunday morning 30 May 1937 despite the eruption at Vulcan Island. This photograph was taken by Sarah Chinnery (1998, 213). She did not have any film of her own, but a roll was found for her by the young missionaries at Vunapope. The towering Vulcan eruption cloud can be seen in the background of this image, which was digitally enhanced and provided by the National Library of Australia.

Many New Guineans who arrived at Kokopo, either on foot from the devastated area west of Vulcan or from the *Montoro*, found accommodation in nearby Tolai villages. Father Benda later wrote to a colleague in Germany telling him that his

> former parishioners from Paparatava distinguished themselves especially. They took in the people from Tavuiliu station and looked after them without payment. How much taro was thus consumed you may work out for yourself. To feed several hundred guests for two whole weeks is a great achievement. (MSC 1937; Arculus and Johnson 1981, 38)

New Guineans who could not be accommodated in villages stayed in or around Kokopo, in the administration labour compound or, in fact, anywhere they could find. Tents were sought and found by administration staff, and many more arrived later on relief ships, swelling the tent town that grew at Kokopo (Figure 3.32). Asians, New Guineans and Europeans were segregated and watched closely by Department of Public Health officials in case of any outbreak of disease.

Exactly how many people were fed and, in many cases, housed at Vunapope is not known. Father Hepers thought as many as about 10,000 (MSC 1937; Arculus and Johnson 1981, 12). This seems quite feasible if, as Brett Hilder (1937, 54) considered, 6,000 people alone were on board the *Montoro*, though not all may have passed through Vunapope. The mission, by Sunday, was 'overflowing', 'a vast army camp', 'buildings simply packed out with people' (MSC 1937; Arculus and Johnson 1981, 12–13). Patients evacuated from Namanula Hospital filled up Bethania Hospital at Vunapope, including the convalescent ward. European women and their children were accommodated in the sisters' quarters: the Daughters of Our Lady cooked for between 60 and 70; the Hiltrup Sisters for 42 white women and children. Father Murche was in charge of a building for mixed-race children who boarded at the mission, and 240 Chinese were accommodated there. The mission houses used as quarters by the fathers and brothers were similarly occupied by milling refugees. Mr Hoogerwerff came off the *Montoro* and found a place to stay in an old store at Kokopo, but, through the kindness of Brother Nattebride, he was able to sleep instead in a boxroom at the mission (Hoogerwerff 1937). The mission kitchens, and others set up outside, served meals for thousands. Bernard Ryan remembers vividly the great piles of bread and butter and huge bowls of hard-boiled eggs (Ryan 1980–84).

The mission even made available to Gordon Thomas its printing facilities, and a special 'Volcano Issue' of the *Rabaul Times*, dated 4 June, was published at Vunapope. Thomas wrote in the first paragraph:

> [R]ight here we wish to record the wonderful work done by the Mission all through this harrowing period and congratulate His Lordship Bishop Vesters, and his untiring staff, on the excellent organisation and display of Christian spirit. (Thomas 1937a, 1)

In this special issue, Thomas warned of

> commenting on the possibilities of the future ... Whether the capital of the Territory is to remain at Rabaul or be removed to some other locality is not a question to be finalised at this juncture. Both the volcanoes and our mental state must be allowed to cool off. In a month or so—perhaps more—we can the better decide such a momentous question. (Thomas 1937a, 1)

Such commentary on the future of Rabaul was present, and inevitable, in the community from the end of May onwards; however, the matter would not be settled for some years. Meanwhile, people were trying to come to terms with the personal and family situations that had resulted from the disaster.

Father Laufer of Rakunai station had to be brought into Vunapope because of his exhaustion (MSC 1937; Arculus and Johnson 1981, 14–15). He had arrived at Vunalama, near Tavuiliu, on Saturday at midnight after fleeing the Vulcan eruption cloud, and on Sunday he carried on towards Vunadidir. Father Hepers at Vunapope had set out from there on Sunday morning to meet him at Rakunai, believing the station was unaffected, but he could proceed no further than Taliligap. He therefore drove on to Vunakanau where he met Father Bogershausen and four Hiltrup sisters from Tavuiliu:

> The sisters had wandered about the whole night in the rain of mud and ash; they were covered with dirt from head to foot and had only saved what they had on them. Fr Bogershausen equally! (MSC 1937; Arculus and Johnson 1981, 14)

Father Hepers heard from some local people that Father Laufer was on his way south to Vunadidir on foot, and he found him about halfway there, staggering through the wet kunai grass in shirt and torn trousers and with completely worn-out shoes. He had nothing else with him ... and had discarded his soutane on the way. (MSC 1937; Arculus and Johnson 1981, 15)

Father Laufer was brought to Vunakanau and he and the sisters were driven to Vunapope.

Others needed rescuing too, particularly those Rabaul evacuees now stranded along the north coast, and the administration made arrangements for their return to Kokopo and Vunapope by boat. The *Asakaze* was sent to the north coast carrying Dr Champion Hosking of the Department of Public Health in case medical relief was required. Rev. Laurie Linggood, with the Trevitt–Chaseling wedding party on the north coast, had no intention of waiting for rescue. Anxious to reach his wife and child, he left Vunairima at 5.15 am on Sunday to extricate his truck and drive to Raluana, first south then eastwards around the southern edge of the devasted area. Rev. Linggood eventually succeeded in reaching his family and the safety of Raluana (Wayne 1937).

Rev. Albert Jones and Ron Wayne at Vunairima took tools and a team of New Guineans back to the abandoned cars on Sunday morning, and cleared logs from the road, including some that had fallen after Rev. Linggood had been through. 'When we reached the vehicles we had to shovel inches of pumice off them before we could open doors and the bonnets ... Rain was falling all the time and still contained mud', Wayne (1937, 7–8) noted, but he and Jones were able to bring the cars back to Vunairima at about 3 pm. They heard during the day that Tavurvur had started activity of its own, and eruptions from Vulcan continued into and through the night. The next day, Monday, Rev. Jones, Wilfred Pearce (business manager of the Methodist Church) and Ron Wayne and his wife, Helen, attempted to get back to Kabakada. Jones wrote later:

> We had to cut a way through to a village 5 miles from K'da, when we were forced to stop as the road had been carried away. I left the car and Mr Pearce and went on with the Waynes. We forded the first stream, half of the road being there. When I returned a few hours later [on his way back to Vunairima] the whole of the road was gone leaving a hole 12 ft deep. Another few miles and we came to a big river that had come down ... [cutting] a huge track 40 ft deep and about 35 ft across. It had been a raging torrent. Though much water was gone it was still racing out to sea. It was a twisted

mass of fallen palms and pumice stone ... One of the boys sat Mrs Wayne on his shoulders with her legs around his neck and carried her across the fallen palms ... We pushed on, passed dozens of cars that had been abandoned by the road side, till we came to K'da ... Our beautiful garden was no more and everything was covered in inches of volcanic ash. The house was thick with it, and the tables still set on the back verandah. There were the place cards still in their places. (A.S. Jones 1937a, 4; see also A.S. Jones 1937b, 5)

Ron Wayne noted that Kabakada was not so seriously affected by the fallout compared with Vunairima, or with Kabaira plantation through which they had passed:

Kabaira ... suffered most, all of the palms being broken, to say nothing of the palms that had fallen. The fronds had not only bent down beneath the weight of mud but the midribs had broken at the bend—a most unusual damage for a leaf to suffer. It was heartbreaking to look along lines and lines of thousands of palms and see only the topmost uncurled frond sticking up like a spearhead. (Wayne 1937, 10)

Rev. Jones's party heard that the schooners *Asakaze* and *Induna Star* had that morning picked up many people, including the large group stranded at J.O. Smith's plantation at nearby Vunawutung (Wayne 1937). The group included Methodist chairman Rev. Lewis and his wife, who were transferred to Kokopo and then to accommodation at Vunapope. This news was later received by Hazel Jones at Vunairima: 'Mrs Lewis is with the Roman Catholic Sisters in a Nunnery, and our Rabaul Minister is in a monastery. We think it is a great joke' (H.L. Jones 1937, 10). Virgil King was also one of those taken by boat to Kokopo from Vunawutung and, like many others, was deeply impressed by the hospitality afforded him by J.O. Smith: 'That big-hearted planter ... was amongst us continually with the perpetual question, "Is there anything else I could do for you?"' (King 1937, 6; Lewis 1937a). Evacuation by sea seems to have been the only feasible way out for many people on the north coast, as the road eastwards to Rabaul was cut by fast-flowing streams similar to those crossed by the Jones party to the west. Rev. Howard Pearson, who was with the Lewises at Vunawutung, had attempted to return towards Rabaul, but he was unable to cross a rushing stream and, on his return, encountered a new stream that had just formed.

Ron and Helen Wayne also wanted to leave the north coast so that they could return to their home at Ulu Mission on the Duke of York Islands, but the relief boats had evidently departed and evening was fast approaching (Wayne 1937). Then, from the beach, they saw the Vunapope vessel *Theresa* moving along the coast towards them. It had come from Kokopo and was in the charge of Patrol Officer C.B. Bates. The *Induna Star* had reported that all people from the north coast had been evacuated, but Bates was out looking for any people still stranded there. The Waynes boarded the *Theresa*, together with Mrs Lulu Miller, joining other evacuees on board. Mrs Miller had managed to reach Watom Island; she had left her pet birds in the charge of the father there, and she now wanted to get back to her home. Instead, she was taken to Vunapope, where, against her wishes, she was obliged to stay for the next three weeks: 'Oh, I asked and asked; they wouldn't let me go back to my place' (Miller 1980, 35).

Damage to the crops of commercial plantations, especially along the north coast, was extensive. No less than seven members of the Planters' Association of New Guinea were offered immediate assistance, as reflected in the ultra-formal language of the association's annual reports, which also referred to the mandated expropriation arrangements that had been set in place following the Treaty of Versailles:

> Application for suspension of payments of Principal and Interest instalments due in respect of affected Expropriated Properties, was made to the Custodian who has granted a suspension of the payments due 1st July, 1937, and assured the Association of sympathetic consideration based on the merits of individual cases in respect of future payments. The extent of the damage done to these plantations cannot yet be fully assessed. It is certain however, that production will be very seriously affected for the next eighteen months to two years, and that consequent loss to owners will total many thousands of pounds. (Planters' Association 1937, 4; see also 1938)

3.9. Reduced Eruptions and Return of the Administrator

The District Office at Kokopo and the staff of Vunapope Mission were, by Monday, coping well enough. However, the flood of refugees from the surrounding evacuated areas was stretching them to their limits and, clearly, the support they were providing would not last indefinitely (Thomas 1937a; Robson 1937; MSC 1937; Arculus and Johnson 1981). More supplies would be needed to supplement those off-loaded from the *Montoro*, held at Vunapope, and being transferred from Rabaul. The supply of roof-catchment water and the water tanks of the *Montoro* were being rapidly drained, and fresh supplies would be needed urgently. Further, the prevention of serious outbreaks of disease in the crowded buildings and camps at Kokopo and Vunapope would have to be tackled—particularly infectious diseases such as dysentery caused by poor sanitation, and also malaria, an ever-threatening reality that could reach epidemic proportions if mosquito nets were not made available, or if open bodies of water were not drained or treated. Dr T.C. Backhouse and, later, Dr Champion Hosking were in charge of medical arrangements at Kokopo, assisted by three other doctors, including Dr Watch, a private practitioner, who had come to Kokopo after assisting overnight at Namanula Hospital, as well as staff from the Department of Public Health, including Roger Davies.

On Monday, District Officer Waugh sent from the *Montoro* a telegram to Canberra detailing the situation, his requirements and the names of supply ships that could help with the relief effort:

> RABAUL POPULATION SUCCESSFULLY EVACUATED TO KOKOPO SUNDAY EXCEPT JUDGE PHILLIPS TOWNSEND MELROSE BALL AND FEW OTHERS STANDING BY AND HUNDRED ODD ON NORTH COAST. NO KNOWN CASUALTIES. ADMINISTRATOR SALAMOA EXPECTS FLY RABAUL THIS MONDAY MORNING. ESTIMATED TWO DAYS RATIONS KOKOPO. MONTORO RUNNING SHORT WATER ARRANGE MORESBY SECURE SUPPLIES THURSDAY ISLAND. ESSENTIAL MALAITA ARRIVE AS SOON AS POSSIBLE SUPPLIES VERY SHORT ENDEAVOUR SHIP MALAITA 50 TONS RICE 500 CASES MEAT 12 OZS TWENTY TONS FLOUR THREE TONS SUGAR FIVE CHESTS TEA THREE TONS FINE SALT 25 CASES CONDENSED

MILK 50 CASES IDEAL MILK FIVE CASES LACTOGEN TEN CASES HAM 100 GROSS EGGS 1000 LBS BACQN THREE TONS POTATOES TWO TONS ONIONS 15 CASES TINNED BUTTER 20 DOZEN HURRICANE LANTERNS 100 CASES KEROSENE 4 CASES MATCHES 50 CASES SOAP 200 TENTS BEDDING FOR 600 PERSONS THOUSAND MOSQUITO NETS 1000 BLANKETS 1000 TOWELS 120 FIVE GRAIN QUININE CAPSULES 60 ASPIRIN TABLETS SIX DOZEN CHLOROFORM SIX DOZEN ETHER 1000 EACH SHIRTS SHORT TROUSERS SANDSHOES ASSORTED SIZES 60 DOZEN EACH TOOTH BRUSHES TOOTHPASTE. WAUGH OFFICER IN CHARGE KOKOPO. (Waugh 1937)

HMAS *Moresby* was surveying the Gulf of Carpentaria and proceeded at once to Thursday Island where it picked up supplies, reaching Kokopo on Saturday 5 June. The Burns Philp vessel *Malaita* at Brisbane was chartered to take supplies to Kokopo, arriving there on Sunday 6 June. Water supplies were brought in by the *Island Trader* the same day and, later, by other vessels.

Short-term relief problems at Kokopo and Vunapope were being dealt with expediently, but many refugees remained concerned about the future of Rabaul. Would the town recover from the effects of the volcanic eruption? If so, how soon before they would be able to return? Was Simpson Harbour now blocked from the open sea? Would Rabaul be built elsewhere? Would Rabaul remain the capital of the Mandated Territory? These questions would soon be addressed by Administrator Ramsay McNicoll, who, on the morning of Monday 31 May, was on his way back from the Morobe Goldfields by air via Lae and was expected at the Taliligap (Vunakanau) Airfield at lunchtime. McNicoll had heard about the volcanic eruption at Rabaul by way of the Wau radio station:

WAU RADIO PICKED UP FROM UNKNOWN SHIP BARE NEWS ERUPTION AT RABAUL AND RESIDENTS EVACUATING. AM RETURNING FROM WAU TO RABAUL BY AIR. RABAUL RADIO REPORTED OUT OF ACTION WILL SEND FURTHER NEWS BY ANY MEANS POSSIBLE. MCNICOLL. (McNicoll 1937a)

The administrator radioed Canberra and set off, accompanied by Director of Public Works C.R. Field, on board a three-engine Guinea Airways plane piloted by Captain A.J. Turner (Thomas 1937a). They arrived over Rabaul about midday, circling the harbour area—including Rabaul town—and obtaining, for the first time from the air, an impression of the extent of the damage to the Blanche Bay area caused by the two volcanoes. The tropical-garden town had been stripped of its fine foliage. The streets, near-deserted and branch-strewn, were shrouded with grey ash and pumice. Much of the waters of Simpson Harbour and Blanche Bay were covered by floating pumice that must have seemed solid enough to land on. To the west, where formerly had been villages, gardens and forest, they saw only devastation. And, instead of low-lying Vulcan Island, a new peninsula jutting out from the western side of the bay and dominated by a new and still growing volcanic cone, Vulcan, was observed (Figures 3.34–3.35). The aircraft circled the town and then landed 'perfectly' at 12.30 pm on 'the new 'drome at Taliligap' (Thomas 1937a, 3) outside the area of devastation. By that time, Tavurvur volcano had, apparently, ceased producing the volcanic mud that had blanketed Rabaul town the previous day and possibly during part of the night.

The degree to which the two volcanoes were still active on the afternoon of Monday 31 May should be addressed, given that Captain Turner was able to circle Rabaul without apparently any interference from suspended volcanic materials and to land safely at Taliligap. Four photographs are of interest regarding the explosive eruptions then taking place at Vulcan (Figures 3.34–3.37). They were all taken at a time when the Vulcan cone was at, or nearly at, full height—likely on Monday 31 May. The 'phreatoplinian' and 'plinian' eruptions had ceased less than two days after the initial sea floor outbursts near Vulcan Island. By 31 May, the style of eruptive activity was 'vulcanian'—a general term used widely by volcanologists for what is a common type of explosive activity, including for many volcanoes in Papua New Guinea—and subaerial, although the extent to which the eruptions were still being affected by seawater deep within the cone is unknown.

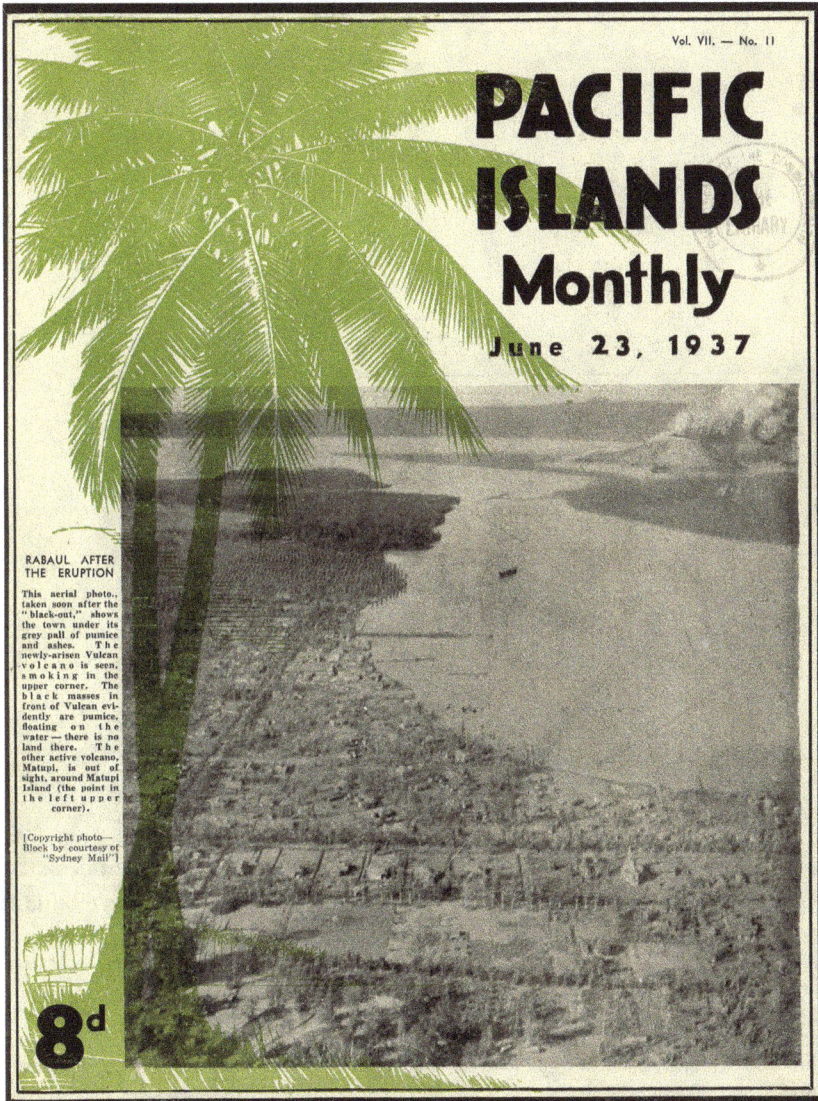

Figure 3.34. Aerial view of Rabaul town and Vulcan on front cover of *Pacific Islands Monthly* magazine.

The new Vulcan cone is seen in explosive eruption in the top right-hand corner of this aerial photograph that featured on the front cover of an issue of *Pacific Islands Monthly*. The photograph was evidently taken during the flight that brought the administrator to Vunakanau Airfield on 31 May (see also Figure 3.35). Rabaul town is shown in the bottom left of the photograph, covered in volcanic mud. The caption read: 'Rabaul after the eruption. This aerial photo., taken soon after the "blackout," shows the town under its grey pall of pumice and ashes. The newly arisen Vulcan volcano is seen, smoking in the upper corner. The black masses in front of Vulcan evidently are pumice, floating on the water — there is no land there. The other active volcano, Matupit [Tavurvur] is out of sight, around Matupit Island (the point in the left upper corner).'

Figure 3.35. Aerial view of the new Vulcan cone and small eruption clouds.

Newly formed and 225 metres high, Vulcan cone is seen from the air in this photograph taken by pilot Jack Turner on Monday 31 May 1937 (Fisher 1939a, 61). This was the day when Administrator McNicoll was flown by Turner into the airstrip at Taliligap/ Vunakanau. Lightly laden ash clouds emerge from the central crater, as well as from a small subsidiary vent on the south-south-western side of the new volcano. This small vent was evidently caused by a blow-out on the flanks of the main cone, and it released vapour and entrained ash for many days after the main activity had ceased. Ravines apparently cut by floods of water from the Vulcan eruption cloud are clearly visible on the extreme left and on the new shore in the right-hand corner where there is no sign of any floating pumice on the waters of Simpson Harbour (see also Figure 3.37). GA negative references M2579-1-12-2 and M2579-11.

Figure 3.36. Emissions from both Vulcan and Tavurvur as seen from Taliligap.

Taliligap provided an excellent vantage point from which to view and photograph the changes to Simpson Harbour and Blanche Bay caused by the 1937 volcanic eruptions. This photograph appears to have been taken on Monday 31 May, perhaps by E.A. Hawnt from Taliligap (compare this photograph with a similar undated one taken by Hawnt and published by Fisher [1939a, 61]) or by Captain Turner on his way by road from the Vunakanau Airfield to Kokopo. Vulcan cone (left of centre) has reached almost its full height and is in moderate or weakening eruption. Ash is falling out of the cloud to the left. Tavurvur (extreme right) appears to be emitting only water vapour and gas. Floating pumice covers much of the bay and harbour. Compare this photograph with the one taken from Taliligap by Mrs Sarah Chinnery on Saturday 29 May just before the Vulcan outbreak (Figure 3.7). GA negative references GB2792 and M2447-33A.

Ash is seen falling from all the Vulcan eruption clouds shown in Figures 3.34–3.36. Figure 3.37 is rather different in that pyroclastic materials can be seen flowing down the sides of the cone. The origin of these pyroclastic flows is uncertain, but they may represent the result of what some investigators at other volcanoes have called 'shallow-pocket' explosions (Perret 1937; Taylor 1958). Such explosions take place at a shallow depth, just beneath the crater, and are fairly weak; the crater fills with a massive, convoluted cloud of fragmental debris and gas that has little tendency to rise. The heavy, buoyant, 'fluidised' mass of pumice and ash then overflows the lower parts of the crater rim. This type of explosion is quite different to the stronger eruptions of a day or two earlier that originated deep within the volcanic conduit, causing a stronger upward thrust to the surface and pyroclastic flows that originated from the collapse of a high, weakening eruption column.

Figure 3.37. Small pyroclastic flows running down the flanks of the new Vulcan cone.

Clouds of hot pumice and gas cascade down the sides of the newly formed Vulcan cone towards the end of the 1937 eruptive period when the overall activity was periodic and waning. Note that the waters of Blanche Bay in the foreground on this occasion appear to be free of floating pumice. The high volcanic cloud shown here appears to be vapour-dominated but ash can be seen falling out of the clouds on the left. The date of the photograph is uncertain but is most probably in the period 31 May – 2 June. GA negative references M2579-1-12-4 and M2579-9.

McNicoll, by the time of his arrival, was seeing both volcanoes in a much less threatening condition compared with people who had endured, or had fallen victim to, the effects of the eruptions—particularly Vulcan—over the previous two days. McNicoll was met at the airstrip by Clive Meares, who briefed him on events while the rest of the party, which included C.R. Field, the director of public works, drove down to Vunapope (Meares 1980; Thomas 1937a). McNicoll then boarded the *Montoro*, where he spent a few minutes with Mrs McNicoll, who had been evacuated from Namanula, before leaving by the *Induna Star* for Nodup and Rabaul (Figures 3.38 and 3.39). That night he sent from the *Montoro* his first report to Canberra:

> LANDED RABAUL DROME TWELVE THIRTY TODAY REACHED KOKOPO THREE OCLOCK PROCEEDED BY SCHOONER TO NORDUP AND ROAD TO REMAINS OF RABAUL. WIDESPREAD DEVASTATION. VOLCANO AT VULCAN ISLAND STILL ACTIVE ALSO OLD CRATER NEAR MATUPI. FINE ORGANISATION AT KOKOPO WHERE WHOLE POPULATION WHITES CHINESE NATIVES INCLUDING THOSE PREVIOUSLY SHELTERED ON NORTH COAST NOW CONCENTRATED. DISTRICT OFFICER WAUGH ASSISTANT DISTRICT OFFICER MCMULLEN AND ADMINISTRATION STAFF ASSISTED MOST ABLY BY GREAT MANY PUBLIC VOLUNTEERS HAVE KOKOPO MATTERS WELL IN HAND. AT RABAUL PHILLIPS BALL MELROSE GREGORY MCCARTHY WITH WHITE AND NATIVE POLICE AND MANY VOLUNTEERS ARE CONTROLLING ISSUES AND ATTEMPTING CLEAR ROADS. IMPOSSIBLE OVERPRAISE WORK OF STAFF AND VOLUNTEERS AND DEMEANOUR OF PUBLIC INCLUDING WOMEN CHILDREN CHINESE AND NATIVES. LARGE AND GROWING CONICAL HILL AT VULCAN ISLAND. HARBOUR RENDERED IMPASSABLE BY THICK LAYER OF FLOATING PUMICE. WILL REVIEW WHOLE SITUATION TOMORROW AND REPORT. MCNICOLL. (McNicoll 1937b)

An important matter for Ramsay McNicoll was assessing the state of Rabaul Harbour. Had the eruptions so modified the sea floor of Blanche Bay and Simpson Harbour that Rabaul could no longer be used as a port town? Captain Michie was no less interested and, at 10 am on Tuesday 1 June, he set off from Kokopo in the *Montoro*'s launch, together with others of the crew, including Second Mate Hilder, in an attempt to get into the harbour through the normal steamer passage:

> About one mile from Matupi we ran into a sea of pumice which blocked the entrance from Vulcan Island to Matupi and right up Matupi Harbour. Vulcan ... was belching forth stones and lava in thousands of tons. Our launch was only able to penetrate the pumice barrier for about 100 yards, when the engine got too hot [pumice had entered the intake of cooling water] and the launch almost came to a standstill. The attempt to get through was abandoned, and we returned to the ship. (Michie 1937, 2; Figure 3.40)

Figure 3.38. Ash-covered steps at the Rabaul Hotel.

Brigadier-General Ramsay McNicoll (right) and pilot Jack Turner pose for a photograph at the Rabaul Hotel presumably shortly after they had reached Rabaul via Kokopo after having flown into Vunakanau Airfield on Monday 31 May. Ash or dried-out mud mantles the steps. GA negative reference GB2594.

the A.W.A. men were driven out by sulphur fumes and ash, Mr. Faulkner's radio and special messengers were sent out in all directions, instructing the people (800 Europeans, 1,000 Chinese and over 5,000

BRIG.-GENERAL W. RAMSAY McNICOLL   JUDGE F. B. PHILLIPS

Figure 3.39. Combined portrait photographs of McNicoll and Phillips in *Pacific Islands Monthly* article.

These photographs of the administrator and judge were published in an issue of the *Pacific Islands Monthly* (Robson 1937, 10). This digital copy was provided courtesy of the National Library of Australia.

Brett Hilder tried to pull up a bucket of water from over the side but obtained one full of pumice. They also saw, while in Blanche Bay, a large launch returning to Kokopo—on board, a police officer and 'two old friends of ours … respectable Europeans' who had, finally, been extricated from their drinking binge and pub crawl of Rabaul's hotels and clubs (Hilder 1961, 57).

Tuesday 1 June was also the day on which Ramsay McNicoll issued the first of a series of 'Circular Despatches' (CD) from the central administration at Rabaul (McNicoll 1937c). His assessment of the situation was not encouraging:

> Rabaul will be untenable for many weeks, the water supply has become polluted and no sanitation arrangements are at present possible. The present bad situation will become intensified when rain comes as the roofs of all houses are thickly covered with volcanic dust which will become thick mud directly it becomes wet. Under the circumstances it is necessary to restrict the number of persons that may be permitted to visit Rabaul. Only those officials on duty

and others with urgent business to attend to will be permitted to proceed from Kokopo and land at Nodup. It is essential that written authority be first obtained ... [and] extremely desirable that at the very first opportunity as many women and children as possible proceed to Australia. (McNicoll 1937c, CD1, 1)

Subsequently, many European women who were still on board the *Montoro* sailed with her when she left Kokopo for Lae on Wednesday 2 June.

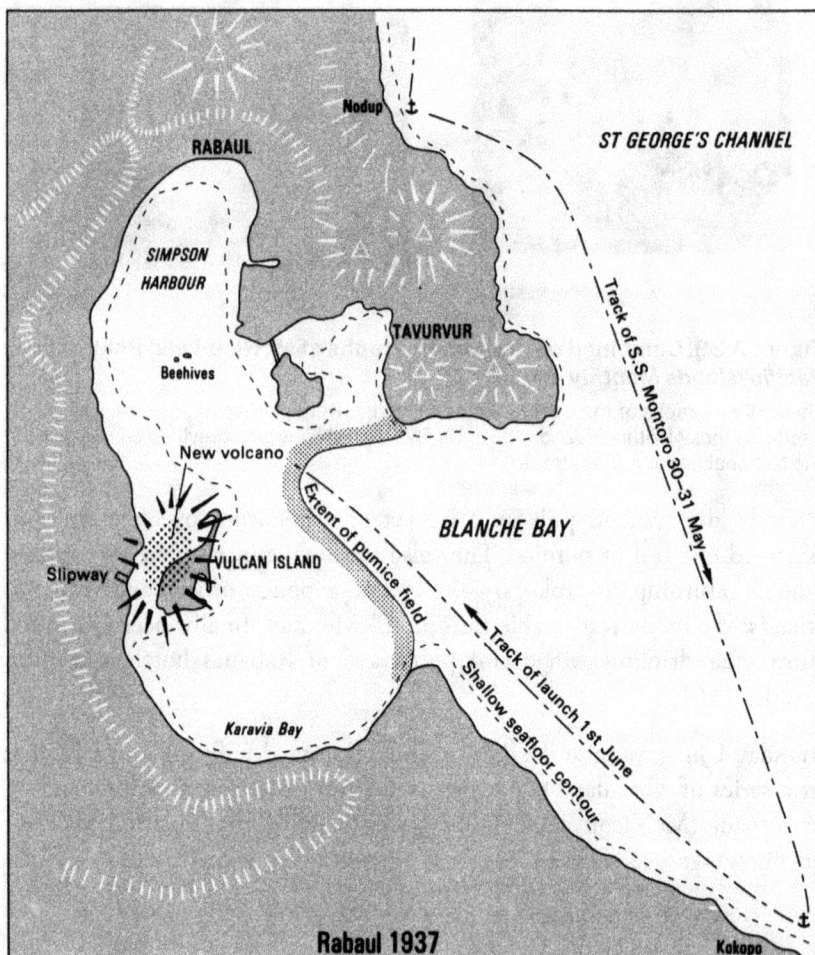

Figure 3.40. Brett Hilder map of Blanche Bay pumice field plus launch and ship tracks.

Brett Hilder drew a rough sketch map of the Blanche Bay area showing the extent of the floating pumice field on Tuesday 1 June 1937; the approximate position of the new volcano, Vulcan, in relation to what used to be Vulcan Island; and the tracks of the *Montoro* and its launch. The diagram shown here is adapted directly from Hilder's sketch.

Figure 3.41. Rescue vessels at Kokopo and new Tavurvur emissions.

Vessels are shown anchored off Kokopo in this photograph dated 31 May 1937. They include, on the left, the *Golden Bear* used in the evacuation from Nodup (see also Figure 3.30). Tavurvur volcano is mildly active left of centre in the background. Note especially that water vapour — and apparently no ash — is emerging from two different sources on the volcano, as described by Olsen (n.d.) when the *Golden Bear* returned to the Rabaul area on 12 June 1937. GA negative reference GB2597.

The McNicolls dined on board the *Montoro* on the Tuesday night. Captain Michie was thanked 'for the assistance rendered', and the old ship left anchorage at 1.14 pm the following day (Michie 1937, 2; Figure 3.41). However, the *Montoro* had not quite finished her role in the Rabaul evacuation; she would be back at Kokopo to take on board more Europeans en route to Australia on 9 June.

3.10. Coming Back to a Damaged Town

The 'garrison' of administration officials and volunteers at Rabaul had spent Monday attempting to tackle the priorities of sanitation, road clearing and police work (Figures 3.42–3.54). Clearing the road over Namanula Hill to Nodup of fallen trees was a priority so that the increased population at Kokopo could be supplied with goods and produce. Other tasks would be added to these, and the arrival of the administrator meant that the central administration at Rabaul could begin organising an integrated relief effort in cleaning up the town (McNicoll 1937c, CD 1–6; Phillips 1937c; Thomas 1937a, 1937b; Radio Roundsman 1937). Tavurvur was still active on Monday, but at a reduced level compared with the previous day; there are no known reports of Tavurvur ash falling on Rabaul that day. Pumice was still being discharged, at times in large volumes, from Vulcan. This ongoing activity did not deter the garrison from immediately beginning clean-up operations in the town. Judge Phillips remained in charge of Rabaul, while his counterpart, Don Waugh, assisted by a committee of officers and residents, was officer in charge at Kokopo. Assistant District Officer H.A. Gregory was placed in charge of the beach at Nodup, R. Melrose of stores at Rabaul and the medical situation continued to be supervised by Dr R.W. Cooper (McNicoll 1937c, CD1).

There were hungry people in Rabaul as early as Sunday morning, and Doug Joycey, assisting Dr Cooper, heard from the police that

> you could help yourself at Burns Philp and W.R. Carpenters' stores … so I made a parcel of tinned ham, chicken, salmon, crab, asparagus, chocolate, and some cigars for the Doctor. Nothing but the best. (Joycey c. 1937, 2)

Figure 3.42. Mango Avenue damage.
The trees of once shady Mango Avenue, Rabaul, were stripped of many of their branches as a result of ash overloading. The photograph was taken at the intersection with Court Street. Rabaul's chemist shop (pharmacy) can be seen on the corner of the intersection on the left, together with W.R. Carpenter's store behind it. GA negative references GB2785 and M2447-34A.

Figure 3.43. Malaguna Road damage.

Branches weighed down by ash on the raintrees running down the centre of Malaguna Road, Rabaul — and probably at its eastern end where the damage was most severe — have collapsed into both lanes of the road. GA negative reference GB2606.

The decision to open the Rabaul stores was made by administration officials but, at least for the Burns Philp's store, company staff were not consulted, nor was a tally kept (Burns, Philp & Co. 1937). Stores were loaded on vehicles once a trafficable road had been cleared and driven to Nodup where they were sent on to Kokopo. There, at least, Burns Philp employees were able to check what was landed. Both trading companies, and others, were quick to set up offices and stores at Kokopo.

Bank officials were concerned about their records, cash and silver. Virgil King, manager of the Commonwealth Bank, reached Kokopo after his north-coast rescue on the *Induna Star* at about 2 pm on Monday, and left on the schooner that same afternoon—with Ramsay McNicoll on board—for Nodup, where there 'were cars everywhere, parked by refugees' (King 1937, 7). They next went to the Rabaul bank where records were retrieved and brought back to Kokopo. King and a group of bank volunteers returned again on Wednesday afternoon for the cash and silver:

> The sea was very rough ... Box after box (Australian silver £5000) was loaded into the row boats and the last issue reached the schooner at dark ... barefooted [bankers] with trousers rolled up above their knees ... and the boats bobbing like corks. (King 1937, 8)

Figure 3.44. Yara Avenue damage.
Ash-laden branches of the casuarinas bordering Yara Avenue in Rabaul have fallen and blocked the roadway. GA negative references M2579-1-12-3 and M2S79-10.

The drama of the bankers' rescue of the cash and silver was increased by the fact that activity was continuing at both Vulcan and Tavurvur, though at a much-reduced level. Indeed, by the following day, Thursday 3 June, the administrator reported that: 'Both volcanoes are now reduced to a small quantity of lazy smoke' (McNicoll 1937c, CD3, 1). The major force of the outbursts had evidently been expended. Vulcan's major activity had, in fact, diminished markedly by the previous Sunday night; by Monday morning the activity at both volcanoes was much less than it had been in the few hours after their initial outbreaks.

Figure 3.45. Damage along Rabaul frontage near radio station.

Foreshore damage near the Rabaul radio transmitting station (the rigging of masts and wires can be seen in the right foreground) was caused by tsunamis on the evening of Saturday 29 May 1937. Tavurvur emits vapour in the background, and trees are visibly damaged by the Vulcan and Tavurvur ash fallout of 29–31 May. The view is south-eastwards towards the main business part of Rabaul town. GA negative reference GB2797.

On 4 June, Gordon Thomas reported in the *Rabaul Times* 'the welcome news'—received on Wednesday 2 June—'that the Father volcano in Nakanai [on the north coast of central New Britain] has broken out again' in eruption (Thomas 1937a, 6). The news was 'welcome' because of a belief among some people that Rabaul and Ulawun (the Father) volcanoes were somehow connected. Indeed, an Australian newspaper had published a report on 2 June that referred to Ulawun as 'Rabaul's safety valve' and noted that eruptions had taken place at Rabaul because of a period of volcanic inactivity at Ulawun (Anonymous 1937a). The idea that active volcanoes in the New Guinea region could be related in this way would be addressed by volcanologists in future years, particularly in the 1950s.

Ramsay McNicoll, presumably, would have considered the possibility of further eruptions taking place in the days ahead and the effect they would have on the already beleaguered town. However, he must have dismissed the possibility because, in his fourth CD, released on Friday 4 June, he advised that:

> Arrangements are being made to select from the Administration officials now at Kokopo twenty (20) officers who will proceed to Rabaul for the purpose of assisting actively in the work of clearing and cleaning the town. (McNicoll 1937c, CD4)

The policy would evidently be carried out irrespective of what may, or may not, happen with the volcanoes.

Administration officials in Rabaul soon recognised the need to remove the ash that had accumulated on the roofs of buildings. Much of the town had collected less than 5 centimetres of accumulated ash from each of the volcanoes, and the total of 10 centimetres was insufficient to cause most roofs to collapse, although the roof of the printing shed of the *Rabaul Times*, containing four printing presses, had fallen in (Hoogerwerff 1937). Such examples were few and, in general, most buildings were still intact, and the town had not yet been greatly damaged. However, the damp, mud-like Tavurvur ash soon baked hard in the sun; a visiting radio reporter said that he 'picked up one piece of crust about 2.5 feet square and it weighed about sixteen pounds [7 kilograms], which gives you an idea of the weight on the ordinary roof' (Radio Roundsman 1937, 2). Future rains would soak into the ash, substantially increasing its weight, and roof collapses might easily take place. Further, the ash would be washed into rain tanks, polluting the water, if gutters were not first disconnected. Officials, therefore, supervised the cleaning of roofs by gangs of New Guineans who were brought into the town as labourers for a range of clean-up activities.

About 400 New Guineans were at work as early as Thursday 3 June (Nicholls 1937c, CD3); two days later there were 660 of them (Phillips 1937c). Piles of ash soon began forming around buildings as the gangs pushed tons of material off the stressed roofs into the streets. The removal of ash from the roofs was not an easy task because of the formerly damp, mud-like Tavurvur ash having been baked hard in the sun. It had to be broken up before being dumped on the ground, and it gave off clouds of fine abrasive dust. Vehicles also churned up clouds of the dust in the streets, and traffic restrictions and speed limits were imposed. Rain tanks that had been open to the ash falls were treated with chlorine, and broken tanks were repaired. Clearing the roads of ash and fallen branches was also a major task, and vehicles were requisitioned for this and other purposes. Some fallen branches were later chopped up and stacked in the yards of residences for householders' use as firewood.

Figure 3.46. Department of Lands building and palm damage.

The umbrella-like folding back of coconut fronds caused by ash loading is clearly illustrated in this photograph of the Department of Lands building in Rabaul. Some ash has been cleared from parts of the roof, and the mud-covered road in the foreground seems to be fairly passable. GA negative references M2579-1-12-7 and M2579-6.

Figure 3.47. Central administration building and ash-covered road.

The central administration building at Rabaul is seen mantled by volcanic ash. Trees behind the building are damaged and ash covers the road. GA negative references M2579-1-12-11 and M2579-2.

Figure 3.48. Rabaul Hotel and compacted mud-ash deposit.
Proprietress Kathleen Bignell is photographed with pilot Jack Turner (left) and customs official Jack Marshall at the Rabaul Hotel soon after the end of the Tavurvur volcanic mud fallout. Roof, courtyard, seats, rails and table are all covered by the material. Note the deep footprints caused by the strong compaction of the mud-ash deposit. GA negative references M2579-4 and M2579-13.

Figure 3.49. Roof collapse caused by ash loading.
Ash loading has caused corrugated iron roofing to collapse onto a parked vehicle. The tree is stripped of leaves and branches. GA negative reference GB2626.

Figure 3.50. Ash being removed from roof.
Brushing the ash from Rabaul roofs and clearing gutters were undertaken after the 1937 eruptions in case rain increased the weight of the ash and caused the roofs and gutters to collapse. There was also concern that the ash would turn to mud and flow into the rain tanks that were the main source of domestic water in Rabaul. GA negative reference GB3285.

Figure 3.51. Clean-up operations in ash-affected Rabaul.

Ash on the roof of the *Rabaul Times* building is being pushed off by a team of New Guinean labourers in this post-eruption photograph taken by J. Hewett of clean-up operations in Rabaul. Hewett was one of the 'barefooted bankers' who helped Commonwealth Bank Manager Virgil King remove the bank's cash and silver from Rabaul to Kokopo via the Nodup beach. GA negative reference GB2957.

Police patrols were a deterrent to looters. Relatively few looting incidents were reported within the town, and those caught were dealt with summarily—for example, with 10 cane strokes across the offender's rump. A *doctor-boi* (New Guinean medical assistant) was found among a group of looters in the Rabaul clinic that Doug Joycey had opened up on the Monday morning. The *doctor-boi* had helped himself to trade goods from Chinatown (including a bicycle, clothing and torches), and had also done 'a good job in the Clinic. He was on his toes the whole time—he could not sit down for a week' after his punishment (Joycey c. 1937, 3). The two Europeans that Brett Hilder had seen being taken into Kokopo by launch on the Tuesday had also been charged with looting, but apparently not caned, having helped themselves to liquor long after supplies were officially put under administration control. Some homes beyond the town limits and not patrolled by the police were plundered, including, for example, Mrs E.S. Garton's home and tearooms at Ravuvu out on the Kokopo Road.

Figure 3.52. Burns Philp store after clearing ash from roof.
These piles of volcanic ash outside the Burns Philp store in Rabaul represent at least part of the ash shovelled off the roof of the building. GA negative reference GB2609.

The sanitation problem in the town had grown serious by Tuesday 1 June. 'Rabaul began to stink', reported Doug Joycey bluntly (Joycey c. 1937, 3). Three days had elapsed before sanitary pans, ice boxes and perishable foods could be properly attended to. Joycey was assigned the task of organising squads of New Guineans who could 'get the stuff buried in back gardens'. Some of the rotting material and waste was dumped in the sea, but this had to be limited, as Judge Phillips pointed out, because the

> floating pumice in the harbour prevents a proper scour … The result in certain states of the tide and the wind may be left to your imagination, especially when I add that dead cats, dogs, fowls, fish, birds and flying foxes added their quota to the effluvia. (Phillips 1937c, 1)

Sanitation arrangements also had to be made for the growing numbers of helpers who were being brought into the town. Central lavatories were constructed, and Ramsay McNicoll 'earnestly requested that no other conveniences be used' because the administration could not 'cope with conveniences in scattered parts of the town' (McNicoll 1937c, CD2, 2).

Floating pumice on the harbour not only prevented a proper scour of the stinking effluvia from Rabaul, but also was a problem for shipping. Smaller vessels, in particular, found ploughing through the pumice fields especially difficult, including the trapped *Desikoko*, which managed to escape the harbour on Tuesday 1 June despite stalling because of a pumice-filled engine (Anonymous 1937a; Thomas 1937a). On the following Thursday, the administrator sailed from Nodup on board the *Nereus* into the harbour and discovered

> an unimpeded entrance between Vulcan and Matupi estimated to be one mile wide. The vessel was stopped and a lead line failed to reach bottom. The Harbour as far as examined, that is from Old Wharf to Main Wharf, appears unaltered. (McNicoll 1937c, CD3, 1)

The SS *Bopple* steamed into the harbour early on the morning of Friday 4 June, followed by the *Induna Star*. A survey by HMAS *Moresby*, which had arrived over the weekend, confirmed that the harbour floor had changed little as a result of the volcanic eruptions. Fears that Rabaul would no longer have seaway access to St Georges Channel were therefore groundless. However, the pumice continued to be troublesome: it abraded paintwork on ships' hulls; could be pumped with seawater into ships' cooling systems causing blockages and engine damage; and could make berthing difficult if it became compressed and jammed between ships and wharves, as happened when the *Polzella* came into the harbour to load copra on 8 June (McNicoll 1937c, CD5, 2). Gradually, however, over the following days and weeks, the pumice dispersed, either by sinking or by being driven out of the harbour by currents and winds during the north-west season.

Tensions, differences of opinion and ill feeling between the administration in Rabaul and the community at Kokopo had, for several reasons, developed by the first weekend of June. First, the business community that had transferred to Kokopo was not appreciative of the way in which the administration had seized foodstuffs and other supplies from their Rabaul stores without consultation. Second, sections of the community at Kokopo were concerned by the heavy-handed way in which townspeople were restricted from re-entering Rabaul to retrieve possessions and records: after obtaining a permit, a place on one of the small vessels plying to and from Nodup had to be secured, then visitors were challenged at Nodup beach to produce the permit, before finally being given a strict time limit to see to their affairs in Rabaul. This led, at times, to heated words being exchanged on the beach. Other people were critical of the way in which the administration was setting about cleaning the town. Why, for example, could not more helpers be allowed into Rabaul? Judge Phillips countered this criticism in a memorandum to Mr Waugh in Kokopo on 5 June, pointing out that there were insufficient facilities to accommodate an indefinite number of people, that 660 New Guineans were already toiling in the town and that accommodation had become available for 380 more (Phillips 1937c).

Dr Champion Hosking was not completely satisfied with the health arrangements in Rabaul, particularly the measures being taken to prevent a malaria outbreak. In a report to the administrator, he noted that the ground drainage system at Rabaul was completely disorganised as a consequence of drains filling with impervious volcanic mud and tree branches, and that, after heavy rains, the town and surrounding area would become a mass of puddles, which, particularly in shaded areas, would last longer than the life cycle of mosquito larvae, causing an outbreak of malaria (Champion Hosking 1937). Champion Hosking acknowledged that the task of levelling the ground would take a long time, perhaps months. McNicoll, not having the same high opinion of Champion Hosking as of Dr Cooper, did not receive his report favourably (McNicoll 1937d). Nevertheless, Champion Hosking's recommendation that Sir Raphael Cilento, professor of tropical medicine at the University of Queensland, Brisbane, be invited to review the medical implications of the Rabaul eruptions was subsequently accepted.

But the greatest reason for discontent among Kokopo residents was the decision, reached by McNicoll over the first weekend of June, that Rabaul should be reoccupied as soon as possible, and that the townspeople should return from Kokopo without unnecessary delay. In his fifth CD, dated Tuesday 8 June, McNicoll wrote:

> The Chinese community is returning to Rabaul commencing from Wednesday the 9th June. This will enable stores, restaurants, bakeries, etc., to function, and those European residents who desire to do so may return and take up residence in Rabaul, commencing from Thursday the 10th June. (McNicoll 1937c, CD5, 1; Figure 3.55)

This decision was not received favourably in Kokopo. Many felt that Rabaul was still too dangerous a place for reoccupation—if not because of the possibility of the volcanoes breaking out again in eruption, then due to the acknowledged health hazards in the town (Purefoy Fitzgerald 1937).

Rabaul was perceived as an unsafe place, so much so that the administration itself was defraying the cost of sending European women and children back to Australia (the *Montoro* left Kokopo on 9 June for Australia carrying many such passengers), yet the administrator was recommending an almost immediate return of both the Asian and the European communities from Kokopo to Rabaul. His viewpoint appeared to many to be a complete turnabout from the statement made the previous Tuesday that 'Rabaul will be untenable for many weeks'. Businessmen who had just completed setting up temporary stores and offices in Kokopo were not in favour of the move. The dispute was publicised in Australian newspapers and different opinions were expressed. Sir Walter Carpenter, for example, chairman of W.R. Carpenter & Co., said in Sydney: 'The Administrator may think that people will go back at once, but nobody else does' (Anonymous 1937b, 17). McNicoll stated that the return to Rabaul was voluntary, but there were obvious compulsions—in particular, the instruction that:

> All free issues (except to indigent natives) shall cease at midnight on Wednesday the 9th. Pending a return to normal conditions goods may be obtained on purchase from the Government Stores at Kokopo and Rabaul. This is an emergency provision only and shall cease at the earliest possible moment. (McNicoll 1937c, CD5, 1)

Figure 3.53. Fallen tree branches on roadside.
This road has been cleared of fallen branches and is trafficable again, but an abandoned car is partly covered by other fallen debris. Road clearing was given high priority in the task of cleaning Rabaul during the first week of June 1937. GA negative references M2579-1-12-8 and M2579-5.

There is no evidence that McNicoll, at this time, revisited Albert Hahl's original decision to establish Rabaul as a town and capital, despite its now obvious vulnerability to volcanic activity. McNicoll cannot have excluded the likelihood that in the years ahead similar or even more destructive eruptions would take place in Blanche Bay. Evidently, in the immediate situation, he had no alternative but to order a reoccupation of the town, which, after all, and despite its covering of ash-mud, could be restored satisfactorily. Ben Sullivan, the Australian Broadcasting Commission's 'Radio roundsman', talked with a young businessman who had been evacuated to Kokopo. The young man suggested, rather prophetically, that the transfer of the capital to outside Simpson Harbour, while keeping the existing harbour and wharves at Rabaul operational for business purposes, might be the way forward (Radio Roundsman 1937). Moving the whole town back to the old German capital Herbertshöhe, now Kokopo, would have been too expensive, particularly during the global economic downturn of the 1930s. Further investigations of the town's future as a capital would require involvement of the Australian Government in Canberra.

Figure 3.54. Recovery of coconut palms.
Coconut plantations in the volcanic fallout zones suffered greatly as a result of the breaking of palm fronds caused by ash loading, although they appear to have recovered somewhat by the time this photograph was taken. The road is clear of debris and is trafficable. GA negative references M2444-1-3 and M2447-32A.

Figure 3.55. Return of evacuees to Rabaul.
Chinese tradesmen and artisans are here seen returning to Rabaul in the second week of June 1937. Collapsed branches from the partly stripped trees litter the side of the ash-covered road leading up from the wharf where these townspeople — among the first of the evacuees to return to Rabaul — have landed. GA negative reference GB2443.

Another cause of contention was the apparent reluctance of the administrator to acknowledge that hundreds of New Guineans had been killed by the Vulcan eruption. Father Zwinge reported in a letter written on 6 June from Vunapope that: 'Many natives perished. Fr Bogershausen had to date a list of 273 names' (MSC 1937; Arculus and Johnson 1981, 19). Similarly, pilot T. McDonald, managing director of North Queensland Aerial Surveys, told reporters on his return to Brisbane from Rabaul that:

> I cannot understand the denials of the number of casualties. One Catholic German missionary told me that 281 natives had perished. Natives reported that there were many bodies in the valley near the new volcano—so many that they could not continue to live there ... There can be no question that hundreds have perished. (Anonymous 1937c, 15)

Further, Rev. Lewis reported from a Methodist standpoint in a letter dated as late as 26 June that 'there has been a consistent attitude of the Administration to belittle the loss' (Lewis 1937b, 4). Lewis estimated that the losses were: four churches, one teacher, 327 people, plus land that was no longer suitable for resettlement.

However, McNicoll believed that most of the missing New Guineans had probably fled and would return. Indeed, according to the *Pacific Islands Monthly* of 23 June 1937, the administrator, on 12 June, estimated the number of dead to be only 18 (Robson 1937, 72). Government officials subsequently investigated how many had been killed. McNicoll, in a memorandum dated 20 July 1937 to the secretary of the Department of the Prime Minister in Canberra, set the number at 424, but even then added that some of the missing 'may have fled to distant places and will return later' (McNicoll 1937g, 1). The exact number of fatalities will never now be known, but the most accurate number is probably that given by N.H. Fisher, who reported a later administration estimate of 505—plus Costner and Elworthy—in his volcanological account of the eruption (Fisher 1939a).

3.11. Vulcan, Floods and Next Steps

The two most seriously affected villages near Vulcan were Tavana and Valaur (where the *tubuan* had been held) and their total dead accounted for well over a half of all casualties. There, along the Kokopo Road close to what had been Vulcan Island, the Vulcan ash was thick—deeper than 10 metres at Valaur—and its burial effects were the most devastating (Figures 3.56–3.60). Several European houses along the road were buried (but no Europeans), including those of Mr and Mrs Furter, the Wallace family and A.R. Reed, owner of the Rabaul Dairy whose herd of milk cows perished in the eruption. The devastated area west of Vulcan held more dead, as well as survivors many of whom were stunned and shocked by the events they had endured. Administration officers were more concerned initially about the condition of Rabaul than the welfare of the survivors west of Vulcan, but missionaries returned to the area to see what support they could offer, including Rev. Jones, who received pitiful ... letters from my teachers up in the hills as to what they should do ... [T]heir gardens had been utterly destroyed, having been covered to a depth of 3 ft with this volcanic ash. Their houses had collapsed, and they were at their wit's end as to what to do or where to go. (A.S. Jones 1937b, 5)

Jones set out on Wednesday 2 June on a long walk into the hills:

[W]e passed hundreds of natives fleeing to the coast. They had their belongings in bundles on their heads. They had caught various birds for food, and one man had a wallaby ... So tangled were fallen trees and palms, so thick was the deposit that I could not recognise the villages. The gardens were a wreck, the ground appeared as though a huge flood had covered it leaving a deposit of mud and stone. (A.S. Jones 1937a, 2)

Jones visited several wrecked villages and talked to teachers and villagers: 'The teachers have all stood firm and stuck to their flocks, not going to their own villages, as it is natural for them to have done.' A wind was blowing, disturbing the ash: '[I]t is most disagreeable, as it cuts into the eyes worse than any sand storm that I have been in' (A.S. Jones 1937a, 6).

Figure 3.56. The *Durour* on the slipway in front of the new Vulcan cone.
The profile of the new cone of Mount Vulcan, or Kalamanagunan, looms up behind destroyed palm trees and a pumice-covered *Durour* stranded on its slipway that, formerly, had been at the shoreline across from Vulcan Island. This photograph was taken by W.B. Ryan along the track that led to the slipway. R.W.J. Collection 30A, Folder 4, Sleeve 51.

Administrator McNicoll would have realised that many people had perished west of Vulcan had he visited the area during the first week after the eruptions, but he was heavily involved in administrative matters in Rabaul, and, by the second week of June, was evidently under pressure from many sections of the community. His telegram of 12 June to Sir George Pearce, Minster for External Affairs, in Canberra reads: 'Human element at Kokopo a big burden—false rumours, vilification, abuse, which I request you to disregard' (McNicoll 1937d, 1). Two days later, he wrote:

> This show has brought out the best and the worst in people. There are some surprising cases of nervous prostration. I am trying to disregard the unpleasant and unhelpful things that are written to me, and bandied about—attributing them to the nervous conditions of the offenders. (McNicoll 1937e, 2)

Sir George made his support for McNicoll's actions at Rabaul publicly known, and told McNicoll he appreciated 'the prompt manner in which you have dealt with the situation and the steps that you are taking to restore Rabaul' (Pearce 1937, 1). Criticism in Rabaul and Kokopo of McNicoll's plans and actions gradually fell away, and, as the town began to recover and no further volcanic eruptions took place, the administrator's strong stand became more accepted.

Figure 3.57. Vulcan ash covering the Baden-Jones's home.
Several European expatriates lived in houses out of town along the Kokopo Road, some of which were quite close to Vulcan Island. All of these Europeans escaped the Vulcan eruption, but their houses were either severely damaged or completely buried. This home was occupied by the Baden-Jones family. GA negative reference GB2937.

Figure 3.58. House collapse caused by volcanic ash from Vulcan.

Amid general devastation near Vulcan, the roof of this house has collapsed under its loading of pumice. GA negative reference GB3310-13.

Figure 3.59. Vulcan cone and probable salt encrustations.

Extreme devastation is seen in this photograph taken by Dr H. Champion Hosking just north of the new cone of Vulcan at the end of its eruptive activity in 1937. The white encrustations in the foreground are salts presumably dried out from the seawater that was flung out during the explosive eruptions. GA negative references GB3309 and GB3309-11.

Figure 3.60. Excavating an occupied shelter buried by Vulcan ash.

A team of New Guineans excavates a native-materials dwelling buried under pumice near Vulcan. They have unearthed the corpse of a New Guinean woman (right-hand photograph, point a) who had evidently attempted to shelter there. These photographs appeared in Sir Raphael Cilento's medical report on the after-effects of the 1937 Rabaul eruptions. GA negative references GB3312 and GB3312-15 and 16.

McNicoll reported on 15 June that the 'work of clearing the town is proceeding apace and good results are being achieved' (McNicoll 1937c, CD6). Electric light and power were being supplied, and extensions to the electricity network throughout the town were being made daily. Requisitioned vehicles had been returned to their owners, and administration departments had again opened their offices in Rabaul. Evacuees, in time, began returning from mainland New Guinea. However, the environment of the town remained uncomfortable, on account of the volcanic dust that was swirled into the air by breezes and passing traffic. The dust remained a nuisance for several weeks, particularly during dry periods. It irritated eyes and people with respiratory problems must have found it particularly unpleasant. Face masks were ordered by the administration, and the task of providing 2,000 of them fell to Doug Joycey. He made a few hundred masks from cotton wool pads soaked with hypo (sodium thiosulphate; a design based on gas masks used in WWI), but he ran out of hypo, and they were never used.

Meanwhile, villagers and missionaries in the west were attempting to resume a normal life after the devastation caused by Vulcan, dust irritations notwithstanding. Valaur and Tavana villages, which suffered the greatest human death tolls, were rebuilt by survivors away from the shoreline on the western foot of the new Vulcan cone—a vulnerable place should Vulcan break out into eruptive activity again (Figure 3.8)—but on the eastern side of a new unsealed road linking Rabaul to Kokopo, meaning there was at least a possible evacuation route southwards for any explosive eruption in the south-east season, as in 1937. A memorial to those killed in 1937 was later constructed at Tavana as a constant reminder of what could happen again (see Figure 7.9).

Figure 3.61. Flooding in Rabaul.

Heavy rains caused flooding in Rabaul in the few weeks after the 1937 eruption, especially where the impervious ash of Vulcan and Tavurvur covered the normally porous soil of low-lying areas. These photographs were taken by Miss Carol Coleman, who returned to Rabaul on 16 June, and who noted in her diary that there was 'a bad storm on 30th June which turned everything into a sea of mud again'. GA negative references GB2933 and GB2935.

Figure 3.62. Erosion gullies near Vulcan.

Deep ravines were gouged out near Vulcan by rains that fell during and after the 1937 eruption. This photograph was taken near Valaur where Vulcan pumice accumulated to many metres thickness. Layers of pumice can be seen in the ravine walls. GA negative reference GB3284.

Figure 3.63. Erosion on Namanula Hill Road.
Extensive gullying on the side of the Namanula Hill Road has been caused by increased run-off of rain that, normally, would have percolated through the tropical vegetation and porous soils at Rabaul. Volcanic ash has not only stripped away much of the vegetation, but also covered the soils, so that the rain runs over exposed surfaces eroding them. GA negative reference GB2630.

The first of the rains began in the second week of June, and these helped to dampen the dust. The rain, wrote McNicoll (1937c, CD6), 'greatly improved the appearance of the town and greenery is again evident on all sides'. A restriction on the use of vehicles in the streets was lifted, but a 15-miles-per-hour speed limit was enforced. Flowers such as frangipani and hibiscus started to bloom in the still drab town. However, heavier rains, when they came, did not help the situation in Rabaul. Rains could no longer percolate through foliage and normally porous soils, but rather ran off surfaces covered by the relatively impervious Tavurvur ash. Torrential floods developed down slopes, and especially down steeper streets such as the Namanula Hill Road (Figures 3.61–3.63). These torrents swept mud into the town, gullied hillsides, eroded the edges of roads, blocked drains and caused flooding, particularly in the lower-lying parts of Rabaul such as Chinatown. Doug Joycey recalled that after one heavy rainstorm: 'Dozens of Chinese began to arrive on the old burnt wharf, with huge bundles on their heads, for a second trip to Kokopo.' The waters drained away quite quickly, and no serious malarial problems eventuated, but the Department of Public Works subsequently developed a system of deep open drains across Rabaul to cope with any further floods (Figure 3.64).

Figure 3.64. Construction of drains in Rabaul.
A New Guinean team works on lining deep drainage channels that were built shortly
after the 1937 eruptions to cope with the possibility of future floods in Rabaul.
GA negative reference GB2622.

Sir George Pearce informed McNicoll in his cablegram of 11 June that:

> [T]he Government feels it cannot concur in the retention of Rabaul as the permanent capital of the Territory without the fullest investigations as to its safety and suitability from all aspects. It is therefore proposed that such investigation should be made from the scientific, commercial, navigational, Administrative and medical viewpoints. I have caused inquiries to be made with a view to securing a qualified scientist to proceed to Rabaul and report upon the volcanic and seismal dangers of the area from the point of view of the safety of the inhabitants and the Administrative and commercial activities. [Further] a Committee should be appointed to review the questions from all angles, [and which would] consist of persons who could give expert advice from the medical, shipping, commercial, transport and aviation and Administrative point of view. (Pearce 1937, 1)

Figure 3.65. Cartoon of visiting volcanologist on volcanic cone.
J.K. McCarthy produced this cartoon of a nonchalant visiting volcanologist carrying a magnifying glass. Its caption reads: 'Eminent scientist (probably American) examines the crater — Bulletin No. 6 "Cute but not acute".' 'Bulletin 6' presumably refers to the administrator's Circular Dispatch No. 6, so McCarthy is here poking fun at McNicoll for his bracketed comment 'probably from America'. A photocopied collection of some of McCarthy's cartoons is catalogued in R.W.J. Collection 30B, Folder 8, Sleeve 50.

The administrator included news of the proposed investigations into the suitability of Rabaul as a capital in his sixth CD of 15 June 1937, adding his own detail 'that the Commonwealth Government is making enquiries with a view to obtaining the services of a qualified Scientist (probably from America) who will visit Rabaul' (McNicoll 1937c, CD6, 1). McNicoll's view of qualified volcano scientists—American or otherwise—was lampooned in a cartoon drawn for wider distribution (Figure 3.65).

Ramsay McNicoll was the target of criticism from sections of the Rabaul community, and he in turn was critical of his detractors. Yet the administrator was clearly impressed with the actions of many Rabaul citizens during the volcanic crisis, because on 26 June he submitted to Sir George Pearce:

> [I]n view of the meritorious work performed by a number of officers and private individuals during the recent volcanic disturbance in Rabaul ... the names of 10 people who had been particularly outstanding. (McNicoll 1937f, 1a)

At the top of the list was Judge F.B. Phillips. All 10 nominees, and one other, would eventually receive appropriate British Empire awards bestowed by King George VI. McNicoll also informed Sir George that he had a reserve list of another 40 or 50 people who had 'performed work of a sterling character'. Apart from Judge Phillips, Acting Superintendent Ball, Dr Cooper and Acting Inspector Prior were included in the 10 nominated, and so too were Dr Watch (for his medical work at Kokopo), Messrs L.W. Heinicke and E. Hopkins (for driving between Rabaul and Nodup and for other services), Mr Barrie (for his work at the electricity station), Mrs Bignell (for the supply of meals and accommodation arrangements) and the Honourable John Charles Mullaly of Natava Plantation, who 'rendered wonderful assistance in the matter of caring for the natives of the devastated areas of the North Coast'.

Gordon Thomas resumed normal publication of the *Rabaul Times* on 23 July 1937: '[T]he linotype is clicking again and the whirr of the presses make a welcome sound.' In an editorial entitled 'Nearing Normal', he wrote that:

> Large areas of pumice still float in the harbour [Figure 3.66]; many of the streets are lined with large heaps of volcanic mud which must be cleared away by the mechanical conveyors [brought especially from Australia] and large gangs of native labourers shovelling them onto an ever-moving stream of motor lorries; roads and drains require

attention; carpenters and plumbers are busy effecting repairs; street lights are still absent and light and power have not been re-installed in all residences yet … furtive glances are still directed at Matupi and Vulcan as they come into view and nerves are still a-tingle if the verandah shakes from the scratching of a dog. It is correct, however, to say we are approaching normal times again. (Thomas 1937c, 8)

In the issue, Thomas acknowledged the cartoons of J.K. McCarthy— who had 'the power to illustrate the humorous incidents at times of stress which does so much to relieve the tension of the moment'—which had been printed in the previous 'unofficial numbers' of the newspaper. The issue also contained news and notice that a ball, organised by Mrs Bignell, would be held at the Rabaul Hotel on Friday 6 August during the visit of HMAS *Australia* (Figures 3.67 and 3.68). Another announcement was that fresh milk, from newly imported and tested Holstein cattle, was available from Reed's Rabaul Dairy as of 18 July. Social life had resumed soon after the town's reoccupation and, like the frangipani flowers that had so impressed the townspeople, Rabaul soon bloomed again in its former style. The Frangipani Ball was inaugurated to celebrate the rebirth of the town, and this continued as a regular postwar event in the Rabaul social calendar.

Figure 3.66. Pumice floating on Simpson Harbour near wharf.
The *Montoro* leaves main wharf, Rabaul, in July 1937, through a field of floating pumice. Photograph supplied by B. Hilder. GA negative reference GB3302.

Figure 3.67. HMAS *Australia* visiting Blanche Bay.

HMAS *Australia*, Dawapia Rocks (Beehives) and Vulcan in Blanche Bay (Fisher 1939a, 66, lower image). Stehn and Woolnough (1937a, 8–9) reported that HMAS *Australia* undertook deep soundings in the harbour on 10 August, including to a depth of 150 fathoms at the entrance to Blanche Bay in the south-east.

Figure 3.68. Mrs Bignell on the rim of Vulcan cone.

Kathleen Bignell, proprietor of the Rabaul Hotel, and her daughter Margaret climbed Vulcan one day in mid to late June 1937 while the volcano was still emitting vapour and gas (Clarence 1982). Mrs Bignell was awarded the British Empire Medal for her work at the hotel in providing meals for the men who remained in Rabaul during the arduous first few days after the May 1937 eruptions. Margaret took this photograph of her mother at the top of the climb on the rim of the new crater. Margaret Bignell was evacuated from Nodup to Kokopo where she worked as a typist for Gordon Thomas helping in the preparation of special issues of the *Rabaul Times* at Vunapope. GA negative reference GB3289.

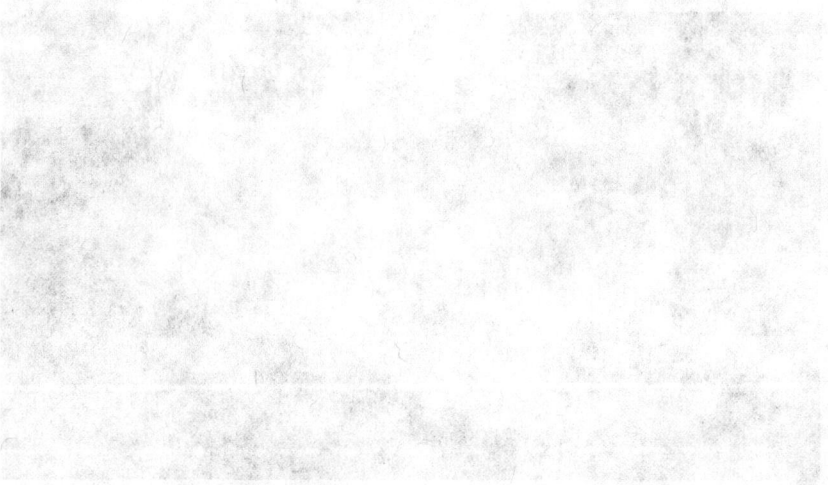

4

Results of Geological and Volcanological Investigations

4.1. Rabaul Investigated and Awards Announced

McNicoll remained in Rabaul as administrator during most of the four years following the 1937 eruption and was at the centre of the controversial discussion on the future of Rabaul. No further volcanic activity of any significance took place during this time, and Rabaul gradually recovered as roads and gardens were cleared, and trees and shrubs regenerated, providing welcome shade. However, the bleak forms of Vulcan and Tavurvur stood at the entrance to Simpson Harbour like sombre monuments to what had happened in May 1937, as Rabaul was investigated and recommendations were made on its suitability as a capital.

Sir Raphael Cilento arrived at Rabaul on 10 July 1937 and began his investigation of the 'question of all precautions of a medical nature that might be necessary either immediately or in the future' as a result of the volcanic activity, including malaria, dysentery and dust hazards such as pneumoconiosis and eye irritations (Cilento 1937a, 1). Cilento supported Dr Champion Hosking's earlier medical recommendations; however, in his comprehensive report, he noted that the medical problem was only one aspect that had to be considered by the administrator in deciding to move the population from Kokopo, where medical risks were already evident, back to Rabaul. Malaria was a primary threat; therefore, the management of surface water where mosquitos might breed was important. Cilento was also concerned about the water-rich nature of the blue-grey volcanic mud from Tavurvur that had fallen on Rabaul:

> The mud ... is of a gummy consistence, very readily flowing; it appears
> to be hygroscopic, specimens have remained moist for weeks without
> the addition of water; and it has lined all the pits and depressions
> in the soil with a plaster-like coating which definitely prevents ready
> absorption or run-off of the contained water. (Cilento 1937a, 16)

Cilento supported McNicoll's controversial decision to occupy Rabaul
immediately, noting that the psychological effect

> was, generally speaking, markedly favourable. The reinstitution of
> orderly routine activity had had a profound effect upon the confidence
> of the individual, and had gone far towards the re-establishment of
> that sense of security which is an important psychological factor.
> (Cilento 1937a, 20)

Sir Raphael concluded that, while there was a malarial risk, this could be
'completely controlled within fifteen months, and may, if circumstances
are favourable, be controlled within six months', assuming that strenuous
preventive measures continued to be adopted (Cilento 1937a, 26).
Addressing the problem of the future of Rabaul from a medical standpoint,
he argued that the European population should be segregated from other
races by living at 1,500–2,000 feet (460–600 metres) above sea level—
and thus away from the 'native and Chinese areas'. This view today can
be judged as distinctly racist; indeed, some people consider Sir Raphael to
have held strong right-wing views (Fisher 1994). Cilento noted in his report
that the expansion of a segregated European settlement would depend on
the response of the Australian Government to the results of an official
volcanologists' inspection that was then being carried out in Rabaul.

Three geoscientists were involved in the volcanological investigations at
Rabaul. The principal investigator was German-born Dr Charles Edgar
Stehn—an experienced volcanologist and director of the Netherlands Indies
(now Indonesia) Vulcanological Survey (van Bemmelen 1949). His services
were requested by the Australian Government through the Government of
the Netherlands East Indies (Press Release 1937). Stehn was a good choice
as he had volcanological experience from studies at East Indies volcanoes,
in particular that of the major 1883 eruption at Krakatau and the related
formation of a caldera, which could be applied to assessing the situation at
Rabaul volcano. Stehn was away from his headquarters in Bandung, Java,
from 8 July until 25 October 1937. He arrived in Rabaul about 24 July
accompanied by geologist Dr Walter George Woolnough, geological adviser
to the Australian Government, who was to be his associate. Stehn wrote later
that he was 'astonished at the exceedingly close analogies which may be drawn
between Krakatau and Blanche Bay' (Stehn and Woolnough 1937a, 5).

Stehn and Woolnough were joined in Rabaul by the young Australian geologist Norman Henry Fisher, who, since 1934, had been the administration geologist at Wau where he undertook geological studies on the Morobe Goldfields (Fisher and Branch 1981; Wilkinson 1996). Fisher was assigned by the administration to give all possible assistance to the two, more senior, visiting investigators at Rabaul. The three scientists undertook fieldwork together, Stehn and Fisher focusing on building up an understanding of the history of the Rabaul volcanoes (Figure 4.1). Fisher would play a major part in the aftermath of the 1937 Rabaul eruption and, indeed, would pioneer the establishment of instrumental monitoring of the Rabaul volcanoes. He produced, in 1939, what turned out to be the standard volcanological reference work of the 1937 eruptions and of the Rabaul area in general up to that time (Fisher 1939a). He even found time to publish a popularised article on Rabaul's recent volcanic eruptions in an Australian magazine (Fisher 1939b). Fisher and Stehn also travelled to the north-central coast of New Britain where there were many young volcanoes, triggering in Fisher a broader interest in the volcanoes of the whole Territory of New Guinea (Fisher 1939c). They climbed the 2,220-metre volcano of Ulawun, or the Father, on 13–15 August 1937, possibly becoming the first Europeans to do so (Fisher 1937).

Stehn and Woolnough had been asked to comment not on maintaining Rabaul as a commercial centre, but rather on its suitability as the main administrative centre of the Mandated Territory. They were uncompromising in their official report, dated 29 September, which the Prime Minister of Australia received in November. They reported that they could

> see no reason whatever for believing that volcanic activity may not recur here at almost any time … [and] the possibility that such outbreaks may occur closer to, or even within the limits of the town area itself (as, for instance, at Sulphur Creek), cannot be eliminated. We are therefore forced to the conclusion that reasonably early evacuation of Rabaul as the main administrative centre of the Territory must be seriously considered [authors' emphasis] … Though it is not improbable that the intensity of vulcanism at this point may have passed its maximum and may now be in its declining phases, such a conclusion cannot be stated with any confidence. [There was also] the possibility that the whole of the capital invested in town and harbour may be jeopardised or wiped out of existence in a few hours by another and more serious eruption taking place under conditions not so extraordinarily favourable as those of the recent phenomena. (Stehn and Woolnough 1937a, 25)

Figure 4.1. Ash-covered *Durour* on slipway and Dr Stehn on foot track.
Volcanologist Charles Stehn returns along the well-worn track made by inquisitive visitors to the pumice-covered *Durour* at the former slipway near Karavia. The *Durour* was left stranded hundreds of metres inland from the new shoreline formed after the May 1937 Vulcan eruption, and was later written off and sold as scrap. The photograph was taken by N.H. Fisher in July 1937. R.W.J. Collection 10, Folder 2, NHF-envelope 5.

Among the 'extraordinarily favourable' conditions referred to by Stehn and Woolnough was the fact that the eruption had started at Vulcan, more distant from Rabaul than Tavurvur, Rabalanakaia or Sulphur Creek; that the outburst began in bright sunlight, rather than at night, so the 'terrors of darkness were not added to those of surprise'; that it took place on a Saturday afternoon when 'families were united in their homes, and motor transport was concentrated in an extraordinarily convenient way'; that the sea at Nodup was calm, allowing the large ships to stand in close; and that the organisation and equipment of the Sacred Heart Mission at Vunapope were available and adequate to meet the emergency (Stehn and Woolnough 1937a, 17–18). In their view:

> The fortuitous concurrence of so many favourable factors could scarcely have been anticipated, and it must not be taken for granted that such a combination must necessarily exist when next Rabaul has to be evacuated at short notice. (18)

Few people in Rabaul at the time of the release of Stehn and Woolnough's report were aware that the two authors had different opinions about the future of Rabaul and about the volcanology of the area. Woolnough, who was not a volcanologist, had returned to Australia after only a few days of field investigations, and had completed the report recommending the withdrawal of the administration from Rabaul while Stehn and Fisher were still in Rabaul undertaking more extensive studies over a period of six weeks. Both Stehn and Fisher believed that Rabaul could remain as the capital so long as a volcanological observatory was established to provide warnings of eruptions. Stehn sent his changes to Woolnough in Australia, but not all of these were incorporated in the report. Consequently, the jointly authored report reflected the opinions of Woolnough more than those of Stehn. Woolnough was the sole signatory of the report's appendix, which suggested the New Guinea mainland as a suitable area for the new capital, and he also wrote the final paragraph:

> The spectacular development of the gold industry in New Guinea and the very high probability of discovery of oil-fields there, together with possibilities of great expansion of timber and agricultural industries, suggest that, in the event of a decision to remove the seat of government being arrived at, selection of a reasonably central site on the New Guinea coast should receive serious consideration. (Stehn and Woolnough 1937b, 158)

That Stehn and Woolnough reached different conclusions as a result of their joint investigations would not have helped decision-makers in the Australian Government. Stehn, based on his experience with active volcanoes and instrumental monitoring in the Netherlands Indies, believed that a well-equipped volcanological observatory would be capable of providing warnings of impending eruptions sufficiently far in advance to enable the Rabaul population to be removed to places of safety. Conversely, Woolnough stated that 'the advantages of retention of the capital in its present site and the provision of elaborate warning systems should not be entertained' (Stehn and Woolnough 1937a, 28). He concluded that the required scale of such an observatory, the uncertainties of eruption prediction, the generally small industrial and commercial development in the town and the fact that the harbour might become blocked by future eruptions were all reasons why an observatory should not be constructed at Rabaul and why the town should be abandoned as the capital.

Instrumented observatories were, by 1937, a well-established feature of some of the world's best-known volcanoes, the idea deriving from the much earlier concept of astronomical observatories where instruments were directed at the stars, rather than at the internal and surface behaviour of a volcanic earth. The first volcanological observatories were built at Vesuvius, Italy, in 1841–45, and at Etna in 1878–81. The Americans had established an observatory at Kilauea volcano, Hawaii, in 1912, and Japanese observatories had been built on Aso by 1928 and at Asama by 1933. Geologist E.R. Stanley had drawn the attention of the Australian Government to the need for volcanological observatories in the Territory of New Guinea after his attendance at the Pan-Pacific Scientific Conference in Honolulu in 1920 (Stanley 1923). Stehn and Fisher were, therefore, building on something of a volcanological tradition.

The British Empire honours for the 10 Rabaul people recommended by the administrator in June were announced towards the end of 1937— around the time of the release of the Stehn and Woolnough report—and the list was published in some Australian newspapers (*Border Watch* 1937). An eleventh awardee, and top of the list, was Ramsay McNicoll himself, who received a knighthood—a decision that, predictably, served to fuel further uncomplimentary remarks about the administrator in some quarters. The list is as follows:

- Brigadier-General W.R. McNicoll: KCBE (Knight Commander of the Order of the British Empire)
- Judge F.B. Phillips of the Supreme Court of New Guinea: CBE (Commander of the Order of the British Empire)
- Honourable J.C. Mullaly, plantation owner: OBE (Officer of the Order of the British Empire)
- W.B. Ball, acting superintendent of the New Guinea police administration: OBE
- Dr R.W. Cooper, medical officer of the New Guinea administration: MBE (Member of the Order of the British Empire)
- Dr N.B. Watch, medical doctor: MBE
- W.B. Prior, acting inspector of the New Guinea police: King's Police Medal.

The following received the BEM (Medal of the British Empire, Civil Division):

- Mrs Kathleen Bignell
- L.W. Heineke
- John Barrie
- Eric Hopkins.

Bishop Gerard Vesters, a Dutchman, was also appointed OBE in recognition of the services provided by the Catholic mission at Vunapope, but only after the Netherlands Government had approved his acceptance. There were no awards for New Guineans—on the contrary, the new land that had been formed at Vulcan and taken over for gardens by local Tolai people was claimed by the administration as 'government land'. Such differences between European and traditional land laws resulted in long and bitter clashes and Tolai protests went on for years. District Commissioner Keith McCarthy, after WWII, supported Tolai claims to Vulcan and their appeal was taken to the Supreme Court, but another district commissioner declared Vulcan an industrial area and then issued a licence for the construction of a racecourse there. However, the administration eventually agreed to hand back Vulcan to the people.

4.2. Calderas and Their Origin

Stehn's, Fisher's and Woolnough's findings were reported comprehensively. Included among their findings was discussion of the origin of the Blanche Bay caldera as a whole. Fisher promoted the agreed view that a huge volcano, perhaps 'at least 9,000 feet [2,700 metres] in height', once covered the area now occupied by the Blanche Bay caldera—or else, 'as is quite likely' in its later history, a group of somewhat lower cones (Fisher 1939a, 12). In Fisher's view, it was likely that a 'tremendous outburst or series of outbursts … blew most of the mountain to fragments', scattering pieces of the central volcano around the countryside (Fisher 1939b, 13); subsidences around the periphery of the resulting explosion crater then formed the caldera, a large eruption finally producing huge thicknesses of 'pumice ash' (Figure 4.2).

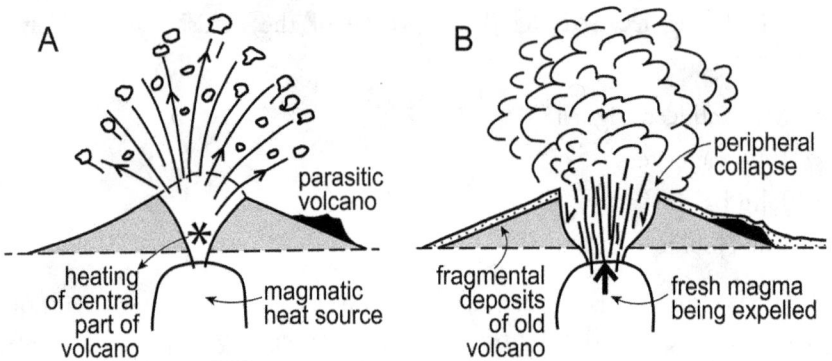

Figure 4.2. Caldera formation by outward explosion.

Simplified cross-sections of the formation of calderas by (A) initial heating and outwardly directed explosions of the central part of a volcano, followed by (B) explosive eruption of fresh magma and then by peripheral collapse of the caldera margins.

The smaller but conspicuous Rabaul volcanoes of Watom, Kabiu, Tovanumbatir and Turagunan were regarded by Fisher (1939a) as 'parasitic' or satellite volcanoes on the flanks of the proposed great ancestral mountain. Fisher noted, however, that the four volcanoes at Rabaul were aligned in a north-west–south-east direction almost at right angles to the central axis of New Britain island as a whole. He referred to Stanley's geological description of the Willaumez or Talasea Peninsula chain of young volcanoes that ran northwards into the Bismarck Sea from the central, east-west coast of New Britain (see the M–D to Garbuna volcanoes shown in Figure 0.1 in the Introduction). Control by major geological structures for both volcanic alignments was proposed by Fisher. However, he went even further, drawing attention to the fact that the north-westward–trending Watom–Turagunan Zone at Rabaul (see also Stanley 1923) paralleled the trend of the strongly elongated New Ireland to the north-east. This structural relationship would be considered again by a later generation of geologists looking into the tectonic setting and origin of this highly complex area of the world.

This interpretation for Blanche Bay by Fisher, and by Stehn and Woolnough (1937a, 1937b), was not, however, the mechanism of formation of the Krakatau caldera proposed earlier by Stehn's colleague, R.W. van Bemmelen (1929; Figure 4.3 left). Nor was it the mechanism for the general model of so-called Krakatau-type calderas that would be promoted later, and internationally, by H. Williams (1941; Figure 4.3 right). Rather, it reflected the view still held by many geologists at that time that calderas were the result of volcanoes that 'blew up', meaning exploded outwards, rather than by collapsing wholesale into underlying magma reservoirs.

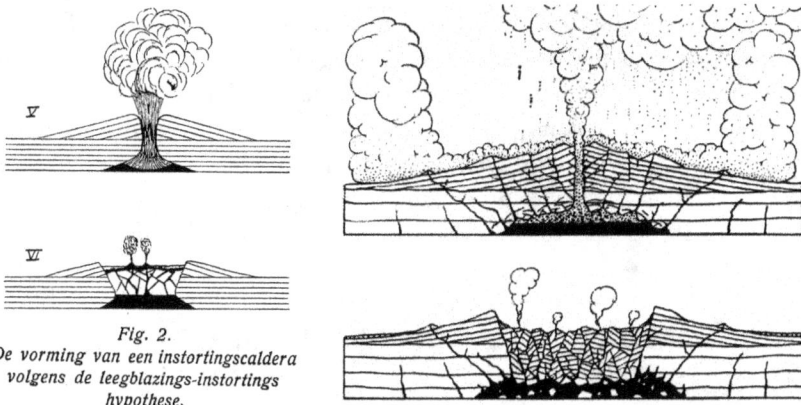

Fig. 2.
De vorming van een instortingscaldera
volgens de leegblazings-instortings
hypothese.

Figure 4.3. Caldera formation by inward collapse.

Two versions of the 'collapse' origin of Krakatau-type calderas are shown here. Left is a detail from van Bemmelen's 'Fig. 2' (1929). The right-hand diagram is from a paper on volcanoes by Howell Williams (1951, 51) based on his own fieldwork on Crater Lake, Oregon, United States. Magma in a shallow reservoir shown in the upper cross-sections of both diagrams is erupted rapidly as a result of a powerful volcanic eruption. The eruption shown on the right is producing pyroclastic flows that cascade down the flanks of Crater Lake volcano. This evisceration leaves the roof of the magma reservoir unsupported, and it collapses and disintegrates, as shown in the two lower diagrams. Subsequent geological studies of many eroded calderas, however, have not revealed such strong disintegration of the roof rocks as shown in these two interpretations.

Fisher, Stehn and Woolnough did not use the term *ignimbrite* for any volcanic deposits at Rabaul, 'ignimbrite' being the important volcanic rock type that had been identified by Marshall (1935) in New Zealand. These rocks or deposits were produced by large-volume lateral flows of pumiceous materials during large explosive eruptions—that is, by *pyroclastic flows* (see Figure 4.3 right). Stehn, however, did refer to 'the eruption of incandescent clouds "fire avalanches" in 1934', but at Merapi volcano in Java (Stehn and Woolnough 1937a, 16). Further, recognition of *nuées ardentes* or 'glowing clouds' at Mount Pelée, in the Caribbean, in 1902, was well recognised internationally (Fisher had access to a copy of the report on Mount Pelée by Perret [1937]) so the three investigators certainly were aware of the pyroclastic-flowage phenomenon in principle. Much later work by others would reveal the deposits of large pyroclastic flows at Rabaul and, indeed, a much more complex geological origin for Blanche Bay.

The 'pumice ash' noted by Fisher was almost certainly the deposit of large-scale pyroclastic flows. Non-use of the term 'ignimbrite' can also be readily understood, as the older New Zealand examples were of the 'welded' type—that is, formed where the great thicknesses of 'pumice ash' of the pyroclastic flows become strongly compacted and flattened after deposition, forming a distinctive lens-like pattern in rock exposures. *Welded* ignimbrites, we now know, are much rarer at Rabaul. Further, Marshall's paper had been published just two years previously and its significance and importance were not yet recognised internationally. Fisher and Stehn did not make any reference either to smaller pyroclastic flows during the eruptive activity at Vulcan in 1937.

4.3. Mapping the Active Volcanoes of Blanche Bay

Stehn, Fisher and Woolnough, during their assessment of Rabaul's volcanic safety, appear to have directed special attention, quite naturally, to the volcanoes immediately south-east of Rabaul township—namely, Rabalanakaia and Sulphur Creek, which were closest to the town, and to Tavurvur, further south-east (Figure 4.4). All three volcanoes had conspicuous geothermal manifestations—fumaroles, hot springs and solfataras—both before and at the time of their visit. We can here refer to the whole area as the Greet Geothermal Field. The north-eastern margin of the field coincides with the south-east-facing caldera wall defining this north-eastern side of Blanche Bay. Rabaul town is immediately adjacent to its north-eastern margin. The significance of the quasi-permanent Greet Geothermal Field is that it must be underlaid by a long-lived heat source, most likely a body of magma or else hot, partially molten rock.

Rabalanakaia had grown not only within the older Palangiangia crater but also directly on the projected trace of the caldera wall. The south-western side of Palangiangia may have collapsed when the caldera itself was created. The flat floor of Rabalanakaia has a distinctive ring or circle of fumaroles marked by mounds of mud and sulphur (Fisher 1939a, upper photograph on p. 56; see also the aerial-photograph image in Figure 7.6). Its previous eruption may have been in 1767, as shown in Figure 1.4. The long, narrow and thermally active Sulphur Creek on the harbour shore trends eastwards towards Rabalanakaia. Fisher believed it may have originated by 'a fissure type eruption of volcanic eruptions' (Fisher 1939a, 39). Three extinct craters were mapped east of the head of the creek, the more south-easterly of which had a freshwater pool in which the Department of Public Health kept its supply of gambusia (fish) because of their reputation for mosquito control.

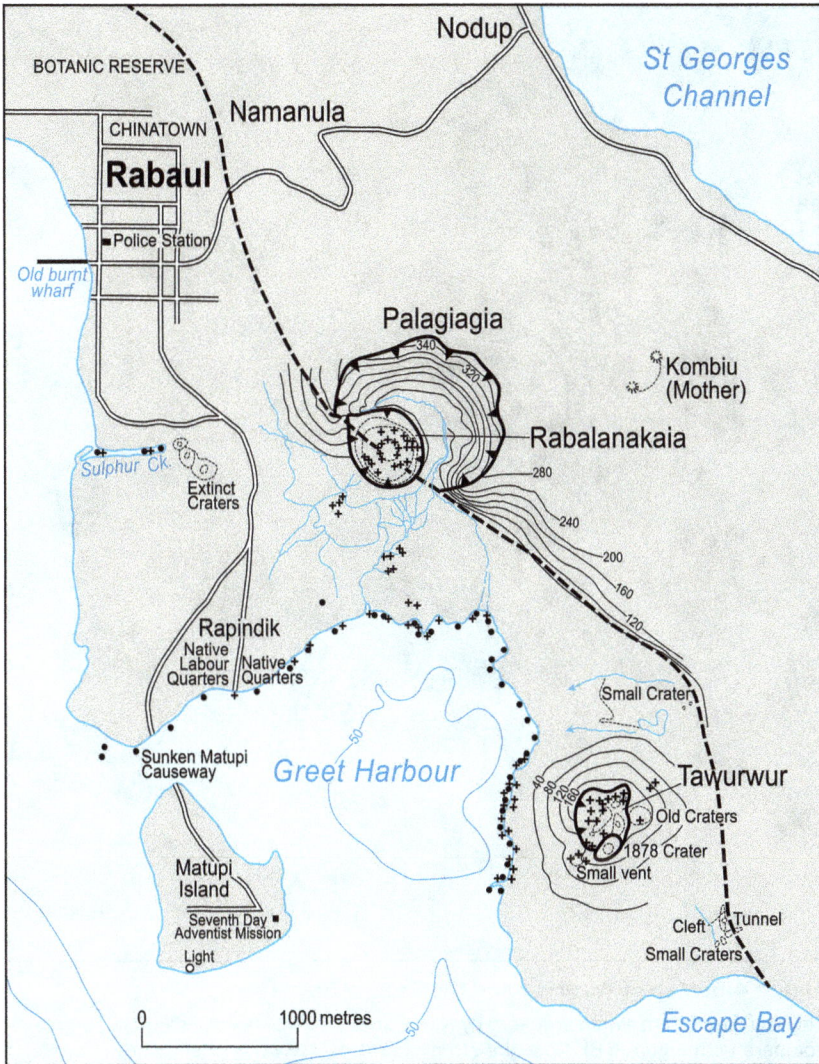

Figure 4.4. Map of the Greet Geothermal Field and Rabaul town.

Features of the Greet Geothermal Field are shown in this detail adapted from the chart published by Fisher (1939a, Plate A1). The bold dashed line refers to the trace of the south-west-facing caldera wall, drawn as if it were at sea level, as envisaged by Fisher (see also Figure 4.5). Note that the small craters near Tavurvur are very close to the trace of the caldera. The spelling is the same as used by Fisher. Contours are in 40-metre intervals (post-eruption) and isobaths are in 50-metre intervals (pre-eruption). Fisher mapped in considerable detail the distribution of 'steam' emanations (crosses) and 'gas ebullitions' around parts of Greet Harbour (filled circles).

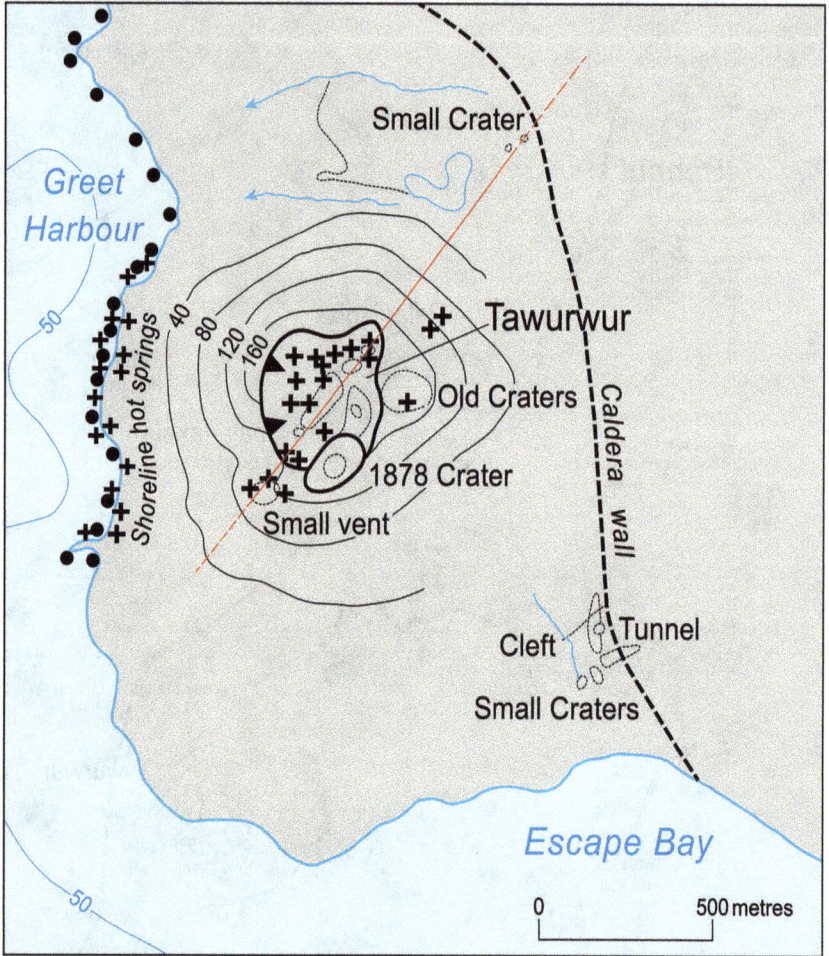

Figure 4.5. Map of Tavurvur and the Escape Bay area.

This is a detail from a map prepared by N.H. Fisher (1939a, Plate A1; see also Figure 4.4; compare with Figure 6.8). Note the strong north-easterly alignment of the craters of Tavurvur (red line) and the proximity of the two 'Small Crater' areas to the trace of the caldera wall.

The most conspicuous geothermal areas seen after the 1937 eruption were on Tavurvur itself and along the eastern shore of Greet Harbour, extending westwards along the southern edge of Rabalanakaia past Rapindik. Fisher described the new craters of Tavurvur in some detail, pointing out that they lay a little towards the north-east of the 1878 crater (Figure 4.5). The recent eruption at Tavurvur had taken place in a narrow central fissure that ran south-westwards beyond the summit area to a small vent on the slope of the cone.

Two other crater areas active in 1937 were of interest too, despite their small size, because of their proximity to the caldera trace. They serve as evidence for minor eruptions beyond the crater area of Tavurvur itself but within the Greet Geothermal Field as a whole. One of the areas was near the shore of Escape Bay and may have been active in 1878 too. It consisted of three shallow craterlets together with a curious east-facing 'tunnel' that apparently represented a lateral 'blow-out' from beneath an old lava flow (Figures 4.5 and 4.6). Fisher noted later that he

> had the unpleasant experience of descending into the cavern south-east of Tavurvur before it was realized that a thick gas existed in the bottom. The gas was undoubtedly mostly carbon dioxide, but at least some of the after effects corresponded to those following on a small dose of carbon monoxide received once before in an abandoned mine ... On returning some months later with the intention of analysing the gas, it was found to have all disappeared and no fresh gas was being given off. (Fisher 1939a, 32)

Salt water was evidently emitted from these small craters in great quantity at the time of the Tavurvur eruption,

> as a flood of water was observed by [local people] at the time, and a heavy flow of mud and sand with a considerable number of huge boulders was left between the craterlets and the shore. (Fisher 1939a, 23)

The second area of craterlets was to the north-east of Tavurvur and apparently they did not exist before the 1937 eruption (Figure 4.5). They were still warm and, again, both held evidence for the presence of the colourless, odourless and heavy volcanic gas of carbon dioxide. A considerable amount of carbon dioxide had evidently been given off at one of them during the 1937 eruption as 'numerous birds, insects and several pigs were found dead in the immediate vicinity' (Fisher 1939a, 38). Accumulations of carbon dioxide remain a potential threat to life in the Tavurvur area, including to local people digging for megapode eggs in the thermally active parts.

Figure 4.6. Volcanological fieldwork at Escape Bay.
Volcanologists C.E. Stehn (left) and W.G. Woolnough are seen here undertaking field work at Escape Bay, south-east of Tavurvur volcano, in late July 1937, close to where a series of minor explosion craters had developed during the Tavurvur eruption. R.W.J. Collection 10, Folder 2, NHF-envelope no. 5. High-resolution copy courtesy of the National Library of Australia.

Figure 4.7. Water-filled 1878 crater near the new Vulcan cone.

What used to be Vulcan Island was all but concealed by the pumice laid down during the 1937 Vulcan eruption. However, the 1878 crater of the island was still clearly identifiable after the 1937 eruptions as a water-filled depression, as seen in this photograph taken by N.H. Fisher probably in about July 1937. Fisher's New Guinean assistant shown here is Sokopo. The slope of Vulcan cone can be seen in the top right-hand corner and the western caldera rim in the top left. R.W.J. Collection 10, Folder 2, NHF-envelope no. 4. GA negative reference GB-2551.

Construction of a new volcanic cone on the western side of Blanche Bay as a result of the 1937 eruption represents the most conspicuous change to the volcanic landscape in the Rabaul area. The cone, which was variously given the name Vulcan, Baluan or Kalamanagunan, grew from a new vent about 1.5 kilometres north-west of the old 1878 crater, filling the gap between the former Vulcan Island and the shoreline (Figures 3.8 and 4.7). This position was closer to the caldera wall on the western side of the bay. The highest point on the north-west side of the central crater rim was, according to Fisher (1939a), in August 1937, 742 feet (226 metres) above sea level, although the general elevation of the crater rim was about 650 feet (198 metres). The crater measured 2,070 x 1,830 feet (631 x 558 metres), being slightly elongated from east to west. Thermal activity was still taking place on the lowest part of the crater floor, as well as on the south-western outer slopes including at the small blow-out craterlet. An inlet or embayment at the northern foot of the cone appears to represent a flank vent that was breached by the waters of Blanche Bay (Figure 4.10). Thermal activity was taking place here too, but, together with all the other thermal activity on the new cone, it would die away within months. This is in marked contrast to the Greet Geothermal Field where thermal activity would continue indefinitely.

4.4. Winds, Ash Falls and Planning

Newsreels were shot of the devastation at Rabaul and the two active volcanoes, some of which were shown in theatres in Australia (NFSAA 1937a, 1937b, c. 1939). One of the films, photographed in 8-millimetre colour from HMAS *Moresby*, shows the shoreline of the affected town and wharves, as well as ash being cleared from town roofs and minor explosive activity taking place at Vulcan (NFSAA 1937a). The *Moresby* undertook a bathymetric survey of parts of the harbour floor and N.H. Fisher (1939a) used the results to estimate the amount of new material on the floor of Blanche Bay. He also mapped ash thicknesses on land throughout the affected area for both the Vulcan and the Tavurvur eruptions of 1937 (Figures 4.8 and 4.9), estimating the volumes of tephra for each one. The total volume for Vulcan, including the cone itself, was calculated to be 400 million cubic yards, or about 300 million cubic metres, although there is some uncertainty about whether this also included the amount of ash that fell over a broad area of the Bismarck Sea to the north-west. Measuring the thicknesses of the Tavurvur mud-ash in the town area cannot have been straightforward because of the flowage of the mud and its removal from roofs by hand or rain, but Fisher estimated the total volume to be 3–4 million cubic yards, or about 2.3–3 million cubic metres—that is, a hundred times smaller than the volume for Vulcan. These results, although only approximations, provide clear evidence that the Vulcan eruption of 1937 was far larger than the Tavurvur eruption, which is not surprising given that Vulcan started its activity well before, and continued after, the relatively short-lived eruption at Tavurvur, creating more devastation, including the loss of hundreds of lives.

Another useful aspect of the isopach map in Figure 4.9 is the way in which the lines of equal ash thickness appear to have been influenced by different sets of winds above the volcanoes. The Vulcan isopachs have a wide divergence, spreading downwind almost at 90 degrees to each other. This was caused by different winds at different times and at greater heights, and presumably also by creation of the volcano's own local 'weather'—including rain and lightning—as the heated water vapour clouds penetrated to higher and colder levels. The dominant trend of the isopachs for Vulcan, however, is west-north-westerly, probably reflecting a combination of both the lower-level south-easterly winds and higher east-to-west winds above 4 kilometres. In contrast, the simple, elongated form of the Tavurvur isopachs reflects the dominant impact of the strong low-level south-east trade winds at this time of year. The distribution of some ash from Tavurvur could have been affected by the higher east-to-west winds, resulting in deposition mainly in Simpson Harbour, as discussed above with reference to Figure 3.27. Alternatively, it could have been deposited on land to the north-west in layers too thin to be measured easily, or been mixed with Vulcan ash before it was deposited on the ground.

Figure 4.8. Two ash layers exposed on a roof in Rabaul.
The preserved thicknesses of ash on this corrugated iron roof of a Rabaul building are shown clearly where parts of the ash layers have slid off onto the ground. Note, particularly, that there are two main layers of ash: the lower, light-coloured one was produced by Vulcan on the evening of 29 May 1937, and the upper one by Tavurvur beginning in the early afternoon of 30 May. The two layers are of similar thickness, meaning that this building was in the main part of Rabaul town. Measurements at different places assisted N.H. Fisher in building up the isopach map shown in Figure 4.9. Note also the damaged trees in the background. GA negative reference GB2934.

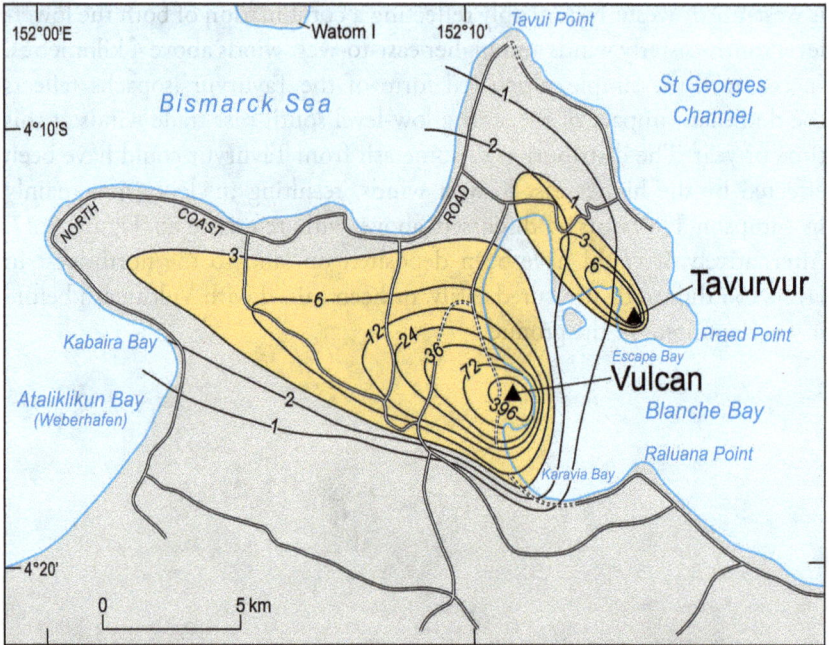

Figure 4.9. Map of contoured thicknesses of Vulcan and Tavurvur deposits.

The isopachs or 'contour' lines shown here have been taken from a comprehensive chart drawn originally by volcanologist N.H. Fisher (1939a, Plate C1). Fisher's chart was then adapted (and reduced in size) by Johnson and Threlfall (1985, 40). The isopachs provide an indication of the thicknesses, in inches, of the ash and pumice deposits produced by the 1937 volcanic eruptions at Vulcan and Tavurvur. Colour has been added here for those areas where Vulcan ash is 3 inches or more in thickness and where Tavurvur ash is 1 inch or more, in order to emphasise the differences between the main axes of ash dispersal for the two volcanoes (the Vulcan axis being more westerly). Note, however, that there is overlap between the isopachs for the two volcanoes corresponding to depths of less than both 3 and 1 inch, respectively. Fisher originally plotted names of villages and their respective number of fatalities but these have been omitted in this simplified adaptation.

The exodus of townspeople from the Rabaul area up Namanula Hill and Tunnel Hill roads in May 1937 was a spontaneous evacuation triggered by the sight of the volcanic fallout. However, most of the land west and north of Blanche Bay was affected by the volcanic fallout from both volcanoes (Figure 4.9). No contingency plans for evacuation had been drawn up by the administration, but the two roads were obvious exit routes for the trapped evacuees. Both roads were the nearest and most effective means of escape by land from within Simpson Harbour, even though they did not lead to nearby refuge points suitable for the accommodation of thousands of people. A lesson to be learnt from the 1937 Vulcan eruption is that the north-coast road would again be made unserviceable by a similar eruption during the south-east season. Those people who used Tunnel Hill Road in 1937 and then turned westwards along the north-coast road fared much worse than those who headed north-eastwards towards Tavui Point at the northern tip of Crater Peninsula where the ash fall was very slight (Figure 4.9). The coastal strip between Nonga, Tavui Point and Nodup is less vulnerable to Vulcan ash falls in any season than other areas around Rabaul, but accommodation was lacking there for large numbers of people in 1937. The thousands of evacuees at Nodup on St Georges Channel were fortunate that the sea was calm, enabling ship embarkations to be made even in the absence of port facilities.

Dr Stehn, before leaving Rabaul, compiled a list of 14 phenomena that might precede any further volcanic outbreaks at Rabaul (Stehn and Woolnough 1937a, 1937b; Fisher 1939a, 45). These ranged from felt earthquakes (number one in the list) to seeing large quantities of dead fish (number 14). Among other recommendations, he highlighted the importance of taking measurements of any temperature changes at existing thermal areas on land and shore, collecting and identifying volcanic gases, and keeping an eye open for any new gas upwellings in places like Greet Harbour and near Vulcan (Figure 4.10). N.H. Fisher used the list after Stehn's departure for his own, ongoing, volcano-monitoring program, commenting on its usefulness in his 1939 report, although he was limited by the availability of suitable instrumentation—he had no seismograph at this time, for example. Nevertheless, Fisher eventually was able to collect some vapour and gas samples along the water's edge by adapting and using a portable gas-analysis apparatus. The dominant emissions were found to be water vapour and carbon dioxide, although Fisher was fully aware that during the eruptions themselves other important gases such as hydrogen sulphide, sulphur dioxide and hydrogen chloride had been identifiable by their distinctive smells. All of these results, he said, 'point in the same direction, namely, that the volcanic activity has reached a fairly decadent stage' (Fisher 1939a, 32).

A suite of seven samples from the 1937 eruptions at both volcanoes was sent to Australia by the Department of Agriculture for analysis at the Waite Agricultural Research Institute in Adelaide. The results were intended for agricultural purposes, but they also had some volcanological interest. Calcium sulphate (gypsum) was detected in the analysis of soluble salts extracted from the samples and was attributed to a volcanic fumarolic origin. Sodium chloride was also common and was 'undoubtedly due to contamination with sea-water' (Hosking 1938, 375; see also Fisher 1939a, Appendix 2). The acidity and alkalinity of the liquids extracted from the seven samples, as measured by pH values, ranged from 5.0 to 7.9.

Figure 4.10. Measuring temperatures at the northern foot of Vulcan.
Volcanologist Norman Fisher (left) and Frank Anderson (a visitor from Sydney) measure temperatures in the embayment or inlet on the northern side of Vulcan cone in August 1937. GA negative reference GB2552.

4.5. Comparing and Contrasting the 1878 and 1937 Eruptions

The fact that two volcanoes, Vulcan and Tavurvur, 6 kilometres apart, were in precisely simultaneous eruption for parts of the eruptive activity in both 1878 and 1937, is a notable if not remarkable feature of Blanche Bay volcanology. Finding examples like these elsewhere in the world is a most challenging if not futile exercise. There are, however, some important differences between the eruptions of the two years, not least being that the 1878 eruption took place during the north-west monsoon season whereas the 1937 activity was in the 'dry season' of south-east trade winds. This meant that the 1937 eruptions were the most destructive to life and property north-west of the volcanoes. Rabaul town, of course, did not exist in 1878, and there is no way of knowing the numbers of people who in 1878 lived in the area destroyed by the 1937 fallout. Much more information is available as a whole for 1937 compared with 1878, so making accurate disaster-management and volcanological comparisons between the two years is also a difficult exercise.

One important difference appears to be that in 1878 Tavurvur was the first of the two volcanoes to break out in eruption and the last to cease its activity. The reverse was the case in 1937 when the Vulcan eruptions lasted about three or four days and Tavurvur for less than a day—ignoring days when post-eruption vapour and gas emissions were observed. Tavurvur in 1878 could have been in activity for upwards of a month, impacting the area towards Praed Point but depositing its pyroclastic material out to sea to the south-east. The exact duration of the Vulcan eruption in 1878 is unclear but is assumed to have been only a few days—that is, of similar duration as in 1937. Another important difference between the activity of 1878 and 1937 at Vulcan is that, in 1878, the eruption was mainly submarine, producing large amounts of floating pumice that were blown out of Blanche Bay into the surrounding sea. Only late in its 1878 activity did the eruption vent break the water surface, creating Vulcan Island and a new subaerial crater.

The volcanic products emitted from both volcanoes in 1878 and 1937 were also different. Vulcan, in both years, produced pumice, lighter ash and larger blocks of lava, all of which represented fresh volatile-rich magma erupted explosively from an underlying reservoir and influenced too by the ingestion of seawater. This contrasts sharply with the 'blue-grey gummy mud' that was expelled from Tavurvur and that fell on Rabaul town on 30 May. Fisher (1939a) called this type of volcanic activity a 'steam explosion', meaning a hydrothermal eruption caused by the heating to boiling point of water held within the volcanic cone of Tavurvur. The water was both groundwater accumulated from seasons of tropical rain and volcanic water vapour being emitted from a magmatic source beneath the Greet Geothermal Field. The water was turned to steam, which expelled the chemically altered materials of the old volcano, and forced them out as if clearing the vent. This activity could have taken place also in 1878, but the records are insufficient to be sure. However, in 1878, Wilfred Powell mentioned in his descriptions of Tavurvur a 'fiery crater … [from which] enormous stones, red hot, the size of an ordinary house' were thrown up, corresponding to strong incandescence and to eruption of molten materials—that is, to freshly erupted magma, not mud, and, therefore, to non-hydrothermal volcanic activity at that time.

Some of the descriptions of the eruptive activity from Tavurvur on the night of Sunday 30 May—such as those of Brett Hilder—refer to incandescent rocks and to an eruption column much higher than the one that deposited the mud on Rabaul. This raises the possibility that the type of eruption at Tavurvur changed at some time during the evening or overnight from a hydrothermal, vent-clearing one to a full or part magmatic eruption, and that the fallout of ash at this time was more to the west over Simpson Harbour when the column reached higher into the atmosphere, rather than over Rabaul (Figure 4.9). The total volume of material emitted from Tavurvur in 1937 as measured by Fisher was still much lower than that produced by Vulcan in 1937, which is not necessarily the case for the longer-lasting Tavurvur eruption of 1878 whose volume could not be measured.

4.6. Discussions on Shifting the Capital, 1938–40

Public release of the 'official' Stehn and Woolnough recommendations at the end of November 1937, and their publication in the *Rabaul Times* of 3 December 1937, marked the beginning of a long period of argument, counterargument and indecision that was eventually resolved in late 1941. Such discussions had already been taking place and informally immediately after the eruptions of May 1937, but the level of debate increased resolutely in the three years between 1938 and 1940. This is illustrated best by the extensive amount of written and published material held on government files in the National Archives of Australia in Canberra, including discussions in the Australian Parliament, as reported in Hansard; government memoranda and reports; newspaper and magazine articles and editorials; lobbying by various business interests both in Rabaul and in the short-listed towns; and letters to government officials, particularly in Rabaul and Canberra. (A selection of this material is presented in R.W.J. Collection 8.)

Administrator Sir Walter McNicoll informed Minister for External Affairs Billy Hughes in Canberra that 'severe tremors' had been felt in the Blanche Bay area at 1.25 am on 8 January 1938, the main one according to Fisher's measurements registering 'Intensity 7' on the Rossi–Forel scale (Territories 1938). This information appears to have formed the basis for a ministerial press release in Australia that was picked up by several Australian newspapers. Headlines such as 'New 'Quake Hastens Rabaul Plan' (*Sun* Sydney, 10 January 1938) and 'Rabaul Shaken Again—2 Tremors' (*Herald* Melbourne, 10 January 1938) are examples. Australian Government Geologist Dr Woolnough in response stated on 11 January that, despite the uncertainties of the quoted data, 'it may be concluded that, while there is evidence of active earth movements in the area, there need be no immediate alarm in regard to [an] imminent eruption' (Woolnough 1938, 1).

A capital-site committee led by a former administrator, Brigadier-General T. Griffiths, was appointed by the Australian Government on 4 February 1938. Its members arrived in Rabaul on 20 February at the beginning of an extensive travel schedule throughout the Mandated Territory. Sixteen sites were considered (Griffiths, Thomas and Thornton 1938). The Minister for External Affairs at the time was the irascible former Australian prime minister Billy Hughes, the irascible former Australian prime minister. Hughes had maintained his interest in, and strong opinions about, the Mandated Territory and the separate Territory of Papua to the south ever since his involvement in the post-WWI treaty discussions and disputes at Versailles in 1919. Griffiths's committee focused on four potentially suitable sites on the New Guinea mainland where there were no active volcanoes but where earthquake and tsunami risk had to be borne in mind. Wau, the only inland town, was rejected because of its relative difficulty of access and because the ongoing goldmining there had a finite life. Madang on the north coast of the mainland also had some disadvantages compared with the two other coastal towns of Lae and Salamaua, both of which were in economically developed Morobe Province in the Solomon Sea. The committee eventually decided on Lae as the most appropriate site (Griffiths, Thomas and Thornton 1938). They recognised, however, that permanent anchorages and a natural harbour—such as developed magnificently at Rabaul—were not well developed at Lae. They therefore proposed the construction of a coastal road to Salamaua where the natural harbour was superior. Salamaua would, therefore, become the port town serving the new capital at Lae.

This proposal by the Griffith Committee was not accepted by the Australian Government in Canberra, and Minister Hughes went even further in proposing that Salamaua itself should become the new capital, and that a new road should be built from it over rugged terrain to Wau. This generated a good deal of criticism from people who knew the Mandated Territory well, and especially from those who knew Salamaua (NAA 1937–39). The conflict and confusion were still unresolved by the end of 1938 in part because of the costs involved but also because of international events that were diverting the attention of those in the Australian Government. The British Government's 'policy of appeasement' towards Adolf Hitler and his relentless expansionism in Europe affected New Guinea because there was the possibility of the Mandated Territory being handed back to Germany. The policy was abandoned, however, in March 1939 when Hitler completed the destruction of Czechoslovakia as a nation in defiance of the

Munich Agreement of September 1938. Japan was expanding militarily too, and its advances were much closer to the Territory of New Guinea than Adolf Hitler's.

Another committee of inquiry was set up in 1939. This one was to address the recurrent question of combining the Australian administrations of New Guinea and Papua into a single territory that would be run by Australia from a new capital. The committee chairman was F.W. Eggleston, a Melbourne lawyer, assisted by H.L. Murray and H.O. Townsend, the administration's treasurer in Rabaul. Murray was related to Sir Hubert Murray, the long-serving lieutenant-governor of the Territory of Papua, who, as his retirement approached, had a strong interest in the future of the Mandated Territory to his north and the prospect of amalgamation (West 1968). By August 1939, the Eggleston Committee had advised against the amalgamation, but it still had to address the matter of recommending a site for the new capital for the Territory of New Guinea (Eggleston, Murray and Townsend 1939). Salamaua was not recommended—a survey had been done that drew attention to its limitations—and the committee opted again for Lae, pointing out that the conditions for off-loading ships of different sizes were not as problematic as had been suggested by the Griffiths Committee (Eggleston, Murray and Townsend 1939; NAA 1941–46).

Rabaul people were becoming used to the stream of recommendations and delays in implementation. There had been no more eruptions at Vulcan and Tavurvur, and volcanologist N.H. Fisher was now on hand to report on the condition of the volcanoes, so residents were coming to the view that life in the town would be continuing as it was before the eruptions of May 1937. The Eggleston Committee had not been instructed to comment on maintaining Rabaul as the administrative capital; Rabaul clealy would continue anyway as an important commercial centre even if the big companies W.R. Carpenter and Burns Philp transferred to a new capital. Nevertheless, the committee summarised the various opinions presented to them on the matter:

> These views may be crystallised as being against the removal of administrative head-quarters at great expense from a suitable site to an inferior site (to which only a small portion of the population of Rabaul would move) in order to avoid a problematical and not extensive damage to government property once every 40 or 50 years. (Eggleston, Murray and Townsend 1939, 43)

Public notice was given in January 1938 that a special committee under the chairmanship of Treasurer H.O. Townsend would be set up in Rabaul to consider the giving of financial assistance to those people who, because of the eruption, were 'unable to carry on their means of livelihood without such assistance' (NAA 1937–50). Nineteen claims (all from non–New Guineans) had been received and partly dealt with by June. Several claims were refused, but loans were given to others, including planter J.O. Smith and A.R. Reed, who had lost his dairy. Mrs Lulu Miller and Mr Furter later received loans of £100 and £500, respectively. Mrs Jane Wallace, however, was one of those refused assistance (Wallace 1938, 1948). She and her son owned a property at Valaur that had been buried by about 10 metres of pumice and from which they had escaped 'merely in the clothes we stood up in, not one instant too soon' (Wallace 1938, 1). In addition, their car had been requisitioned and 'wrecked' by the administration during the week after the eruption. Mrs Wallace continued her appeal to the Australian Government at least as late as 1948 from her home in Nice, France (Wallace 1948). The reason for the refusal is not known, though there is some suspicion that the administration discriminated against the Wallaces because the family was considered too 'pro-native' by prewar European standards.

German forces invaded Poland on 1 September 1939 triggering World War II, and Australia became an ally of Great Britain in the same month. These momentous world events helped to delay a decision on Rabaul's future. The transfer to Lae would require funds and manpower, which were difficult to justify given wartime economic measures. Attention was directed towards the matter of Rabaul's defence should Japanese expansionism threaten the town, and there were internments of Germans (including Mr Furter) and Italian residents in Rabaul. The New Guinea Volunteer Rifles, made up of part-time civilian soldiers, was formed, and many Australian and British residents enlisted for overseas duty. War, rather than volcanic activity, was foremost in the minds of many Rabaul people.

5

World War II and the Tavurvur Eruptions of 1941–43

5.1. Monitoring the Volcanoes

Dr Stehn's proposal that a volcanological observatory capable of providing early warnings of volcanic eruptions should be established at Rabaul began to take effect despite the earlier, reasonable criticisms of the idea by his co-investigator, geologist Dr W.G. Woolnough. Funds were made available in 1939 for building the observatory and equipping it with instruments as part of the administration's Department of Lands, Surveys, Mines and Forests (Fisher 1940a; R.W.J. Collection 10). Norm Fisher, geologist, would be in charge of the observatory; he was sent to Java for further training under Dr Stehn in 1939. Fisher informed the world geoscience community of the construction of the 'Vulcanological Observatory' by preparing a short note for the international journal *Bulletin Volcanologique* published in Italy (Fisher 1940a). The observatory was to be built in a commanding position on a south-western spur of Tovanumbatir on the northern rim of the caldera overlooking Rabaul and all the historically active volcanoes of Blanche Bay. In addition, observation posts would be established near Tavurvur and Vulcan by using tunnels driven into the caldera walls near each volcano, fitted with instruments and gas-proof doors, and connected by telephone to the main observatory.

Figure 5.1. First volcanological observatory to be built overlooking Rabaul town.
This was the first building to be erected on Observatory Ridge for the prewar volcanological observatory run by N.H. Fisher. A residence was subsequently built, together with a tennis court. GA negative reference GB1352.

Fisher's plans were quite ambitious and involved construction of several above-ground rooms: an office, laboratory, darkroom, observation room and a small museum. The building consisted initially of a simple fibrocement-sheet construction over an insulated cellar that would contain most of the instruments (Figure 5.1). Fisher wrote in the final paragraph of his note that 'the instruments have been on order for some time, so it is probable that the station will be in operation before the end of the current year' (Fisher 1940a, 187). The hoped-for instruments included a German-designed Wiechert seismograph and Italian tiltmeters, but obtaining them during global wartime conditions proved impractical. Two seismographs, a tiltmeter and an annunciator had to be manufactured locally in Rabaul (Figure 5.2). Completed in June 1940 by Mr W.E. Jackson of the Department of Public Works, the instruments were 'a great job of work', being 'wonderful exhibits of precise workmanship and delicate adjustments', reported the *Rabaul Times* (1940, 5). Tenders for erection of a residence at the observatory were let in May of the same year.

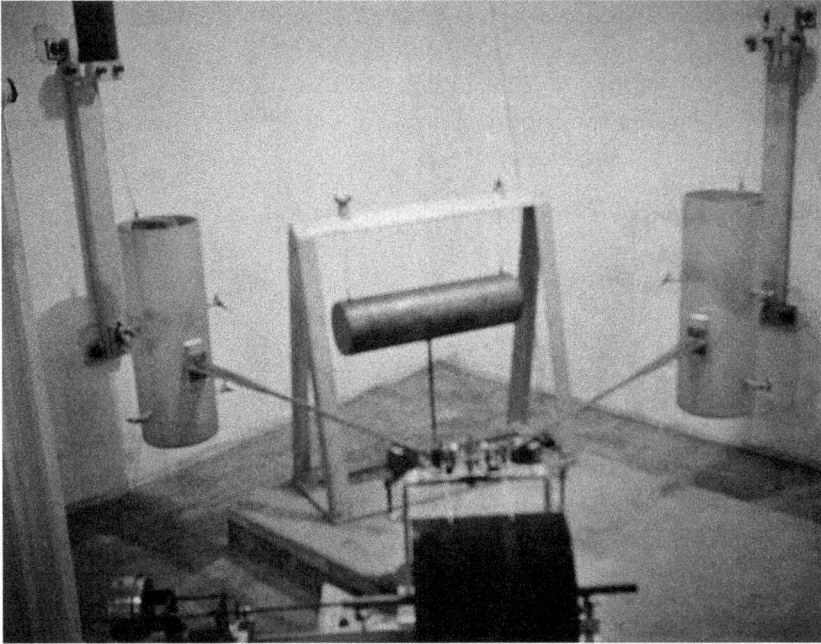

Figure 5.2. Volcano-monitoring equipment in observatory cellar.
Instruments installed in the cellar of the prewar volcanological observatory were designed and built in Rabaul under the direction of N.H. Fisher. Two parts of a seismograph are seen in this photograph (taken by Fisher) hanging on the western and southern walls of the cellar. Arms extend from them to the smoked-drum recorder in the foreground where the earthquake vibrations are traced out. The horizontal cylinder suspended in the trestle-shaped steel frame is an earthquake annunciator made after a design by Dr Stehn. Earthquakes would produce movement of the cylinder that, in turn, would cause a pointer to move across a series of mercury-filled electrical terminals to a point dependent on the size of the earthquake. A bell would ring automatically in the observatory residence, and shutters dropped in a panel in the observatory building, when earthquakes took place. Printed from original negative held by the National Library of Australia. R.W.J. Collection 10, Folder 2, NHF-envelope no. 4. High-resolution digital copy provided by the National Library of Australia.

Staff of the volcanological observatory led by Fisher provided weekly reports of their observations between December 1937 and January 1942. These reports were published in the *Rabaul Times* and were made available to the Australian Government in Canberra, where many were archived officially. Fisher was assisted by L.E. Clout, a draftsman with the Department of Lands; by the end of 1940, the administration had appointed an assistant volcanologist, C.L. 'Clem' Knight. Other contributors to the weekly reports, particularly when Fisher was in Java, were K.L. Spinks, H.J. Badger and L.C. Noakes. These reports and others, including copies of radiograms between the administration and Canberra, are still held by the Rabaul Volcanological Observatory (RVO) and digital copies of them are provided in the RVO information management system (IMS) (RVO 1937–42). Fisher and Noakes also published on aspects of the geology of New Britain as a whole (Fisher and Noakes 1942).

No further volcanic eruptions took place at either Tavurvur or Vulcan during 1938–40 but, reported Fisher in a memorandum to the government secretary in Rabaul:

> [A] series of small steam explosions occurred at Tavurvur Volcano between 9.30 a.m. and 10.15 a.m. Sunday 3rd March 1940. These were apparently due to the choking up of the principal vents on the crater floor by material washed down by the recent [wet season] rains and by slips within the crater … The explosions were witnessed at close quarters by a party of ten or eleven natives, mostly from the Seventh Day Adventist Mission at Palm Beach, Rabaul, who were gathering sulphur at the foot of the vertical face [just] west of the edge of crater. (Fisher 1940b, 1)

There were no casualties and the emitted dust and rocks barely extended beyond the crater limits. Nevertheless, the event was indicative that Tavurvur was still retaining significant amounts of volcanic heat, unlike Vulcan, which had cooled significantly. Further, rising temperatures at Tavurvur had been recorded by Fisher in his reports by the end of the year and had been noted in newspapers. For example, the headline for a short article published in the *Pacific Islands Monthly* for January 1941 was: 'Eruption? Rabaul's Volcanoes Under Suspicion. From Our Own Correspondent. Rabaul, Jan. 3' (*Pacific Islands Monthly* 1941, 9). The 'correspondent' was presumably Gordon Thomas.

A severe earthquake shook the whole of the Gazelle Peninsula and elsewhere at about 2.28 am on 14 January 1941, particularly in the north-east, causing damage to houses and contents, and concern in Rabaul town (Fisher 1944). The seismograph at the observatory was put out of action by the earthquake so few scientific details were available on the nature of the main earthquake, which was followed by many aftershocks. Fisher, however, undertook field work, examining the different directions in which houses had responded to the main shock. He concluded that the epicentre of the main earthquake was about 20–21 miles (32–34 kilometres) south-west of Rabaul, near the village of Wunga. Fisher even produced an 'isoseismal map' for the whole region encompassing New Ireland and eastern New Britain. He showed that the areas of maximum earthquake intensity—measured at 8–9 on the Rossi–Forel scale—defined a broad north-north-west trending zone that was evidently related to a major geological fault running from Ataliklikun Bay across the Gazelle Peninsula to beyond the south coast of New Britain. The earthquake was, therefore, of tectonic rather than volcanic origin.

Dr Woolnough in Australia, early in the first week of January, had sent a memorandum to the Department of the Interior, Canberra, expressing his concerns about the volcanic situation in Rabaul (Woolnough 1941a). He was significantly more anxious about the situation than he had been about the local earthquake on 8 January 1938 (Woolnough 1938). After news of the 14 January earthquake reached him, Woolnough completed an article that was published in the *Sydney Morning Herald* (23 January 1941) and *Rabaul Times* (31 January 1941):

> The news of severe earthquakes at Rabaul, following upon reports that both of the recently active volcanoes have again been emitting steam, focuses attention upon the conditions existing at the centre of administration of the Mandated Territory of New Guinea …
> It is inescapable that the recent increase in seismic activity in so critical a region must give ground for some anxiety as to the ultimate outcome. (Woolnough 1941b, 10)

Woolnough believed that January's seismicity represented 'deep-seated action' that 'must favour the movement towards the surface of masses of molten rock, bringing fresh supplies of heat and energy within the reach of surface waters. This favours explosive activity.' He also expressed his opinion that Vulcan and Tavurvur were connected by a sea floor fracture and that any new opening and eruption along this crack might 'block completely the harbour of Rabaul, and for ever bottle up any ships which might be caught there' (Woolnough 1941b, 10).

Norm Fisher replied to Woolnough's article in the *Rabaul Times* on 7 February, introducing his generally critical remarks thus: 'Dr Woolnough has presented the most pessimistic and unfavourable aspect of the situation in Rabaul' (Fisher 1941, 4). Fisher stressed that the earthquake and its aftershocks had not led to a change in the state of the volcanoes 'apart from subsidences around Vulcan caused by submarine slippage'. He questioned the evidence that Vulcan and Tavurvur were connected by a potentially active submarine fracture. Woolnough had written that the conditions for evacuations during the 1937 eruption 'could scarcely have been luckier'

(Woolnough 1941b, 10). Fisher, however, took the opportunity to point out that circumstances during the 1937 Rabaul evacuation had not been especially favourable (as Stehn and Woolnough had stated in their report), noting that similar eruptions during the north-west season, especially any from Tavurvur, would not have affected Rabaul; that dusk and darkness did interfere with the evacuation; and that families 'were not so reunited but dispersed largely about their various recreations'. Fisher added that another evacuation from Nodup would not be likely to arise as an 'emergency camp has been prepared at Tavui with full facilities for looking after the population of Rabaul' (Fisher 1941, 4).

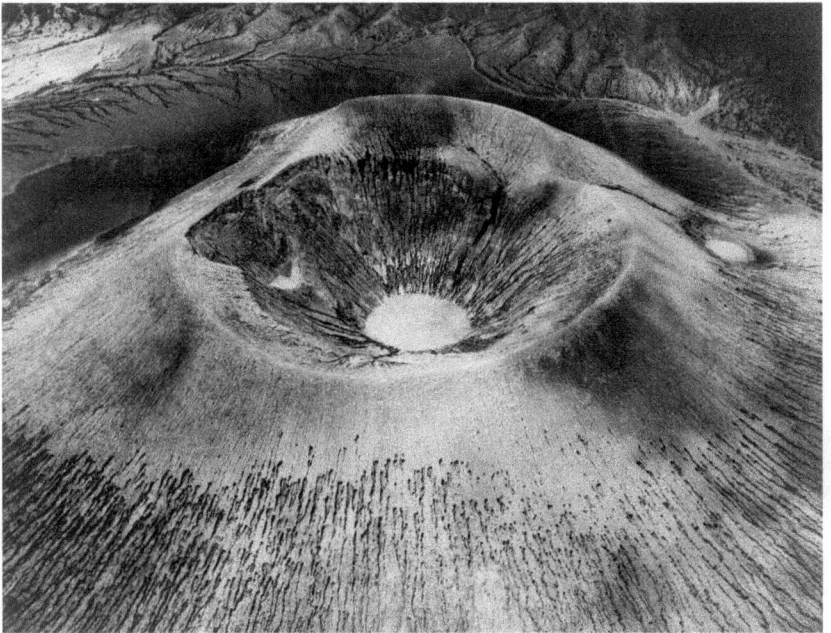

Figure 5.3. Oblique aerial photograph of Vulcan cone in 1941.

Vulcan was eroded by rains in the few years after the 1937 eruption. Numerous small gullies with rounded headwalls were formed, as seen in this photograph taken by Flying Officer R.J. Love from a Catalina flying boat in May 1941. Not much vegetation has taken hold on the new volcano, but by the end of WWII the cone was shrouded in vegetation and further erosion was thus inhibited. The erosion of Vulcan and of the nearby shorelines (including what was formerly Vulcan Island) resulted in pumice being washed into Blanche Bay. Some of this pumice floated and was driven into Simpson Harbour during the south-east seasons until at least 1945, though not in the large amounts seen shortly after the 1937 eruption. Photograph courtesy of Mrs M. Love. GA negative reference GB2381.

The January earthquake had apparently not affected the volcanoes, although Fisher had been observing some important changes to the state of Tavurvur—but not Vulcan (Figure 5.3)—as far back as late August 1940 when new hot ground was noted in one of the craters. He summarised this and later developments in a valuable paper written 35 years later (Fisher 1976b). Temperatures remained steady in the craters at about 100°C until early November when an 11°C rise was noted, followed by a spectacular rise to 224°C in December prior to the 14 January earthquake (Figure 5.4). A general rise in temperature followed over the next few months until the last recorded temperature reached 392°C on 3 June 1941. New vents vigorously discharging gas had also appeared during the previous August, giving off a strong smell of sulphur dioxide. The discharges increased here and elsewhere on Tavurvur until 27 May when gas and vapour were 'issuing with a "roar like a train" and the steam cloud was visibly larger' (Fisher 1976b, 205).

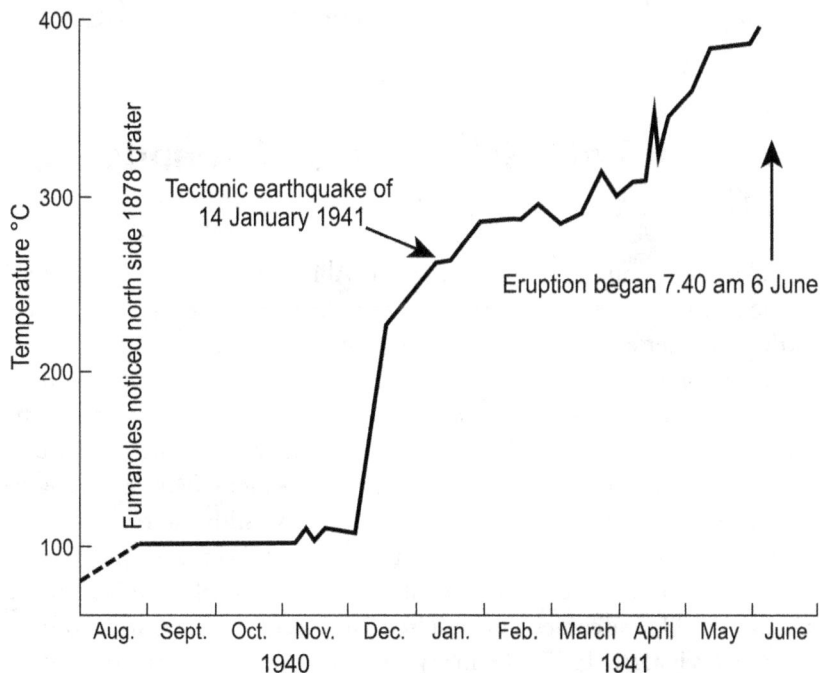

Figure 5.4. Time series graph of Tavurvur temperature measurements, 1940–41.

This rise in temperature of hot ground in the 1878 crater of Tavurvur was detected by N.H. Fisher until a short time before an eruption took place on 6 June 1941 (Fisher 1976b). Adapted from GA drawing, reference 24/B58-2/35.

Sulphur dioxide and hydrogen chloride gases became readily detectable, but no build-up in earthquake activity was recorded on the two seismographs—one at the observatory and the other at a point about 1 kilometre north of Tavurvur—probably because of the low sensitivity of the instruments. The bottom of the main crater of Tavurvur had silted up since 1937, and, in February and March 1940 (the end of the wet season), there had been minor steam explosions caused by the heating of the crater muds—as reported above. Four explosions on 3 March 1940 'hurled mud and small stones up to about the height of the crater rim' and on 12 and 16 March the ejections of 'mud, stones, and later dust were thrown out to a height of a hundred metres or more above the crater rim' (Fisher 1976b, 207).

All these observed changes at Tavurvur were preparatory to the start of a main period of volcanic eruptions that began at 7.40 am on 6 June 1941 and would affect the unfortunate town of Rabaul over the next several months. They would affect not only the usual residents of town, but also, subsequently, the Japanese forces who invaded and occupied Rabaul in January 1942.

5.2. Arrival of 'Lark Force' and Abandoning the Capital

Australia had committed to joining the Allies in the war against Nazi Germany in Europe, but it also had the challenge of guarding its own northern border against possible invasion by the Japanese military, which was expanding resolutely in the Western Pacific. The magnificent deep harbour at Rabaul was an obvious military target for the Japanese; the challenge for Australia was how to defend what was still the capital of a territory mandated to Australia by the League of Nations. More particularly, could enough military resources be assembled to constitute a strong defence of a territory that was, in reality, distant from Australia's continental shores? The fact that the actual military deployments at Rabaul were inadequate and the terrible outcomes that resulted are described in detail in official histories (Wigmore 1957), in many unofficial ones (Hall 1981; Nelson 1992; Lindsay 2010; Threlfall 2012) and in major compilations based on the accounts of individuals and the members of organisations (Aplin 1980; Stone 1995; Downs 1999; PNGAA 2017).

Australian soldiers of the 2/22nd Battalion were among those who saw the 6 June 1941 eruption at Tavurvur. They also saw the succeeding eruptions of what was to be the first phase of volcanic activity that lasted until 29 June. The battalion was part of the 23rd Brigade of the 8th Division, 2nd Australian Imperial Force, and its members were largely young men, in their late teens or early twenties, from Melbourne and rural Victoria. They had volunteered expecting to fight against the Germans in Europe, but had been sent to Rabaul for garrison duty as the defence force for what they discovered was an isolated outpost. An advance party of troops arrived in early March 1941, and, by late April, over 1,000 soldiers had been established in an army camp along Malaguna Road. The battalion was part of a composite force in Rabaul, known as 'Lark Force', that included units of, for example, fortress artillery, anti-aircraft artillery, fortress signals and six women from the Australian Army Nursing Service. Lark Force also had an excellent brass band, some of its members being formerly from the Brunswick Salvation Army Band. They had enlisted as brass bandsmen and stretcher bearers, along with their bandmaster, Sergeant Arthur Gullidge, a talented conductor and composer of band music. Most of the other bandsmen were Salvation Army members from other corps. The Royal Australian Air Force deployed Wirraways a trainer and general-purpose aircraft, in the Rabaul area, but these would prove inadequate in competition with fast 'Zero' fighters of the Japanese Air Force. Members of the already-established New Guinea Volunteer Rifles supplemented the small Australian military force that had been imported to defend Rabaul.

Garrison duties were not especially exciting for the young 2/22nd soldiers, and boredom was a common enemy. However, the Tavurvur eruptions provided some excitement, as few, if any, of the troops had ever before seen a volcano in eruption. Stan Whitty was a member of a fortress signals unit. He recalled:

> One evening, about mid-June 1941, a group of us became aware of some sense of vibrations in the very still air. Then there was this most frightening and terrifying loud explosion. The sensible [people] seemed to be moving out of Rabaul, either in cars or on foot ... The ignorant stupid, like us, just stood and looked in awe ... The mouth of the volcano was spewing out red hot rocks and ashes. A great cloud was rising high into the sky ... The flames belching out of the open mouth were quite plain to see ... Though other belchings and eruptions of rock and ashes took place in the following weeks, that first terrifying explosion was never equalled in sound level. The experience made you realise how puny man is compared with

> nature ... One chap refused to sleep in his tent ... Always slept
> fully clothed—boots and all—in the back of one of our trucks. Keys
> always left in ignition switch. (Whitty 1981, 1)

The first eruptive period consisted of irregularly spaced explosions of
different intensities, the largest of which took place on 17 June, hurling
blocks beyond the foot of the volcano. Dust and rocks were flung out during
the first two days of the eruption, but on the evening of the second day,
red-hot rocks began to be expelled. The ash during this period was blown
towards Rabaul by the south-east winds, and was a considerable nuisance
to the townspeople. Mr Hoogerwerff, who was still manager of the Rabaul
Printing Works, recorded on 12 June that the 'only inconvenience so far is
that the air is full of dust and ashes which make everything very dirty'; two
days later he added that he had 'thought it best to go to bed dressed with
shoes on ... The happenings of the last week do not make one feel too good'
(Hoogerwerff 1941).

Many photographs were taken of Tavurvur eruption clouds in 1941 although
precise dates are rare (Figures 5.5–5.7). These images, plus descriptions of
the eruptions, provide good evidence that most, if not all, of the eruptive
activity at these times at Tavurvur, and probably into 1942, was mildly
'vulcanian' in nature. Similar vulcanian eruptions probably took place at
Tavurvur in 1878 but were not photographed. Further, the short overnight
eruption at Tavurvur in May 1937 also may have been vulcanian in part,
although it was not well described.

'Vulcan' is the name not only of the Roman God of Fire but also of a young
volcanic island in the Mediterranean that was active in 1888. Volcanic
gas and vapour in vulcanian eruptions are exsolved explosively from new,
viscous magma beneath craters, producing ash, dust and blocks that are
commonly incandescent. Vulcanian explosions can also create atmospheric
shock waves. Strong vulcanian eruptions were recorded at the Italian Vulcan
in 1888 but, as the Italian volcanologist Mercalli (1907), noted:

> In the less violent explosions large ejecta were lacking and the jet
> consisted of a dense gray mass of lapilli, sand, and ash that rose
> slowly, taking the form of a great cauliflower or giant mushroom ...
> [T]he cloud expanded in dense globes and volutes, finally building
> up to a height of 3 or 4 kilometres. (cited in Macdonald 1972, 223,
> translated from Italian)

Figure 5.5. Tavurvur in explosive eruption (Hawnt).

E.A. Hawnt took this photograph of Tavurvur in vulcanian eruption in late June 1941. The photograph was supplied by Mr C.N. MacL. Stirling, a former lieutenant in the 2/22nd Battalion. Members of the 2/22nd Battalion and Lark Force Association in Victoria responded most generously to requests for photographs such as this of the 1941–42 volcanic activity at Tavurvur. GA negative reference M2447-11A.

Figure 5.6. Tavurvur in explosive eruption (Hutchinson).

A cauliflower-shaped vulcanian eruption cloud from Tavurvur billows upwards in this 1941 (possibly 1942) photograph supplied by Dr R.C. Hutchinson, a civilian who managed to escape from the Japanese invasion of Rabaul. GA negative reference GB2538.

Figure 5.7. Incandescence in Tavurvur eruption cloud.
Early morning light catches the eastern side of a vulcanian eruption cloud from Tavurvur and is sufficiently dull that incandescence at the rim of the volcano is readily visible. This photograph was supplied to N.H. Fisher courtesy of Mr F. Kollmorgen, a member of the Salvation Army band. Its exact date, other than 1941, is unknown. R.W.J. Collection 30D, Folder 13, Sleeve 13.

Conditions in Rabaul at this time are summarised in a letter dated 20 June 1941 and written by Mrs Mona Anthony to her brother-in-law the Hon. Hubert L. Anthony, a minister in the Australian Government:

> Well, she's still going up with great gusto to a height of 2–3000 feet [600–900 metres], blowing away with great roars and explosions which sound most of the time like a train going over a big overhead bridge. From outside of the town where you can get a good view of it, it presents a wonderful spectacle at night, throwing out great red-

hot boulders and sparks and glowing away threateningly. No one seems particularly nervous now and there is absolutely no sign of panic, though, naturally, at first it caused a certain amount of alarm, particularly at night. Many people, particularly those with children have gone out of town, and I think that it finally decided those who were hesitating about going South to leave when passages became available. You can't conceive the dust and filth which covers the place … Everyone is going about with handkerchiefs over their noses and mouths and coverings over their hair. We eat it, sleep in it and sit in it, and what a mess! … Nearly half of the children are away from school … If it becomes permanent we will not be able to live here permanently. I should think many people are extremely indignant that the powers that be have missed their opportunity of moving the capital and think that it should have been done after the first [1937] eruption. Of course, there is a war on now as an excuse. Also many of them are extremely irate of the seeming complacency of the Administrator and pray daily that all the dust will blow in his direction. Quite heart breaking to see the homes. We thought this morning that we were going to have some heavenly relief with the wind blowing in the opposite direction but, alas, a strong S.E. has sprung up and … [the ash is] pouring down like rain. (Anthony 1941, 1)

Tavurvur remained inactive for a few days after 29 June, but had resumed its explosive activity by 3 July. However, it lapsed again into quiescence by 18 July. There were more explosions on 27 July, but the succeeding eruptions were mild—at least until 9 August, when activity began to increase at the start of what was to be the longest and most continuous period of activity of the whole eruption. Especially violent explosions took place on 22 and 25 August, and the week of activity ending on 27 August was judged by Norm Fisher to be 'easily the most violent' since the beginning of the activity in June (Fisher 1976b, 208).

Lieutenant David M. Selby was in command of an anti-aircraft battery in Rabaul, and had arrived there in August when Tavurvur was in eruption:

It was throwing up a tall column of black smoke and pumice dust. From time to time this would die down, then there would be a roar and a huge cloud of swirling black smoke and dust would burst out of the crater, swelling into a mushroom-shaped cloud … The tremendous force of the gas coming from the crater would lift this up until it became the familiar column, rising to a great height until it bent almost at right angle when the wind caught it and carried it over the town of Rabaul. A curtain of dust and pumice stone would fall from this over the town. Everything exposed to it would be covered with a layer of black pumice. When rain was expected, it was common to see natives on the roofs of houses sweeping clouds of ash off the roof … At times the [eruption] column would be shot with flames and large rocks would be hurled from the crater. A strong smell of sulphur pervaded the town the whole time I was there … Whatever was in the gas … was powerfully corrosive. Unprotected metal was very soon destroyed and this was particularly noticeable on motor vehicles where any scratched or unpainted parts quickly developed gaping holes. We became quite resigned to the fact that everything exposed to the open air was thickly covered with black grit. (Selby 1981, 1–2)

The metal deterioration mentioned by Selby was probably caused by volcanic gases such as sulphur dioxide, hydrochloric acid, fluorine and hydrofluoric acid:

[G]alvanised iron, wirenetting, gauze containing zinc or copper, and many other metals were attacked, mainly by the hydrochloric acid in the atmosphere. The result was that extensive replacements became necessary, and drinking water collected in tanks from roofs became fouled with zinc chloride and unusable. (Fisher 1946a, 2)

Further, Mr Hoogerwerff recorded in early August that:

The dust seems to be harmful to the vegetation … Only casuarina, mango and a few other strong trees do not suffer but otherwise all the trees and hedges have lost their leaves. (Hoogerwerff 1941, 42)

And, on 28 August, that:

[Tavurvur has been] ejecting huge quantities of black dust [which is] badly affecting the vegetation, all ironware and textiles. Dust is everywhere, so, our lungs and eyes are sure to suffer someday. If this goes on, Rabaul will no longer be a place to live in. (Hoogerwerff 1941, 42)

The administration, in fact, had already reached the same conclusion as Hoogerwerff. Sir Walter McNicoll—who used his other Christian name after being knighted—called a meeting of the Executive Council for Saturday 23 August, and the decision—delayed for almost four years after the release of the Stehn and Woolnough report—was made to move the seat of government from Rabaul (NAA 1941–46). The decision was announced in an editorial of the *Rabaul Times* the following week (Thomas 1941). Weeks of enduring the effects of Tavurvur's repeated eruptions had finally removed any remaining doubts about the advisability of leaving Rabaul. Tavurvur ash had even fallen on the administrator's home on Namanula Hill. This was caused by the Tavurvur eruption clouds eddying around Kabiu mountain—no doubt providing satisfaction to those Rabaul people, mentioned in Mona Anthony's letter, who had irreverently hoped for fallout on Government House. Approved by the Australian Government in Canberra, Lae was proclaimed the capital of the Mandated Territory of New Guinea on 9 September 1941, ending Rabaul's 31 years as a colonial capital.

The move of the administration to Lae could not, however, be made immediately. The shift was planned to take place over several months, and the task of crating documents and all the paraphernalia of a bureaucracy took many weeks. McNicoll made several trips between Rabaul and Lae, but, by the end of November 1941, the administrator had transferred his household to Lae and taken up official residence there (*Morobe News* 1941). So ended the line of governors and administrators living on Rabaul's Namanula Hill. The army's hospital was moved into Government House from its tents on Malaguna Road.

Eruptions from Tavurvur continued through September until 7 October. These were generally less severe than those in late August, but, while the south-east season lasted, they nevertheless continued to make life unpleasant for those in Rabaul. The eruptions of Tavurvur helped, however, to alleviate boredom for at least some of the 2/22nd troops. A group of soldiers, wearing masks for protection against volcanic gases, would climb the flanks of Tavurvur while it was active and plant a sign as near as possible to the volcano rim. They would, on their return to camp, then challenge any other group to retrieve the sign, and this game would go on as later groups tried to stick the sign farther up the slopes of the volcano! These daredevil antics were stopped when a commanding officer heard about them. The southern side of the volcano could, in fact, be approached quite closely, with caution, because the eruption point was near the base of a lava plug in the crater formed in 1937, and because the southern wall of this crater was almost vertical. This meant that most projectiles did not fall to the south. Nevertheless, the sign-planting practice could hardly be recommended, especially as, in late October, powerful explosions hurled rocks—that were hot enough to set fire to dry grass—about 1 kilometre from the crater. Visits were also made to the deep gullies on Vulcan (Figure 5.8).

Figure 5.8. Visit to deep erosion gully near Vulcan.
Four soldiers from the 2/22nd Battalion are here on an excursion to the 'Grand Canyon', their name for one of the deep gullies cut into the 1937 pumice near Vulcan (see also Figure 3.62). Mr S.G. Whitty supplied the photograph and marked with crosses those of his mates who died during World War II. GA negative reference GB2675.

A lull in volcanic activity took place from 27 October to 8 November, another from 13 to 15 November, followed by a period of strong activity until 25 November, which was characterised by some exceptionally powerful outbursts. Volcanic bombs were hurled great distances—the farthest yet—on 19 November (Fisher 1976b). However, Tavurvur was quiet for the rest of the year after 25 November, except for short outbursts of activity on 4 December and on 9 January 1942. The north-west season had set in by this time, so ash was not blown over the town. In any case, the people of Rabaul were more concerned with the problems of the war that was about to break over the town. Tavurvur, however, began another eruptive period in the week beginning 19 January (Fisher 1976b).

5.3. Japanese Invasion and the Arrival of Takashi Kizawa

The Japanese Navy had a major base at Truk Lagoon in the Japanese Mandated Territory of Micronesia, which was only about 1,300 kilometres directly north of Rabaul. Rabaul was closer to Truk than it was to Cairns in northern Queensland, where there was no equivalent military base. The Japanese first attacked Pearl Harbor, Hawaii, on 8 December 1941 (Rabaul time), and during the remainder of the month they advanced rapidly on many fronts in the South-East Asian and Western Pacific regions. The attack on Pearl Harbor was the trigger for the US to enter the war as one of the Allies.

Japanese reconnaissance aircraft were seen over Rabaul in December. An attack on the town seemed inevitable (Figure 5.9). There were concerns about the ability of the small force at Rabaul to repel any significant attack, and about the lack of preparation among the soldiers should they have to withdraw from Rabaul; most of the troops had no knowledge of tropical bush craft. The orders from Australia were quite clear. On 12 December, the Australian Chiefs of Staff advised Cabinet in Canberra that there would be no military withdrawal from Rabaul and that the troops should remain there as 'hostages to freedom' (Nelson 1992; Stone 1995). A forward observation line was necessary, they said, but shifting out a large force by sea was dangerous at that time, and there were tasks of higher priority for them to consider. European women and children were compulsorily evacuated to Australia, Chinese stores announced that business would be on a cash-only basis, banks restricted trading hours and the number of social events declined markedly. Christmas was quiet and subdued for many of the all-male households of Rabaul's European community.

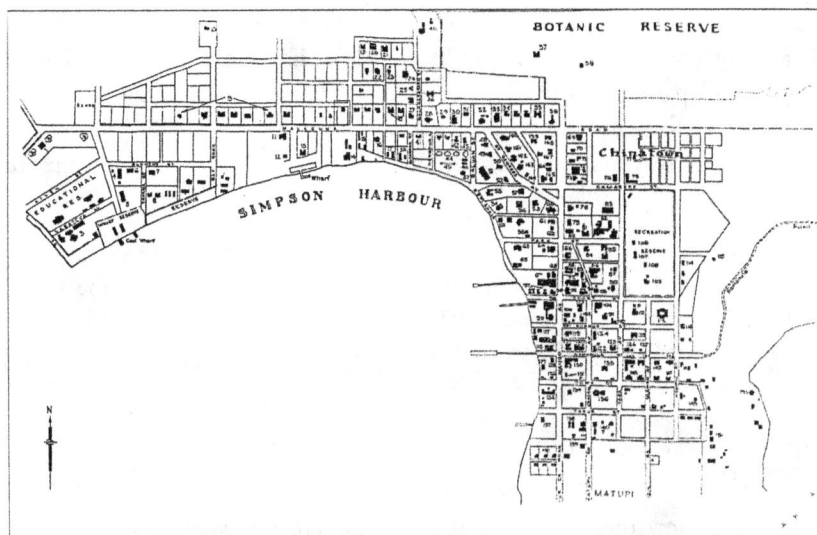

Figure 5.9. Wartime military map of Rabaul.

This is a detail from an original chart of the Rabaul town area that was compiled for Australian military intelligence purposes during WWII (AGS 1944, Map 12; see also Thomas 2012, 9). The original includes an index (not shown here) of 168 placenames compiled by Lieutenant Anthony in April 1943.

Lieutenant David Selby was summoned to headquarters shortly after the news of the Pearl Harbor attack was broadcast on the radio in Rabaul, and was instructed to move his anti-aircraft battery from Malaguna camp to a more strategically advantageous site (Selby [1956] 1971). Selby chose Observatory Ridge:

> Early one Monday morning Dr Norman Fisher, the Government Vulcanologist, was horrified to see two guns lumbering on to his front lawn, accompanied by load after load of ammunition, sandbags and supplies. Slit trenches were dug where he had been patiently coaxing a garden to grow, trees were felled, three times in one day bringing down the electric wires, and troops were quartered in his observatory. Within twenty-four hours a remote and peaceful residence was converted into an important military objective. (Selby [1956] 1971, 11)

Selby's anti-aircraft battery had its first experience of the war on 4 January when Japanese aircraft made a high-level bombing run over the Rabaul airfield. Fifteen New Guineans were killed at the nearby Rapindik Labour Compound. More aerial attacks followed. Tavurvur was active again on 9 January; on 19 January, it began another eruptive period. A major

Japanese invasion force sailed into St Georges Channel on 22 January, and that night—while Tavurvur was in eruption, lighting up the cloud with its glow—Japanese troops were loaded into landing barges and sent into Blanche Bay for a pre-dawn attack on 23 January. Selby ([1956] 1971) was obliged to destroy his anti-aircraft guns on Observatory Ridge in advance of the invasion. He recalled:

> On the night before the invasion the volcano was extremely active with continual flashes and rumbles of explosions. Many of the [Australian] troops were convinced that the invasion fleet, which had been sighted the previous afternoon, had been engaged by an American Fleet and that they were hearing the sounds of a naval battle. (Selby 1981, 2)

However:

> Tomorrow those dim grey ships on the horizon would disgorge their troops by the thousand, the distant warships would blast our positions on the beaches, the aircraft carrier would launch forty or fifty planes to bomb and machine-gun us from the air. (Selby [1956] 1971, 35)

In a tragic case of troops and civilians being sacrificed and abandoned by their military masters for no good reason, the Australian force at Rabaul was routed before 23 January was over. Soldiers and civilians were killed, captured or else escaped into New Britain where many subsequently perished. Others travelled along the northern and southern coasts of New Britain and managed to be rescued or escape back to Australian-held territory and eventually to Australia. Australian soldiers escaping down the eastern side of the Gazelle Peninsula were intercepted by the Japanese at Tol on Wide Bay and approximately 150 of them were slaughtered on about 4 February in what became known, historically and infamously, as the 'Tol Massacre' (Stone 1995).

Norm Fisher, volcanologist, but also a corporal in the New Guinea Volunteer Rifles (NGVR), saw the Japanese landings between Vulcan and Malaguna on 23 January: 'Despite the size of the Australian forces, the Japanese took few early precautions and were laughing, smoking and flashing torches in the darkness of the harbour foreshore' (Stone 1995, 67). This area was later 'subjected to intensive dive bombing. Our mortar received a direct hit and was put out of action' (Downs 1999, 64). Fisher's NGVR party under Lieutenant J.C. Archer retreated to Toma where 'advice was received from commanding officers that further resistance was considered useless'.

Fisher, with others, including assistant volcanologist Clem Knight, retreated from Rabaul rather than surrender to the Japanese. They trudged resolutely southwards along tracks across the central, mountainous part of the Gazelle Peninsula, through Lamingi where there was a mission, avoiding the east coast, but soon reaching Wide Bay—fortunately before the Tol Massacre had taken place. The group there met up with other escapees including Leo McMahon and Bill MacGowan, whose escape stories were later recorded in some detail (Stone 1995; Downs 1999). A small boat was acquired from Kalai Mission on Wide Bay that allowed them to move westwards along the south coast of New Britain, avoiding prowling Japanese ships, to Palmalmal at Jacquinot Bay where they secured another, larger boat. The group of nine, including Fisher and Knight, reached Lindenhafen Plantation, just east of Japanese-held Gasmata, and, after acquiring additional food supplies and gasoline, they escaped at night on 13 February to the relative safety of the Trobriand Islands, then Samarai, then by a Catalina aircraft to Australian-held Port Moresby. Fisher and Knight resumed their careers as civilian geologists on their return to Australia (Wilkinson 1996).

Figure 5.10. Tavurvur in eruption during Japanese occupation.

The volcanic plume being emitted from Tavurvur appears to contain some ash in this colour-enhanced photograph probably taken in early 1942 during the north-west season. Lakunai Airfield and a Japanese aircraft being serviced are seen in the foreground. Photograph supplied courtesy of Dr Yuichi Nishimura.

The Japanese invaders now had to reorganise operations in the Rabaul area to establish the expansive Simpson Harbour as a major centre of mainly naval operations aimed at further military advances southwards against the Australians (*Maru Special* 1984; Figure 5.10). Many prisoners of war (POWs) by now had been incarcerated in Rabaul and they were dealt with by embarkations onto ships for transport to, and eventual imprisonment in, Japan. The *Montevideo Maru* was one of these vessels. Loaded with human cargo at the end of June, it was tragically torpedoed off the Philippines on 1 July 1942 by the US submarine *Sturgeon*. All 1,053 prisoners on board perished (PNGAA 2017; Spurling 2017). Only a few POWs remained in Rabaul for the duration of the war, but they included Gordon Thomas, former editor of the *Rabaul Times* (Thomas 2012). Those who perished on the *Montevideo Maru* included eight of 15 official and non-official members of the Mandated Territory's Legislative Council (Nelson 1995) as well as missionaries such as reverends Laurie Linggood, Howard Pearson and Jack Trevitt (Figure 3.14), and laymen Ron Wayne and Wilfred Pearce (Threlfall 1975), who were mentioned earlier in connection with the 1937 eruption and its aftermath.

The Japanese invaders had inherited from the Australians the problem of how to deal with the volcanoes, especially Tavurvur, which had greeted them in such dramatic fashion. Their commanders were concerned about the effects of the volcanic activity, and especially Tavurvur's ash falls, on the movement of naval vessels and aircraft in and out of Simpson Harbour, and the services of a seismologist were requested from Japan. Takashi Kizawa of the Central Meteorological Observatory in Japan arrived at Rabaul in May 1942 and took charge of a seismological group that monitored the volcanoes for the duration of the Japanese occupation (Kizawa 1961; Kusaka 1976; Nitta 1980; R.W.J. Collection 1 containing correspondence with Takashi Kizawa in 1961–99). Kizawa had not volunteered for the position of Rabaul seismologist; he came as a civilian. A seismological and volcanological observatory was soon built for him by the Japanese Army on the northern edge of Sulphur Creek (Figures 5.11–5.13). The observatory looked across the nearby Lakunai Airfield that was used by Japanese aircraft, towards Tavurvur about 3 kilometres away to the south-east (Figure 5.14). The observatory was equipped with two seismographs—a German-made Weichart and a Japanese Omori—as well as a tromometer for measuring slight shocks. The instruments were housed in an underground shelter (Figure 5.12).

No. 118.　火 山 研 究 所

地 震 観 測 所

(温 泉 ハ リ ー リ)

Figure 5.11. Japanese volcanological observatory at Sulphur Creek.

The eastern side of the Japanese seismological/volcanological observatory at Sulphur Creek is seen in this photograph supplied by Takashi Kizawa. Meteorological equipment was housed in the white hutch in front of the main office. GA negative reference GA9988-5.

Takashi Kizawa saw no volcanic activity from Tavurvur during the first 18 months after his arrival. Soldiers told him that the volcano had been active after the invasion, but the date of the last eruption of the 1941–42 period remains unknown. Kizawa recalled soldiers shovelling Tavurvur ash off a tennis court in Rabaul and heaping it into piles. This ash may have fallen at the beginning of the south-east season just before Kizawa arrived in Rabaul, or else was the remains of Tavurvur fallout from the previous season—that is, well before the Japanese invasion. Norm Fisher wrote later that he thought 'the main eruption probably came to an end about March, although there are reports of some small later outbursts of activity' (Fisher 1976b, 209).

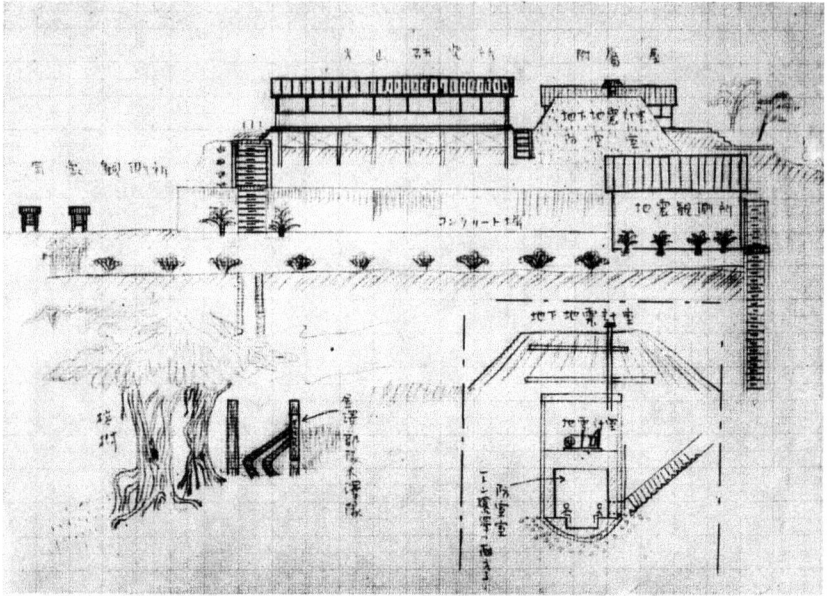

Figure 5.12. Two-part sketch of buildings and bunker at the Japanese observatory.

This sketch of the Sulphur Creek observatory has two parts. The main illustration is of the northern side of the seismological/volcanological observatory, including its entrance in the lower left-hand corner. The inset in the lower right-hand corner is of the air-raid bunker built beneath the room for housing seismological instruments. GA negative reference GA9988-8.

Kizawa made a trip home to Tokyo in the Japanese summer of 1943 for discussions at the Central Meteorological Observatory. Life in Tokyo seemed to be peaceful enough, but Kizawa was troubled by doubts that Japan would eventually win the war. The Allies, led by US General Douglas MacArthur, had been victorious in battles in New Guinea on land and sea, sending Japanese forces into retreat. Kizawa set out his feelings in an emotional article later published under the title 'Death Line' (Kizawa c. 1943). He was concerned in the first place about the prospects of making successful earthquake and eruption predictions in Rabaul, given the extreme conditions of active wartime bombing by the Allies. Another source of anxiety was the task of making choices while in Tokyo about who among his colleagues should return with him to Rabaul. Kizawa made special efforts in Rabaul to have a combined double-basement air-raid/sleeping quarters constructed that was capable of withstanding a direct hit from any large bomb (Kizawa c. 1943; Figure 5.12).

Figure 5.13. Seismograph being used at Sulphur Creek observatory.

Takashi Kizawa is seen here working on seismographic equipment at the Sulphur Creek observatory probably sometime in 1943 (Kusaka 1976; one of several unnumbered photographs between pp. 160 and 161). Digital copy kindly provided by Dr Yuichi Nishimura, Hokkaido University.

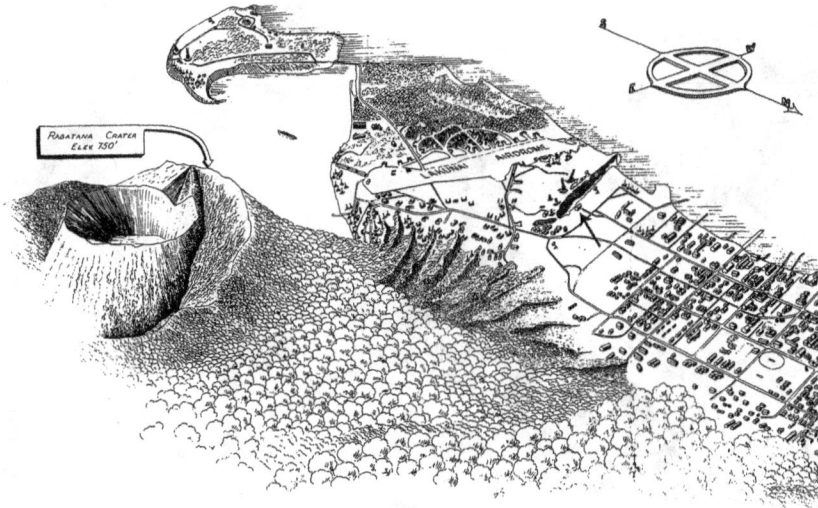

Figure 5.14. Oblique aerial drawing of eastern side of Simpson Harbour.

This is the south-western part of a US military drawing entitled 'Rabaul Harbor looking south-west 18 Nov '43' (CIU, Directorate of Intelligence, Allied Airforces, SWPA, Litho no. 978). The observatory on Sulphur Creek is shown by the arrow. This detailed drawing was evidently crafted from oblique aerial photographs taken after Allied bombing runs on Japanese-held Simpson Harbour on 7 and 11 November 1943 (McAulay 1986). Japanese deployments, including the positions of anti-aircraft guns, are shown in detail around Lakunai Airfield and the geothermally active Sulphur Creek down to the right (north). The southern end of Rabaul town is further to the right. Matupit Island is in the upper-left quadrant and a single ship occupies Greet Harbour to the east of the island. 'Rabatana Crater' is actually Palangiangia volcano in the middle of which is the younger and still thermally active volcano Rabalanakaia. Tavurvur volcano is not shown as it is left of the south-western limit of the diagram.

Figure 5.15. Japanese seismograph reinstalled at Rapindik after WWII.

This is the Omori seismograph used by Dr Kizawa at the Sulphur Creek observatory. The instrument was found after the war and is here shown in its reconditioned state and in use at Rapindik, Rabaul, in the early 1950s. The photograph was supplied by Australian volcanologist M.A. Reynolds, who worked at the RVO at that time (Reynolds 2005, 6).

Instrumental records of earthquakes were obtained by Kizawa after the observatory was established at Sulphur Creek, but he was to have increasing problems with the recording because of the Allied bombing raids on Rabaul. These became more and more intense towards the end of 1943 and into 1944. Nevertheless, Kizawa collected some apparently significant data starting on 11 October 1943 when tremors gradually increased, culminating on 16 October when a felt earthquake took place at 9.30 am directly beneath Rabaul (Kizawa 1951). This was followed by earthquake aftershocks. Kizawa had previously adapted his Omori seismograph (Figure 5.15) as a tiltmeter by slowing the rotation speed of the recording drum. Therefore, he had been able to record, earlier on 16 October, first a rising of the ground in the direction of Tavurvur and then, later in the day, after the 9.30 am earthquake, a down-tilt, as measured at the observatory. No eruption took place immediately—at least not until 24 November when Tavurvur started belching out dark ash at the beginning of about one month of activity (Kizawa 1951). A final eruption was photographed on 23 December and the image later reproduced in a Japanese publication (Kusaka 1976). This late-1943 activity at Tavurvur was the last of the 1937–43 period of volcanic eruptions at Rabaul.

Life in Rabaul for the Japanese troops became more and more unbearable as the Allies found ascendancy in their war efforts (Figure 5.16). Buildings in the Simpson Harbour area were smashed, and rebuilding them and living above ground became pointless. Japanese defence measures then diverted to construction of a remarkable network of tunnels and caves that eventually housed several tens of thousands of Japanese troops and their facilities and equipment (Figure 5.17). The tunnels were dug into the soft pumice deposits that form a mantle around much of the Blanche Bay caldera, and they later proved to be successful in withstanding many of the numerous bombing attacks by the Allies. One set of tunnels directly on the St Georges Channel coastline was used by Japanese submarines (Figure 5.18) for supply and delivery and, even today, is known as Submarine Base, or 'Sub-Base'. The Japanese took full advantage of the steep and deep coastal bathymetry, which, decades later, would be recognised as the near-vertical wall of another caldera named Tavui.

Figure 5.16. Aerial photograph of bombing of Sulphur Creek area.
Allied bombing of Japanese-held Rabaul, such as this on 2 November 1943, eventually caused the destruction of the town. Part of the Japanese Navy fleet can be seen in the upper-left corner. Tovanumbatir and the caldera wall dominate the background. Sulphur Creek is seen in the foreground, and the buildings of Dr Kizawa's volcanological observatory can be made out at the top of the far bank of the creek on the right. Phosphorescent bombs have exploded over the defence positions on the near side of the creek. Published courtesy of the Australian War Memorial, Canberra. AWM 100146.

The seismological/volcanological observatory at Sulphur Creek was hit several times and the damage repaired, but by late 1944 attempts to re-establish the buildings were abandoned and Dr Kizawa was obliged to retreat into caves dug into the caldera wall beneath Tovanumbatir. Kizawa established another observatory near Vunakanau and Latlat village, overlooking Vulcan, in December 1944 – January 1945 (Figure 5.19). It was not bombed directly, but the quality of scientific data obtained there was necessarily of a low standard because of poor instrumentation, and, in any case, no more volcanic activity took place, at either Tavurvur or Vulcan.

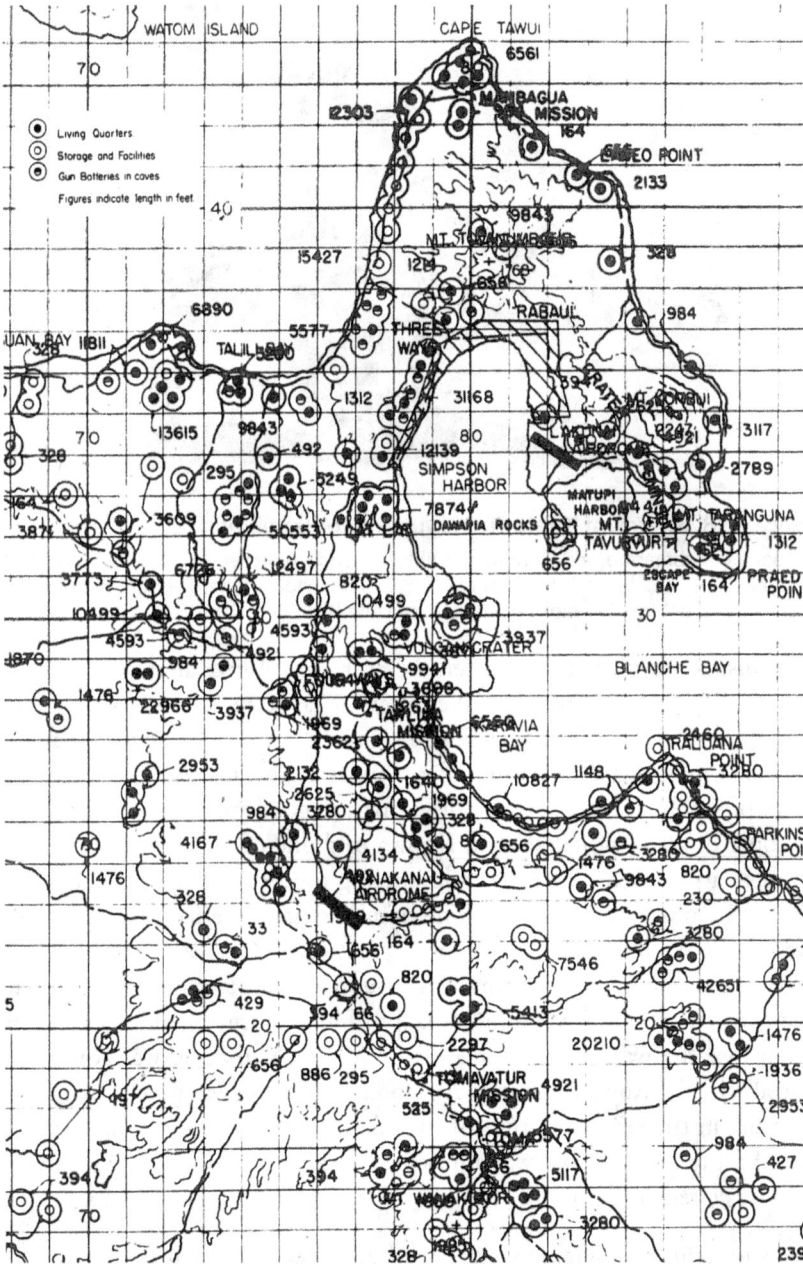

Figure 5.17. Map of Japanese caves and tunnels dug into pumice deposits.

This detail and legend are from a map of Japanese cave and tunnel distribution in the Blanche Bay area (United States Strategic Bombing Survey [Pacific] 1946, Map 19, p. 140). Note that most of the caves have been constructed well away from the seriously targeted bombed site of the former Rabaul town and Lakunai Airfield. Thick pumice deposits are well developed in these areas.

Figure 5.18. Sketch of Kabiu–Tavurvur area plus submarine.
A Japanese submarine and Tavurvur in mild eruption feature in this sketch, looking north-eastwards. Kabiu (the Mother) volcano is in the background. The sketch is part of an advertisement in a Japanese postwar magazine (*Maru Special* 1984).

An unusual story circulated after the war that Allied aircraft attacking Rabaul had dropped bombs into the craters of Tavurvur volcano to trigger an eruption that, hopefully, would be sufficiently large to dislodge the Japanese from their military base. So unusual was the story that many people believed it was probably untrue—a rumour, a myth, a good yarn. But the volcano bombing did take place. Takashi Kizawa told one of us (R.W.J.) of seeing bombers avoiding other targets and circling towards Tavurvur where they released bombs into the craters. The bombing had no effect, and Japanese soldiers considered the event to be a great joke. Kizawa, however, took a different view: that the enemy was inventive and determined enough at least to attempt a new technique. It helped him confirm his view that Japan would lose the war. Norm Fisher also told R.W.J. that he was consulted officially by the Royal Australian Air Force during the war about the use of bombing in triggering an eruption at Rabaul. He dismissed the suggestion outright, but after the war was amused to discover bomb craters within the volcanic craters of Tavurvur.

The thousands of Japanese soldiers in Rabaul were trapped militarily and were almost entirely cut off from military authorities in Japan, as were other Japanese soldiers on islands elsewhere in the Western Pacific (Hiromi 2004). Many of them had agricultural and farming skills that were put to good use in growing crops, thus adding to the evidence of self-sufficiency of the troops now living mainly in tunnels and caves. They were not troubled by any volcanic eruptions; indeed, the whole period of Japanese occupation does not appear to have been affected significantly by volcanic eruptions, most of which (from Tavurvur) seem to have taken place during north-west seasons when winds blow away from the harbour and town towards the south-east.

The Pacific War ended in August 1945 immediately following the American atomic bombing of Hiroshima and Nagasaki and the formal surrender of Japan on 15 August. Japanese forces in New Guinea surrendered formally after a ceremony on board the British aircraft carrier HMS *Glory* in St Georges Channel on 6 September 1945 (Special Correspondent 1945; Nelson 1995; Hiromi 2004). The arriving flotilla of Allied ships, including the *Glory*, had felt 'violent shudders' that morning—an earthquake had taken place 'as if to mark this important and historical day' (Threlfall 2012, 354). The Australian destroyer HMAS *Vendetta* entered Simpson Harbour that same afternoon, followed on 10 September by escort ships and transports carrying the Australian occupying force. A correspondent on board HMAS *Manoora* described the scene on 10 September:

> The peaks of the Mother and Daughter were veiled with fleecy clouds, and at the base of each could be seen some of the extensive gardens laid out by the Japanese. The scenic beauty of Simpson Harbour unfolded as we made our way slowly to our anchorage. Matupi [Tavurvur] had been active, and wisps of steam were rising from the crater, vents and fissures as we passed. Vulcan looked quite serene. His slopes now are about three quarters covered with vegetation. (Special Correspondent 1945, 59)

Virtually all that had been Rabaul itself was in ruins (Figure 5.20). The skeleton outline of streets, lines of trees and foundations remained, but the months of bombing had been so intense that hardly a wall was left standing. Destruction of the town by war had far exceeded the minimal damage inflicted by the 1937–43 volcanic eruptions:

> The town of Rabaul as the old residents knew it has been completely wiped out. There remains the front wall of Burns Philp's store, battered by shell fire and bomb blast, and the concrete entrance to what appeared to be the Rabaul Club. Concrete foundation posts were a mute reminder of the homes that once stood there, together with broken windmills and water tanks ... The whole shore line of

Simpson Harbour is littered with bombed, gutted and burnt out ships, some half on the beach, others submerged, with only their masts and the tops of funnels showing here and there. A very effective job had been done by our Air Force. (Special Correspondent 1945, 60; Figure 5.20)

Japanese still in Rabaul, including Kizawa, now became POWs of the incoming Australian forces until their repatriation back to Japan in 1946 (Figure 5.21). Kizawa closed, then boarded up, the observatory at Vulcan and left the following message on the entrance: 'For the coming generation these machines have wrought each function to the civilised world and progress of the scientific world during the War' (Kizawa 1961, 2; Figure 5.19). After his repatriation to Japan, Kizawa continued a lifelong career in seismology and volcanology—and an interest in Rabaul (R.W.J. Collection 1).

Figure 5.19. Two-part sketch of second Japanese observatory.

This sketch is of the second observatory used by Dr Kizawa in Blanche Bay. He provided the following description: 'This was built in February 1945. Therefore became the last seismograph room. We dug a cave in a cliff at the back of the somma on the west side of Vulcan and set the seismogram. It was in a violent air raid.' The words in the box on the right-hand side of this sketch read: 'For the coming generation these machines have wrought each function to the civilised world and progress of the scientific world during the War.' GA negative reference GA9988-10.

Figure 5.20. Remains of Rabaul town at the end of WWII.

The skeleton of streets in what had been Rabaul is laid out in this photograph taken on 15 September 1945, westwards from Namanula Hill. Published courtesy of the Australian War Memorial, Canberra. AWM 96796.

Figure 5.21. Japanese sketch of the volcanoes of Blanche Bay.

A peaceful Blanche Bay is seen in this Japanese drawing, dated 20 December 1945, of the view looking north-eastwards across Karavia Bay from the western caldera rim. A vapour plume drifts off to the south-east from Tavurvur volcano, and part of Vulcan cone is seen on the left-hand margin. The artist was Mr Akira Shigeta, a member of the AA60 Second Company's camp. English translation of the caption is courtesy of Dr Yuichi Nishimura. A copy of the sketch is held in the Lex McAulay Private Records collection at the Australian War Memorial, Canberra.

5.4. Reconsidering the Wartime Eruptions

Some special consideration is required at this stage—before dealing with the postwar period in Chapter 6 of this book—with regard to the Tavurvur eruptions of 1941–43. Perhaps the most notable feature is that Vulcan was not in eruption at the same time as Tavurvur, although it had been in both 1878 and 1937. There was, in other words, no 'twin' or double eruption in 1941–43. Vulcan had cooled to some sort of ambient temperature soon after its relatively short, but powerful sub-plinian eruption at the end of May 1937, and it was covered in trees and grasses by the end of the war (Kizawa 1951). In contrast, Tavurvur had retained its geothermal activity after May 1937, producing some minor hydrothermal outbreaks in March 1940. Temperatures measured by Fisher in the crater of Tavurvur started to climb later the same year. This conclusion will have some merit in considering below how the volcanic system in Blanche Bay actually operates or 'works'.

Another observation is that the Tavurvur eruptions of June 1941 took place four years after those of 1937. This is a significantly long gap—bearing in mind that the definition of a single 'eruption' used in the definitive global database of volcanoes and eruptions is that any eruption preceded by a three-month or greater gap from the previous eruption at the same volcano should be regarded as a *separate* eruption (Siebert, Simkin and Kimberly 2010). However, there are much longer gaps than three months between some of the eruptions at Tavurvur in 1941–43. The three-month definition used in the global database is a convenient one for compiling information in systematic encyclopaedic listings, but it does not correspond with the conclusion, developed in the following pages, that the time range 1937–43 represents a *single* eruptive period.

The 1941–43 eruptions at Tavurvur are not the first occasion for which the term 'vulcanian' can be used for historical eruptions in Blanche Bay. The vivid description given by Wilfred Powell of Tavurvur activity in 1878, for example, corresponds well with vulcanian activity, and the sketch of the eruption cloud in Figure 1.5 may correspond to vulcanian activity in 1791. Similarly, photographs taken of the later stages of the Vulcan eruptions in 1937 are best interpreted as showing vulcanian activity (Figure 3.36), even though these smaller eruptions had been preceded by 'sub-plinian' activity at Vulcan.

Perhaps of even greater interest is the short duration of the Tavurvur eruption in 1937—less than a day—and the question of whether any true vulcanian activity took place then at all. The hydrothermal eruptions that produced the fallout of blue-grey gummy mud on Rabaul town on Sunday 30 May may not have continued into the night, judging by the descriptions left by observers on the *Montoro* on the Sunday evening, including Brett Hilder (Figure 3.27) and George Clarke from Kokopo. Further, Doug Joycey said that the initial eruption of mud lasted only about half an hour, and Dr Cooper had a photograph of Tavurvur in which the eruption appears to be of a more normal vulcanian type (Figure 3.25). Perhaps speculatively much of the fallout that Sunday night into the Monday morning was vulcanian in character, or else was a mixture of fresh magma and mud. If so, the vulcanian phase of the eruption was shorter than in 1941–43 and presumably than in 1878. The implication is that eruptions at Tavurvur in 1937 cut out early but that the volcanic system retained eruption potential that was realised as the vulcanian eruptions of June 1941.

Seismologists working on the subject of forecasting volcanic eruptions have a particular interest in recording any earthquakes that may precede an outbreak and, therefore, that may be related in some way to 'causing' the eruptions. Fisher, Stehn, Woolnough and Kizawa all had such an interest, but they were restricted in their work by the instrumentation available to them in Rabaul and in the Territory of New Guinea as a whole. Regional and local networks of several or, ideally, many seismographs are needed to determine the epicentres and depths of earthquakes and so identify whether they are nearby and 'local' or further away and 'regional'. Fisher (1939a) drew attention to severe earthquakes felt in Rabaul in 1910 and 1916 (and noted by Stanley [1923]) in the context of the 1937 volcanic eruption, and Kizawa (1951) drew attention to a preceding event in 1906. The large earthquake of 14 January 1941 is also of interest in the context of the June 1941 eruption, but Fisher (1944, 11; 1976b) emphasised that temperatures at Tavurvur had begun rising *before* the 14 January event and so concluded that the two events were only 'in a general way related'. Similar seismic/volcanic relationships would be studied internationally in the years ahead as more earthquake stations were installed globally and in specific volcanic areas such as Rabaul. Interpreting the significance of those relationships, however, remains challenging.

6

Rebuilding Rabaul and Re-Establishing the Observatory, 1945–69

6.1. The Immediate Postwar World, 1945–50

One of the main purposes of this book is to compare the Rabaul eruptions of 1937–43 with the ones that broke out in September 1994, more than 50 years after the last eruption witnessed by Dr Kizawa in December 1943. That half-century was filled with great change, politically, socially, environmentally and scientifically (Threlfall 2012; Johnson 2013). First, the former Mandated Territory of New Guinea in the north and the former Australian Territory of Papua in the south were amalgamated after WWII. This was under a new 'Australian Trusteeship Agreement for the Territory' that was approved on 14 December 1946 by the General Assembly of the United Nations that had replaced the League of Nations (Downs 1980). Port Moresby, and neither Rabaul nor Lae, was declared the capital of the new Australian-administered Territory of Papua and New Guinea (TPNG). Colonel Jack K. Murray was appointed the administrator of the combined territories by the Australian Government, which was then held by the Australian Labor Party. He would have the responsibility of reconstructing the new territory on Australia's behalf during a period of great, accelerating postwar changes.

Another organisation that would face postwar challenges was a new Australian Government geoscience agency that combined the functions of a national geological survey and bureau of mines (Wilkinson 1996). It was created in March 1946 and went under the unwieldy name of the Bureau of Mineral Resources, Geology and Geophysics. Its abbreviation, however, the 'BMR'—or even just 'the bureau'—soon caught on. The BMR was given the authority to carry out geological and geophysical surveys, undertake related research in both Australia and TPNG, and obtain basic earth science data by running geophysical observatories. Geologist Harold Raggatt helped create the bureau and he became its first director in 1946. There were delays in staffing the new BMR—finding suitably trained people, professionals and technicians, in the immediate postwar environment was challenging— but volcanologist N.H. Fisher was appointed to the senior position of chief geologist as early as 1946. The BMR, however, had to wait until 1949 for a resident geologist, A.K.M. Edwards, to be chosen to run a Geological Office in Port Moresby and so commence geoscientific work in the new territory.

Postwar changes included notable developments in the fields of science, engineering and technology, in part stimulated and driven by ongoing competition and conflict between nation-states for political power and domination. The relatively minor subject of volcanology, being a multidisciplinary subject, also benefited on many fronts, including volcano-monitoring instrumentation, new concepts in areas such as the tectonic setting of volcanoes, and more specialist topics such as the origin of pyroclastic rocks and the mapping of volcanic hazards and risk. The Cold War created the so-called space race when the Soviet Union in 1957 successfully launched the first satellite, the *Sputnik*, high above the earth's atmosphere. This led to extensive satellite monitoring of the earth from space, including the mapping of changes on volcanoes and the detection and analysis of eruption clouds.

BMR Chief Geologist N.H. Fisher returned to Rabaul after the end of the war and became involved in reporting on the selection of a more suitable site for the town (Fisher 1946a, 1946b). He noted in early December 1946 while undertaking fieldwork in Rabaul that the volcanoes were quiescent and that no measured temperatures were greater than 100°C, even at Tavurvur where the only detectable gas was hydrogen sulphide, which was depositing crystalline sulphur. The 1937–43 eruptive activity had ended, to all practical purposes.

Fisher, in a final BMR record, attempted to prioritise five areas in the north-eastern Gazelle Peninsula that might be considered for postwar development, at least from a geological and geophysical safety point of view. The geohazards he considered were volcanic eruptions, earthquake ground-shaking and tsunami inundations. Fisher also gave consideration to some of the 'utilities' or assets at each of the five places, namely: water supplies; aggregates for construction purposes, especially roads and building foundations; ready access to existing ports and airfields; and local climate. These assets are what today might collectively be called 'exposure'. He excluded several other factors, such as cost and availability of land, ease of supply of building materials and suitability from a town-planning point of view, all of which were beyond his area of expertise. Fisher also presented what today can be regarded as the first geohazard map of the north-eastern Gazelle Peninsula (Figure 6.1).

Fisher preceded his analysis of each of the five prioritised areas—which are summarised below numerically—with the following statement:

> I have always maintained that while the Administrative establishment existed at Rabaul, the direct danger from volcanic eruptions was not sufficiently great to justify the expense of moving that establishment. Now, however, when it is a matter of starting practically from zero, obviously the sensible thing to do is to re-establish the administrative centre at some place not directly in the path of possible outbursts. (Fisher 1946b, 2)

1. Rabaul, the old town now destroyed by wartime bombing, was still the most exposed of all five areas to the three geophysical hazards. It also had town water supply problems, including the use of bore water contaminated by the latrines used for the Japanese prison camps and native labour compound. Rabaul's microclimate also tended to be hot and windless, the town being hemmed in by caldera walls on two sides. Rabaul scored well, however, in its relatively ready access to a port (Simpson Harbour) and airfield (Lakunai).

2. Nonga–Tavui. This coastal area in the north on Talili Bay and facing Watom Island was judged to be the least hazardous of the five areas. Its main disadvantages were poor access to both a port and an airfield.

3. Kerevat and Kabaira were on the eastern side of Ataliklikun Bay, down on the west coast and well away from the port and main airfields. The area had the advantage of running water from the Kerevat River, but it was judged to be the most susceptible of all the areas to ground-shaking from earthquakes and, being coastal, was potentially prone to tsunami impact.

4. Vunakanau–Taliligap, in the hills overlooking Blanche Bay to the north-east, had the most hospitable climate. Vunakanau Airfield was nearby, and there was virtually no tsunami threat, but it, like Kerevat–Kabaira, was also susceptible to earthquake shaking and had relatively poor access to the port.

5. The coastal Kokopo–Rapopo area west of Cape Gazelle faced northwards into St Georges Channel and was not entirely free from the impacts of any of the three geohazards. This included, however, only light falls of volcanic ash from both Vulcan and Tavurvur and then only in the north-west season (Figure 6.1). There was only some threat from tsunamis; much of inland Kokopo is actually on a raised terrace that protects it somewhat from tsunamis. The area had no major anchorage of its own but it was linked by road to the port at Rabaul. Further, surface water was available from Matanatava Creek.

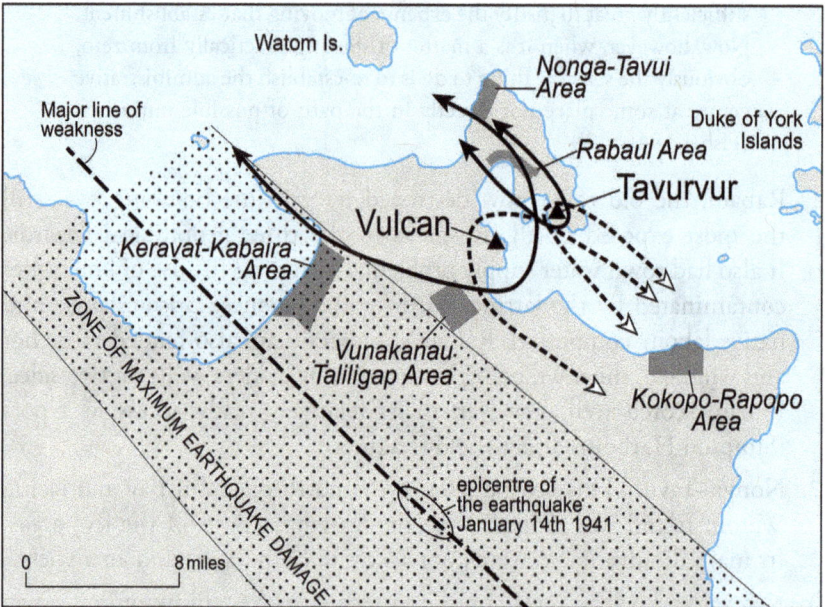

Figure 6.1. Geological hazards map for the north-eastern Gazelle Peninsula.

The five areas considered by Fisher (1946b) in his geohazards analysis are shown here in grey in this adaptation from his original map. They are seen in relation to: (1) ash fallout zones for both Vulcan and Tavurvur volcanoes; and (2) the major earthquake zone that Fisher believed ran south-eastwards across the north-eastern Gazelle Peninsula, based on his fieldwork following the 14 January 1941 earthquake (Fisher 1944). The two curves whose ends point north-westwards refer to ash deposition in the south-east season (trade winds) from both Vulcan and Tavurvur volcanoes, and the dashed curves pointing south-eastwards to deposition in the north-west season (monsoon).

In considering Tavurvur and Vulcan volcanoes as a future threat in the Rabaul District, Fisher wrote the following in his draft report:

> The 1937 eruption is regarded as having exhibited the probable maximum severity … [I]t is the most serious eruption that has occurred for hundreds, probably thousands of years … There is no evidence that volcanic activity at Matupi [Tavurvur] has ever reached the violence of the 1937 outbreak at Vulcan and it is assumed that eruptions at Matupi much more severe than those of 1937 and 1941–42 [sic] are not probable. (Fisher 1946b, 1)

These conclusions are somewhat surprising as Fisher did not mention the 1878 eruptions at both Vulcan and Tavurvur only 59 years previously (see, however, Fisher 1939a). In any case, his quoted conclusions, which are concerned with eruption periodicity or frequency, would have to be amended as more geological data on past eruptions at Rabaul came to hand, and when further eruptions took place at both Vulcan and Tavurvur later in the twentieth century.

Fisher recommended that Rabaul and Kerevat–Kabaira should be ruled out as sites for postwar rebuilding—at least on the basis of relative geohazards threat—and that the order of preference for the remaining three sites should be Nonga–Tavui, Vunakanau–Taliligap and Kokopo–Rapopo. This order would be reversed, however, when considering only the 'utilities' discussed by Fisher—that is, Kokopo–Rapopo, the area where the Germans had first established Herbertshöhe, would be the first preference. The extent to which Fisher's opinions about abandoning Rabaul would be accepted by the authorities appears to have been minimal.

Fisher, in an earlier draft, had stated strongly that 'it is most important that a decision be reached as rapidly as possible, as every day's delay makes the move from Rabaul more difficult and more expensive' (Fisher 1946a, 6), but this statement did not appear in the final version (Fisher 1946b). Importantly, reoccupation of the old town was already taking place at the time of Fisher's visit in December 1946. Populous Tolai villages could still be found in the north-eastern Gazelle Peninsula, including the Simpson Harbour area. Other New Guineans also lived in the same peninsula area. Moreover, Chinatown and Malaytown, including shops, were quickly being re-established by Asian people who had been held captive by the occupying Japanese.

Expatriate Australians and Australian businesses that had owned land and property before the Japanese invasion also returned to the Blanche Bay area, perhaps fuelled by a nostalgia-laden ambition to rebuild Rabaul and its environs, including Kokopo, as it had been in its colonial prime before the war. At the same time, this return was no doubt driven by the commercial advantages of the harbour and its now repaired wharves. Yet Rabaul would no longer be the capital, and its initial civil administration—after the departure of the Japanese and Australian military authorities from the Blanche Bay area—was based initially in the new capital of Port Moresby.

The story of indecision in the years from 1937 to 1942 concerning the future of Rabaul was repeated between 1946 and 1953. The Australian Government in Canberra, the TPNG administration, and the spontaneously growing business population in Rabaul itself all had different views about the future of the town, including the administrator, Colonel Jack Murray, who supported the move to Kokopo. Rabaul town, however, soon became re-established and its population began to grow in size and influence. Another delay related to the TPNG administration not being in an immediate position to employ suitably trained professionals; indeed, it had to wait until new postwar university graduates appeared on the market. Nevertheless, Major J.K. McCarthy—who had witnessed the 1937 eruption in Rabaul as a patrol officer and given outstanding war service—was appointed to the newly created position of District Commissioner for New Britain (McCarthy 1963). Another appointment was G.A.M. 'Tony' Taylor, a newly recruited BMR scientist who arrived in Rabaul in April 1950.

Taylor had enlisted in the Australian Imperial Force in 1942 and, by September 1945, after the Japanese surrender, had moved to Rabaul, where he was able to observe at length the volcanic nature of Rabaul (Fisher 1976a). Taylor was discharged from the army in 1947 and began a science degree at the University of Sydney under the Australian Government's postwar reconstruction scheme. He joined the BMR as a base-grade geologist in March 1950. Taylor's BMR supervisor would be the Canberra-based N.H. Fisher. Taylor's role in Rabaul was to re-establish the volcanological service for the territory administration and to reinstate the Rabaul Volcanological Observatory. It was a logical place for him to be stationed, given Rabaul's now well-known history of volcanic activity, high volcanic risk and volcano monitoring, although he would also have responsibility for the active volcanoes in the entire area of the new TPNG that included the old Territory of Papua.

Taylor's presence in Rabaul had the additional advantage—at least for those people who wanted to remain in and develop the town—of his being able to provide volcanic early warning advice, if needed, as Norm Fisher had done in 1937–42. Taylor found that the instrumental cellars of the old Australian observatory on Observatory Ridge were intact, and he was able to reinstall seismographs from Australia in a harbour network, including the Japanese Omori at Rapindik on Greet Harbour. However, the work of re-establishing the destroyed observatory buildings on the hill would take some time. Additional staff members were required and, from the beginning, Taylor would employ young, talented Tolai men to assist him in the field and office. Leslie Topue was an early recruit.

6.2. Influence of the Mount Lamington Eruption, 1951

Taylor visited Bagana volcano on Bougainville Island in December 1950, only a few weeks after a particularly powerful phase of explosive activity had taken place (Taylor 1956; Bultitude 1976). The activity during his visit was much reduced, but stronger eruptions would recur in the years ahead. Following this visit, Taylor returned to Rabaul to resume 'normal' duties (Figure 6.2), but, in the week beginning 14 January, he began hearing of volcanic unrest at Mount Lamington in the Northern District of distant Papua (*South Pacific Post* 1951a; Figure 0.2). There was some initial speculation that the reported activity might be from Goropu, or Waiowa, a Papuan volcano that had been in eruption in 1943–44. Taylor wanted to investigate the reports by visiting Mount Lamington but there were problems in facilitating immediate travel out of Rabaul. The administrator, Colonel Jack Murray, happened to be visiting East New Britain at that time (*South Pacific Post* 1951a), and Taylor received his approval to fly out with him from Rabaul on the morning of Monday 22 January. They were on their way to Lae when they heard that Mount Lamington had erupted catastrophically at 10.40 am on the previous day, Sunday 21 January 1951. Almost 3,000 people, mainly local Orokaiva, had perished and the government district headquarters at Higaturu had been destroyed, as had the Anglican mission at nearby Sangara (Taylor 1958). The Lamington eruption would have a major impact on perceptions of the extreme dangers of active volcanoes in TPNG, as well as on the international trajectory of Taylor's career as a volcanologist. The Lamington eruption came at a critical time for Rabaul, no official decision regarding the town's future having been made by 1951.

Figure 6.2. Climbing out of the crater of Tavurvur volcano.

Tony Taylor, lower left, and two unidentified expatriate colleagues are seen climbing out of the crater of Tavurvur sometime late in 1950 or in early January 1951. The photographer is unknown, but the photograph was published in the *Illustrated London News* on 3 February 1951 (Anonymous 1951, 169).

Tony Taylor and Leslie Topue became heavily involved in assessing the Lamington disaster and spent considerable time away from Rabaul. Taylor, in fact, would spend nearly two years on the monitoring and study of Lamington volcano, broken at times by other commitments in the territory. A conclusion soon recognised by Taylor at Lamington was that the 1951 eruption had similarities to the catastrophic 1902 eruption at Mount Pelée in the Caribbean, which had inundated the town of Saint-Pierre on Martinique (Lacroix 1904). The Lamington eruption, therefore, was labelled as *peléean*, a type of volcanic eruption not yet recognised in the Rabaul area and one quite different to those of 1937–43. Taylor soon established an effective working relationship with the administration officers—including the administrator himself—all of whom came into the Lamington area as part of the post-disaster rescue and relief effort, as described in greater detail elsewhere by Johnson (2020). BMR geologist John G. Best visited Taylor in the Lamington area but soon moved to Rabaul where he ran the volcanological service in Taylor's absence as part of the administration's resident staff.

Norm Fisher arrived in the Lamington area from Canberra on 31 January bringing considerable experience in volcanological work in the former Mandated Territory of New Guinea and especially at Rabaul in 1937–42. Colonel Murray was in the area too, making Fisher's presence timely, as they were able to discuss not only the disastrous Lamington situation but also the volcanic threats and risks elsewhere in the territory—most notably the ongoing question of relocating Rabaul. Fisher (1946b) had expressed the opinion four years earlier that rebuilding the town in the same place after its wartime destruction did not make sense, but the situation had changed: the town was becoming re-established almost spontaneously. Murray, in January 1951, was under pressure to allow this growth to continue, especially given that the volcanological observatory was being re-established by the BMR for eruption early warning purposes. However, on Friday 26 January 1951, the Port Moresby–based *South Pacific Post* contained the following editorial:

> The time for argument and indecision is long [past]. The matter is no longer a question of comfort or discomfort, financial gain or loss. The people of Rabaul must be removed from the possibility of a repetition of the Higaturu horror … The important and glaring necessity is to get the place moved and get it moved quickly. If the Administration wants to clutter up its routine activities with red tape then it can do so. But red tape where human life is endangered cannot be tolerated. (*South Pacific Post* 1951b, 8)

Figure 6.3. Visit to Mount Lamington, Papua, in May 1951.

A group of Rabaul residents poses in front of Lamington volcano, Papua, in the second week of May 1951. A similar photograph was published on the front cover of the July issue of *Pacific Islands Monthly*, including the statement that: 'On their return to Rabaul all were of the opinion that the sooner that volcano-encircled town was moved to a safer spot the happier they would be.' Geoscience Australia negative reference M/2438-3-1.

Murray still favoured the Kokopo area to the south of Rabaul as an alternative town site. But what was the risk that Rabaul, like Lamington, might break out in catastrophic eruption? Fisher flew to Rabaul and, on his return through Port Moresby, was able to reassure the administrator that there was

> no evidence whatever of impending [eruptive] activity for some considerable time and that in any case the present station he [Fisher] has at Rapindik [near Tavurvur volcano] can give two days notice of major eruption. (Murray Administrator 1951, 1; see also *South Pacific Post* 1951c, 2)

This information was sent by Colonel Murray to the minister, Department of External Territories, Canberra, on Wednesday 14 February—the day that Fisher returned to Australia—together with Murray's opinion 'that without compulsion [it] would be unlikely that majority of total nonnative population would move out of Rabaul township' (Murray Administrator 1951, 1). Nevertheless, in the wake of the Lamington disaster, public safety concerns in Rabaul gained new momentum. District Commissioner J.K. McCarthy cancelled a Rabaul evacuation plan dated 1950 and introduced a new one on 15 February 1951 (McCarthy 1951). Later, the administrator visited the town intent on persuading the community to transfer Rabaul to a safer site—the administration even arranged for a representative party of Rabaul residents to visit the Lamington area for two days, leaving Rabaul on 12 May (*South Pacific Post* 1951d). According to a sub-headline in the *South Pacific Post* (1951f) the visit represented a 'Move by Administrator to Stress Horrors of Eruption' at Lamington—'apparently with the idea of impressing upon them what a volcano can do', as stated in the caption on the front cover of the July issue of the *Pacific Islands Monthly* (Figure 6.3). However, no definite decision about relocating Rabaul resulted from the visit.

Significant political changes were taking place in Canberra at this time. A cabinet reshuffle of ministerial positions was undertaken in mid-1951 by the ruling Liberal government under R.G. Menzies and Paul M.C. Hasluck was appointed minister for territories, a position he held until 1963 (Hasluck 1976; Downs 1980). Hasluck paid his first ministerial visit to TPNG in May 1951, travelling to the Lamington disaster area and to Rabaul, which he said

> was a shambles with the main port facility the upturned hull of a ship sunk in war time. Most buildings were temporary makeshifts. Chinatown was a higgledy-piggledy warren of old iron. Everywhere was the evidence of the bombings by Allied planes when it was in Japanese occupation six and seven years earlier. The chief reason why there had been scarcely any post-war building was the lack of a final decision whether ... the town site should be moved. Rabaul Harbour still had scores of war-time wrecks around its shores and the war-time tunnelling of the Japanese was visible everywhere. (Hasluck 1976, 19)

Hasluck respected the Labor-appointed Jack Murray, calling him 'a good and devoted man' (Hasluck 1976, 15), and appreciated the colonel's outstanding leadership in the aftermath of the Lamington disaster (Hasluck 1976). Yet he believed that Murray's original appointment as administrator in 1946 by the Labor government 'was the wrong one' (Hasluck 1976, 50). Hasluck, therefore, issued a statement on 10 May 1952 announcing that Murray would relinquish his position as administrator from 30 June and that Donald M. Cleland, who had been appointed acting administrator when Murray went on leave in March, would continue in that role. Cleland was appointed administrator early in 1953.

Hasluck and Cleland, but not Murray, officiated at the opening of the Mount Lamington Memorial Cemetery at Popondetta in November 1952, at which Cleland gave out awards for services during the Lamington relief and recovery operation. Tony Taylor received the George Cross and Leslie Topue the British Empire Medal (Civil), awards that helped promote— not only locally but also internationally—the role of the volcanological observatory centred on Rabaul. Fisher (1976a, x) noted later that the Lamington eruption had 'catapulted Taylor, normally one of the most reserved and retiring of men, into public prominence'. An article on Taylor's early volcano-monitoring work in Rabaul had even appeared in the *Illustrated London News* of 3 February 1951 immediately after the Lamington eruption (Anonymous 1951; Figure 6.2).

The rebuilding of Rabaul town continued unofficially and resolutely on its existing site throughout 1951. Then, in June 1952, the Australian Cabinet in Canberra provided official government approval for its ongoing reconstruction (Territories 1952). Rabaul town was to be rebuilt on the same vulnerable site that had been identified by Governor Albert Hahl over 40 years previously, had been invaded by Australia in 1914, had been reoccupied after the 1937 eruptions at Tavurvur and Vulcan, and had been reoccupied after being destroyed in WWII.

Australia was experiencing important changes at this time. While stronger political alliances were established with the US, Australia remained firmly within the British Commonwealth of Nations, which had the new Queen Elizabeth II as its figurehead. Postwar immigration was being promoted, the economy was improving and family homes were being established in new suburbs, many equipped with a new technological gadget: black-and-white television.

6.3. Eruption Time-Cluster and Earthquake Mapping, 1951–59

The period 1951–57 was a particularly busy time for volcanologists at the Rabaul Volcanological Observatory. Eight volcanoes were active in different parts of TPNG during that time, requiring investigations by John Best and Tony Taylor, as well as by a new recruit, geologist Max Reynolds, who started work in late 1953 (Reynolds 2005). The eight volcanoes were:

- Lamington, 1951–52
- unnamed submarine volcano near Karkar, 1951
- Bagana, 1951 and onwards
- Long, 1953–55
- Tuluman, 1953–57
- Langila, 1954–56
- Bam, 1954–55
- Manam, 1956–57 (and up to 1966).

These volcanoes formed an eruption 'time-cluster' that, notably, did not include Rabaul volcano itself; neither did it include young volcanoes in the Dawson Strait and Esa'ala area of eastern Papua, where there had been local, and concerning, earthquake activity in 1953–55. The previous eruptions at Tavurvur and Vulcan in 1937–43 do not seem to have been part of an equivalent time series. However, the 1878 eruptions at Rabaul might have been. Eruption information is not nearly as comprehensive for the 1870s as it was for 1951–57, but the following volcanoes may constitute an eruption time-cluster for 1878–88: Bam, Manam, Ritter, Langila, Ulawun, Rabaul, Bagana, Bamus and Lolobau (Johnson 2013, Table 1). Norm Fisher in 1957 published, internationally, a major catalogue on the eruptions and active volcanoes of Melanesia, including Solomon Islands and New Hebrides (Vanuatu), but the eruption information for the 1956–57 time-cluster is incomplete because of the publication date (Fisher 1957).

A notable feature of the volcanological research carried out in the 1950s and led by Tony Taylor was consideration of the concepts that the eruptions were related in some way: (1) to regional geophysical unrest, or tectonic stress release, that affected the whole of TPNG; and (2), more controversially, to the changes in earth-tide forces caused by the motions of the moon and sun— so-called luni-solar influences—as reflected in several reports of the time (e.g. Taylor 1958, 1960) and dealt with in greater detail elsewhere (Johnson 2013). Instrumental monitoring on active volcanoes remained critical, however, and Taylor recommended that separate, permanent volcanological observatories be established on Manam Island and at Esa'ala in south-eastern Papua. Monitoring at Rabaul, where the territory's volcanological observatory was established and named 'Central Observatory', and where maintenance of the seismographs and regular temperature measurements were undertaken (Figure 6.4), continued unabated. These observatories could be used for the detection of regional earthquakes as well as local volcanic ones.

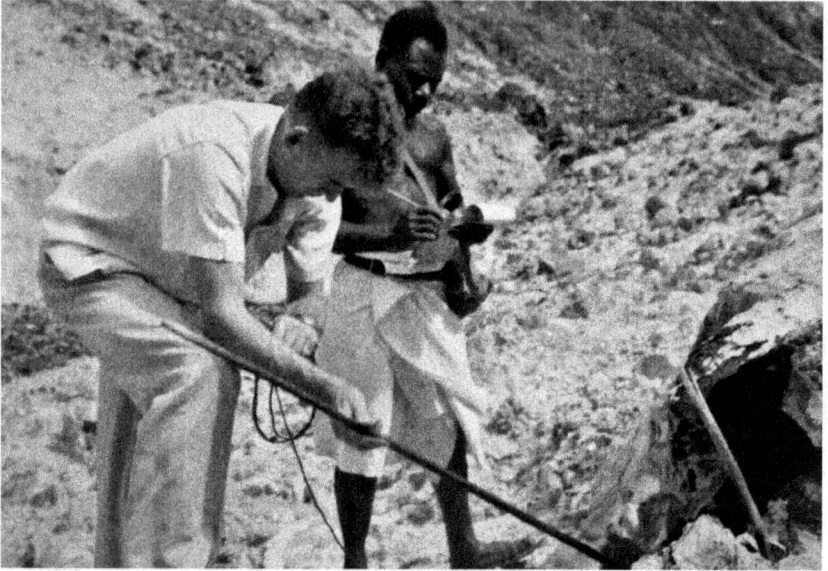

Figure 6.4. Temperature measuring on Tavurvur.

Max Reynolds and Leslie Topue are here seen taking temperatures at a fumarole on Tavurvur volcano, Rabaul, at some time when Reynolds was not investigating other volcanoes during the 1950s eruption time-cluster (Reynolds 2005, cover photograph; see also Figure 6.6).

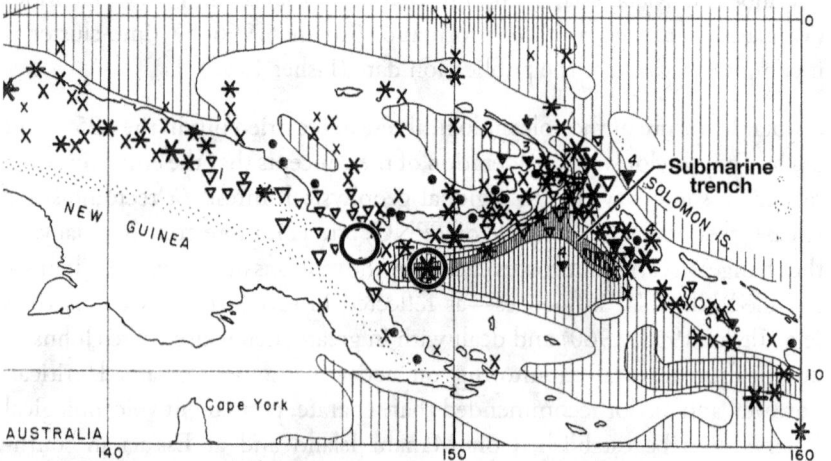

Figure 6.5. Early earthquake epicentres for the New Guinea region.

The strong concentration of earthquake epicentres in eastern New Britain, St Georges Channel and Bougainville Island (shown here as one of the 'Solomon Is.' [sic]) is illustrated in this detail from a map of Western Pacific earthquake epicentres (Gutenberg and Richter 1954, Figure 16). The epicentres are shown in relation to the conspicuous submarine trench that runs along the south coast of New Britain and south-west of Bougainville

Island. The large encircled asterisk at the south-western end of the submarine trench represents the epicentre of the large-magnitude tsunamigenic tectonic earthquake of 14 September 1906 whose effects were reported in Rabaul. This epicentre was determined by Sieberg (1910), but the International Seismological Centre later revised it to a position in the Finisterre Range of mainland New Guinea, as represented here by the large unfilled circle (McCue and Letz 2019).

Figure 6.6. Duke of Edinburgh's visit to the volcanological observatory.

Max Reynolds is seen in the centre, with his back to the camera, hosting a visit to the Rabaul observatory by the Duke of Edinburgh on 13 November 1956 when Leslie Topue wore his Medal of the British Empire (Reynolds 2005, 41; see also R.W.J. Collection 9). The district commissioner John Foldi, on the right, is checking the time.

Considerable progress was also being made in the 1950s in detecting and locating large earthquakes, both globally and regionally. The BMR in 1958 established a Geophysical Observatory in Port Moresby (PMGO) as part of a wider Australian network for the monitoring of earthquakes and local changes in the earth's gravity and magnetic fields (Brooks 1962, 1965). PMGO would work in partnership with the volcanological observatory in distant Rabaul. There had also been major global compilations published of the earth's seismicity, perhaps most notably that of Gutenberg and Richter (1954) in which the high density of earthquake epicentres for parts of TPNG—and especially the area encompassing Rabaul—was illustrated clearly (Figure 6.5; see also Brooks 1962, Figure 2).

6.4. New Instrumentation and Geoscience Surveys at Rabaul, 1960–69

There were staff increases and other employment opportunities in the 1960s at the Rabaul Volcanological Observatory. These included the beginning of overseas recruitment of graduates in geology and geophysics from countries other than Australia—Italy and Britain and later the US and France—as well as local New Guinean staff for technical and office work. A series of different overseas scientists became head of the Rabaul Volcanological Observatory (RVO) up to the 1980s. Tony Taylor maintained his commitment to the work of the RVO but took on more senior BMR positions in Canberra and Port Moresby in the 1960s and early 1970s.

Figure 6.7. Rabaul town and volcanoes as photographed from the observatory.

This view from Observatory Ridge is a famous and oft-photographed one. This shot was taken by observatory volcanologist John Barrie around 1960 or slightly earlier. Kabiu and Palangiangia volcanoes are seen on the left, and Turagunan and the light-coloured Tavurvur are behind them (see also Figure 6.8). Rabaul town hugs the foreshore around to the wharves on the right. Photograph supplied courtesy Mr Barrie.

Figure 6.8. Aerial view of Tavurvur volcano in the early 1960s.

Tavurvur volcano, Rabaul, is seen from the south-west in this oblique aerial photograph taken in the early 1960s. Thermal activity is preventing vegetation from growing on much of the volcanic cone (compare with Figure 4.5). Hot springs discharge into the waters of Greet Harbour in the foreground. St Georges Channel can be seen in the background to the left of Turagunan volcano. Digital copy provided by the State Library of New South Wales. Australian National Travel Association photograph, published in *Walkabout* magazine.

There was some international economic interest in the early 1960s in using New Zealand expertise in assessing the geothermal potential of Greet Harbour for possible electricity generation (Studt 1961; Fooks 1964). Power for the township of Rabaul was supplied by diesel-driven generators but at high cost, so the relative merits of hydro-electric and geothermal power were investigated. The New Zealand surveys, which were inconclusive, were undertaken after a geophysical survey by BMR staff in 1960 of the shallow structure of the Greet Harbour area (Wiebenga and Polak 1962). Much more attention, however, was being paid at this time to improving earthquake monitoring in the Rabaul area.

Seismologists, internationally, had been discussing in the late 1950s the value of establishing a global network of standardised seismographs (plus clocks for accurate time-recordings) to be used in the detection and measurement of earthquakes worldwide. The opportunity for doing so was realised partly as a result of Cold War discussions on the banning of nuclear tests. In the early 1960s, the US funded the establishment of a World-Wide Standardized Seismograph Network, or WWSSN, and the installation of recording stations in many countries (Peterson and Hutt 2014). Significant underground nuclear tests could be detected by the WWSSN as well as natural earthquakes. The RVO received a WWSSN station in 1962, adding significantly—in conjunction with other stations both locally and globally—to the mapping of earthquakes in the region (Latter 1966). There remained, however, the challenge of determining the location of smaller earthquakes within the Blanche Bay area that might be of volcanic, rather than tectonic origin. Could any of these relate to nearby volcanoes in the area? Could they be used for forecasting or predicting volcanic eruptions? Were any of them directly related to earlier tectonic earthquakes? Was there even any causative relationship?

The need for a more effective local seismic network for the detection of small harbour earthquakes was recognised in the early 1960s (Latter 1966). Gordon Newstead, professor of engineering physics at The Australian National University, visited Rabaul in 1963 and, in conjunction with Tony Taylor and others, recommended a five-station network in which seismic signals would be transmitted electronically to the recording room at the RVO (Newstead 1968, 1969; Myers 1976; Cooke 1977). Four of the stations were linked to the RVO recording room by telephone cables and the fifth by radio transmission, and data were transferred to a set of 'helicorders' or direct-writing recording drums (Figure 6.9). The network, however, had not yet been completed when, on 14 August 1967, two major earthquakes of magnitude 5.0 and 5.3 shook the Kokopo–Kabaleo area and nearby villages, causing some damage but no loss of life (Heming 1967; Threlfall 2021a). The two earthquakes, whose epicentres were determined by the WWSSN, took place in St Georges Channel just east of Cape Gazelle and south-east of the Duke of York Islands.

Figure 6.9. Observatory recording room in 1969.
Tony Taylor (left) and Noel Myers examining helicorder records in the Rabaul observatory recording room during the 1969 crustal survey (Wilkinson 1996, l.xii).

Another limitation on the accurate mapping of earthquakes in the north-eastern Gazelle Peninsula in the early 1960s was the dearth of knowledge on how fast earthquake waves travelled through the earth's crust at different depths—their 'seismic velocities'—before being recorded by the seismographs at Rabaul and elsewhere. This deficiency was addressed initially for the Rabaul area in 1966 using preliminary crustal seismic tests (Cifali et al. 1969). Although the results were regarded as unsatisfactory, they triggered two major inter-agency geophysical surveys of the New Britain and New Ireland regions in 1967 and 1969 led by BMR scientists (Brooks 1971; Finlayson 1972; Finlayson et al. 1972; Wiebenga 1973), including a 'seismic refraction' survey of the region. Both surveys were truly ambitious, and resource-intensive, involving ships for letting off explosions at many points at sea, together with numerous receiving stations on land where the seismic waves were recorded. The depth of a seismically layered crust beneath the Gazelle Peninsula, including the Rabaul area, was calculated to be 32 kilometres.

A major aspect of the 1969 survey was the concurrent geological and gravity mapping of New Britain at a scale of 1:250,000, involving large field parties of geologists and geophysicists using both helicopter and fixed-wing aircraft support. Six sheet areas were completed. This included the northernmost Gazelle Peninsula where a Port Moresby–based BMR geologist, Peter R. Macnab, had been field mapping on and off since 1966, traversing the difficult, largely unpopulated and mountainous terrain of the interior (Macnab 1970; Davies 1973). Macnab defined and mapped the major and complex Baining Fault that runs across the Gazelle Peninsula from the south-east coast to Cape Lambert in the north-west; in doing so, he suggested that the January 1941 earthquake south-west of Rabaul and described by Fisher (1944; Figure 6.1) may have been related to movements along it. The outer limits of Rabaul volcano are shown on Macnab's map of regional geology (see also Figure 1.1), but more detailed geological work on the rocks of the volcano itself was being undertaken more or less concurrently by RVO volcanologist R.F. 'Bob' Heming (Heming 1973, 1974).

A major aspect of Heming's research included university laboratory work in California on numerous rock samples collected from Rabaul volcano (e.g. Heming 1973, 1974; Heming and Carmichael 1973). This work involved examining slices of rock using a petrographic microscope, chemically analysing minerals and rock-sample powders, and undertaking thermodynamic calculations in order to quantify magmatic properties. The Rabaul rocks were named basalt, andesite, dacite and rhyolite, and were distinguished serially by their silica content (SiO_2). Miyake and Sugiura (1953) had earlier published some chemical analyses of Rabaul rock samples collected by Takashi Kizawa in 1942. They noted the different chemical compositions between basaltic rocks of Tavurvur and the almost dacitic compositions of the pyroclastic rocks produced at Vulcan in 1937. Relationships between the different rock types were complex, however, and the question remained of how two volcanoes could be in near-simultaneous eruption, as in 1937, yet produce magmas of different compositions. These and other questions would serve as a basis for future studies by others.

There had been global advances in the 1960s in understanding the nature and origin of pyroclastic—'fire broken'—rocks and the explosive eruptions that produced them. This included recognition that the rapid expulsion of large amounts of hot pumiceous material can produce extensive pyroclastic flows that leave behind 'ignimbrite', the rock type first named by Marshall (1935) in New Zealand. Calderas can be produced at the surface where the shallow roof of the now evacuated magma reservoir collapses, as discussed above (Figure 4.3). Bob Heming recognised two such ignimbrite layers at Rabaul and, using radiocarbon measurements, dated them at 1400 and 3500 BP. He related the younger (1400 BP) eruption to a collapse that produced the deep water of a caldera at Karavia Bay in the south, and the older (3500 BP) one to formation of a larger caldera that included Simpson Harbour to the north. One difference between the ignimbrites of New Zealand and those identified by Heming at Rabaul is that the former are commonly 'welded' whereas the Rabaul examples are rarely so. 'Welding' takes place when the still-soft pumice fragments of a thick, hot pyroclastic-flow deposit cool into disc-like blebs that are very distinctive in outcrops and in rock samples.

6.5. Plate Tectonics and Subduction

The late 1960s were an extraordinary period in post-WWII geosciences not just in the New Guinea region but in the world as a whole. Geologists and geophysicists had, for generations, speculated, or hypothesised, on how the islands of the New Guinea region—including New Britain—came to be what and where they are, as reviewed by Johnson (1979). Now, however, a 'new global tectonics' was emerging, based on the notion of sea floor spreading and the earlier concept of continental drift (Le Pichon 1968; Isacks, Oliver and Sykes 1968), soon to be called the theory of plate tectonics. This was a momentous time in geoscience, and the New Guinea region was soon the focus of attention because of its high level of earthquake activity and its complex configuration of both large and minor 'plates' (Figure 6.10). The earth is covered by a few major plates that move according to the principles of spherical geometry—'tectonics on sphere' (Mackenzie and Parker 1967)—some separating from each other, others moving past each other laterally, along well-defined faults. BMR geologists Jack Thompson and Norm Fisher even wrote in 1965—somewhat prophetically—in the language of plate tectonics that the structural development of the tectonically complex New Guinea and Papua region 'has been dominated by lateral movement of the Pacific Plate relative to the Australian Continent' (Thompson and Fisher 1965, 121). The region is now recognised as being characterised by the presence of two or more minor plates in addition to the two major ones (Figure 6.10).

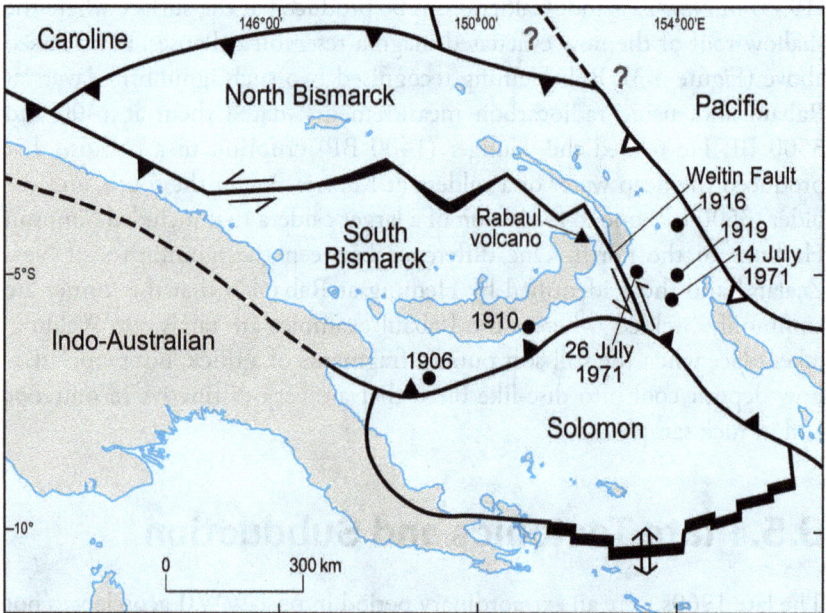

Figure 6.10. Tectonic plates of the region of the Territory of Papua New Guinea.

Three minor plates between the major Pacific and Indo-Australian plates are shown in this schematic map based on the initial study of plate tectonic relationships in the TPNG region published by Tracy Johnson and Peter Molnar (Johnson and Molnar 1972, Figure 2). Directions of plate subduction are shown by the filled arrow heads, and lateral relative plate motion by the pairs of half-headed arrows. Plates diverge at the southern and south-eastern margin of the Solomon plate (see double-stemmed arrow). Dashed lines represent places of earlier plate tectonic activity. Filled circles represent the epicentres of (1) the three high-magnitude (7–8) earthquakes from 1906–19 shown in relation to the epicentres of two others from July 1971 (see text for details). Note that Rabaul volcano is on the eastern edge of the South Bismarck plate (north-west of the Weitin Fault) and that four of the plotted earthquakes are close to the presumed (not shown for the sake of clarity) south-eastern end of the North Bismarck plate.

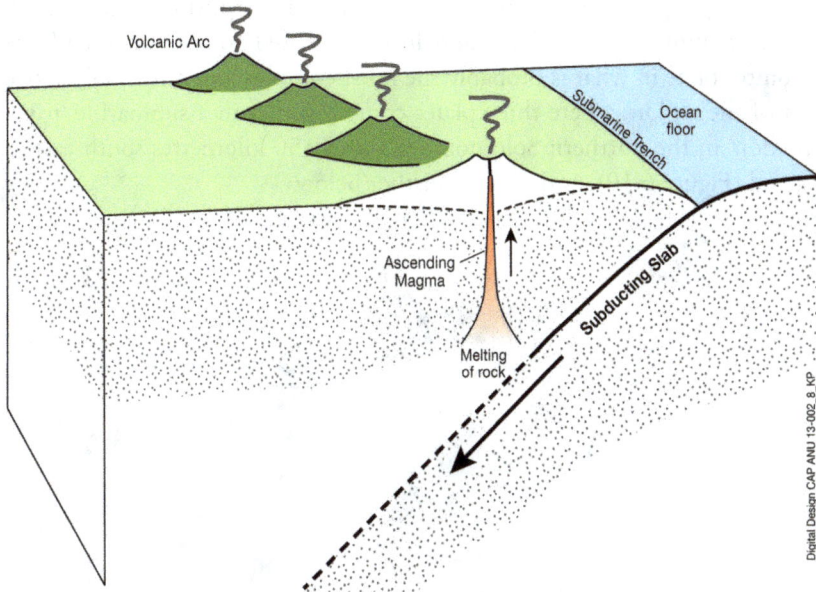

Figure 6.11. Subduction of a tectonic plate.

The process of subduction is shown schematically in this oblique sketch of a tectonic plate on the right (blue) under-thrusting the plate on the left (green) along a submarine trench and creating magmas that rise to form a line or 'arc' of subaerial volcanoes (dark green) at the surface (see also Figures 7.2 and 9.5).

Plates also descend into the deep mantle of the earth in a process that became known as 'subduction' (Figure 6.11). Port Moresby–based BMR seismologist David Denham was already signalling the importance of subduction in the case of New Britain when he published a cross-section through the island arc and plotted earthquakes of increasing depths northwards down to about 300 kilometres (Denham 1969). This represents subduction or 'under-thrusting' of the minor Solomon plate north-westwards beneath the South Bismarck plate along the New Britain–Bougainville submarine trench that had been discovered by the SMS *Planet* (Figure 1.17) during German times (Figures 6.5 and 6.11).

Plate subduction is important in understanding the origin of magmas from volcanoes along the central north coast of New Britain (Figure 0.2). Subduction is thought to drive the hydrated rocks on the Solomon Sea floor to depths of about 100–150 kilometres where they dehydrate or even melt. The fluids so formed rise and trigger partial melting of the overlying upper-mantle 'wedge'. Magmas then rise buoyantly, eventually reaching shallower levels in the earth's crust beneath the active volcanoes, where magmatic gas pressures may increase, ready for final eruption (Figure 6.11). The formation of Rabaul volcano itself is related to subduction of the Solomon plate in what is probably the most complex and seismically active part of the region, where three plates come together in a submarine 'triple junction' in the northern Solomon Sea about 150 kilometres south-east of Rabaul (Figure 6.10), as discussed further below.

7

Geophysical Unrest: Build-Up to Another Eruption, 1970–94

7.1. Demanding Times, 1970–83

The outstanding amount of geoscience-related work on Rabaul undertaken in the 1960s continued resolutely into the 1970s and beyond. The 1970s, in fact, were a particularly demanding time for the staff of both the Rabaul Volcanological Observatory (RVO) and the Geophysical Observatory in Port Moresby (Figure 7.1). The decade began with a large eruption at Ulawun volcano south-west of Rabaul in January–February 1970 (Figure 0.2). Then a magnitude 7.1 tectonic earthquake in October of the same year beneath the coastal ranges of mainland New Guinea—further west along the Bismarck Volcanic Arc—shook the Madang District, causing fatalities and building damage (ACSEE 1973). The seismic energy from two large earthquakes of magnitude 7.9 was released beneath the northern Solomon Sea in July 1971, generating ground-shaking and tsunamis that severely affected the north-eastern Gazelle Peninsula including the Rabaul area. Next, another earthquake series was recorded in January 1972 beneath the coastal ranges of mainland New Guinea, onshore from Manam volcano. Port Moresby–based Tony Taylor visited Manam volcano to inspect its condition in August 1972 but he collapsed and died while undertaking the field work (Fisher 1976a).

Figure 7.1. Rob Cooke and Elias Ravian at a Rabaul Volcanological Observatory staff gathering.

RVO staff are seen in this photograph, which was taken probably in 1977 or 1978 (R.W.J. Collection 14, Folder 2, Folio 13). Rob Cooke is second from the left in the back row with his arms folded. Elias Ravian is in the middle of the back row, standing behind Lesley Topue who is third from the left in the front row. The photographer is unknown.

Six volcanoes in the Bismarck Volcanic Arc—Manam, Karkar, Long, Ritter, Langila and Ulawun (Figure 0.2)—broke out in eruptive activity in 1972–75, constituting what can be considered another eruption 'time-cluster', like the one in the 1950s (Cooke et al. 1976). The long-inactive Kadovar volcano in the extreme west of the arc in 1976 showed signs of volcanic unrest but no eruption took place, and Bagana volcano on Bougainville Island was also in eruption during the same period. However, volcanoes in the Rabaul area were not part of this time-cluster. Karkar was in eruptive activity again in 1979, producing violent explosions that killed RVO staff members R.S.J. 'Rob' Cooke, the senior government volcanologist at Rabaul (Davies 1981), and Elias Ravian, a Tolai from Tavui no. 1 village who worked as a volcano observer (Talai and Pue 1981; Figure 7.1).

The first of the two magnitude 7.9 earthquakes that affected the Blanche Bay area in 1971 struck in the late afternoon of 14 July. Its epicentre was beneath the floor of the northern Solomon Sea, south-west of Buka Island and north-west of Bougainville (Braddock 1973; Everingham 1975; Figure 6.10). This epicentre is remarkably close to a large-magnitude earthquake of 6 May 1919 that was also felt strongly in Rabaul. A large tsunami that reached Blanche Bay just before dusk was generated by the 1971 event. It became trapped in the harbours, slopping back and forth from shore to shore. One of us recalled:

> Buildings shook and shuddered, their timber frames creaked and groaned, trees swayed, water tanks rocked off their stands and burst open. Cupboards fell and the contents of shelves cascaded to the floor. Parked cars rocked on their springs and drivers halted as the roads heaved under them. People running out of buildings found it hard to keep on their feet as successive shock waves made the ground appear to rise and fall under their feet like the deck of a ship in rough seas.

> In other parts of the Gazelle Peninsula there were damaged buildings, burst tanks and roads blocked by landslides. The worst hit area was the United Church centre at Gaulim where sloping land slipped downhill, toppling buildings off their stumps or leaving them standing at a crazy angle.

> Then at about 5.40 pm ... [the] waters of Simpson Harbour and Matupit Harbour receded until the seabed was exposed far below the low-tide mark. Suddenly the sea rushed back and moved over the land to a height of about two metres above the normal waterline. The waters fell and then rose again every few minutes for over an hour, with the height of each rise gradually growing less. (Threlfall 2012, 478–9; see also Threlfall 2021b)

Moderately large earthquakes of magnitude 7.9–8.1 are rare enough in the New Guinea region: those in 1906 and 1919 and referred to above (after Brooks 1965) are probable examples (Figure 6.10). As one of us (R.W.J.) recalls, there was surprise at the Rabaul observatory when another tsunamigenic magnitude 7.9 earthquake took place only 12 days later, on 26 July 1971, this time under the sea off the south-eastern coast of New Ireland and hence closer to Rabaul. More damage was sustained by ground-shaking in the Rabaul area. A notable feature is that the epicentres for the two earthquakes of 1971, as well as for those of 1916 and 1919, are all south-east or east of New Ireland, just to the north of where the New Britain submarine trench changes its north-easterly trend strongly to south-eastwards, down the south-western side of Bougainville Island (Figure 6.10). The implication is that this region must be one of particularly strong seismic-energy release related to lateral bending of the subducting Solomon Sea plate. Further, both of the July 1971 earthquakes were followed by numerous aftershocks that defined in plan a broad, composite arcuate zone in the northern Solomon Sea and mainly to the north of the submarine trench (Everingham 1975). The aftershocks of the main 26 July earthquake in the western half of this aftershock zone contribute to the definition of a down-going subducted slab beneath Rabaul volcano (Figure 7.2).

The tsunamis, or seiches, of 26 July 1971 did further damage to the causeway at Matupit Island already affected by the earlier tsunami, as well as flooding homes on the island's shores and at Rapindik (Threlfall 2012). Matupit islanders were evacuated and those who could not find accommodation elsewhere were sheltered in a tent camp set up by the administration on the edge of the golf course near the Lakunai Airfield (Figure 7.3). Little wonder, therefore, that after this 'double' impact of natural forces there was some talk yet again of abandoning Rabaul and moving to Kokopo (Threlfall 2012).

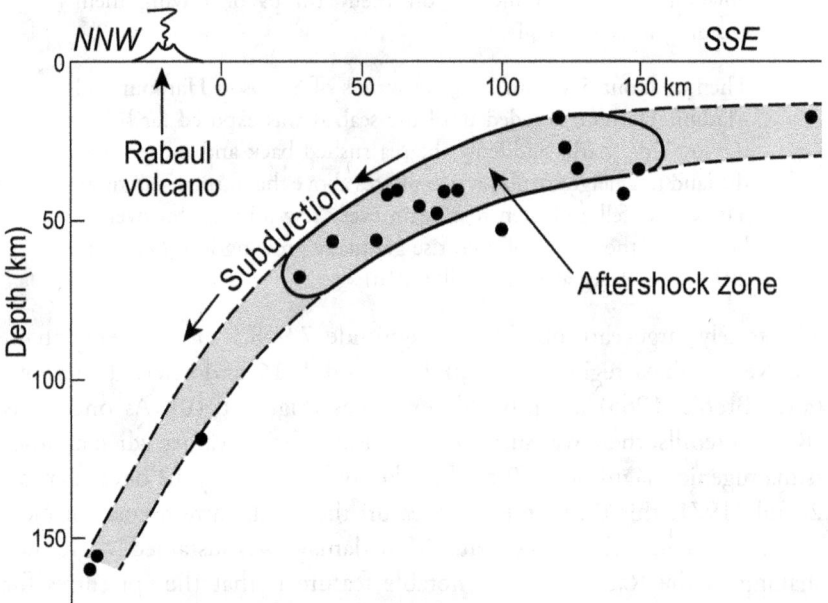

Figure 7.2. Subducting plate beneath the Rabaul area.
The upper part of a northward-dipping, subducting Solomon plate is shown in this schematic cross-section running from the Rabaul area in the north-north-west to the New Britain submarine trench in the south-south-east (adapted from Everingham 1975, Figure 5). The filled circles represent earthquakes that took place between August 1971 and May 1972, and were recorded by more than 10 seismograph stations. The 'aftershock zone' includes numerous events for the two days after the magnitude 7.9 earthquake of 26 July 1971. The shaded area represents only the seismically active upper part of the down-going plate; its base is not shown (see also Figure 9.5).

Figure 7.3. Aerial-photograph mosaic of the north-eastern Blanche Bay area.

Principal features of the north-eastern area of Blanche Bay are shown in this computer-enhanced aerial-photograph mosaic compiled in the early 1980s. The original photographs were taken before uplift of the south-eastern end of Matupit Island became strongly noticeable. The sea is coloured synthetically. Rabaul in the north-east extends down to the golf course and Lakunai Airfield.

The 1970s were a decade of great social and political change in the Territory of Papua and New Guinea, not least in the north-eastern Gazelle Peninsula where the demands for change were more strident, if not more threatening, than elsewhere in the territory (Threlfall 2012). The Gazelle Peninsula Local Government Council seemed to be one of the success stories of the territory but there were Tolai people who opposed its policies and planning, so much so that an opposition Mataungan Association was created in 1969. Protests and street marches ensued. The situation deteriorated considerably when, in August 1971, District Commissioner Jack Emanuel was stabbed and killed. There had been ongoing discussions in Canberra and Port Moresby about the Australian-administered territory achieving independence as a new nation-state, and those New Guineans anxious to see some form of release from expatriate-dominated control did not have too long to wait. Papua New Guinea achieved self-government in 1973 and full independence in 1975. Expatriates continued to lead the scientific and technical work of the RVO after 1975, but a measure of change was evident when Benjamin Talai, from the Duke of York Islands, graduated with a degree in geology from the University of Papua New Guinea, Port Moresby, in 1973, the first local staff member to do so. Talai had joined the RVO staff in 1967.

Small earthquakes in the Blanche Bay area began to be noticed and measured, starting in November 1971 (Mori et al. 1989), just four months after the two magnitude 7.9 earthquakes. The harbour's seismic network of five stations began recording them, but the earthquakes were found to be mainly in the Karavia Bay and Matupit Island area rather than beneath the Simpson Harbour area to the north. Three additional stations therefore were added to the network south of the bay to improve the mapping of the epicentres (Myers 1976; Cooke 1977; McKee et al. 1984). The initial results were quite surprising as the epicentres defined a continuous zone of points between two roughly concentric lines, or what became known informally as the 'seismic annulus', stretching across the entrance of Blanche Bay (Figure 7.4). Different shapes for the annulus would be mapped in the years ahead.

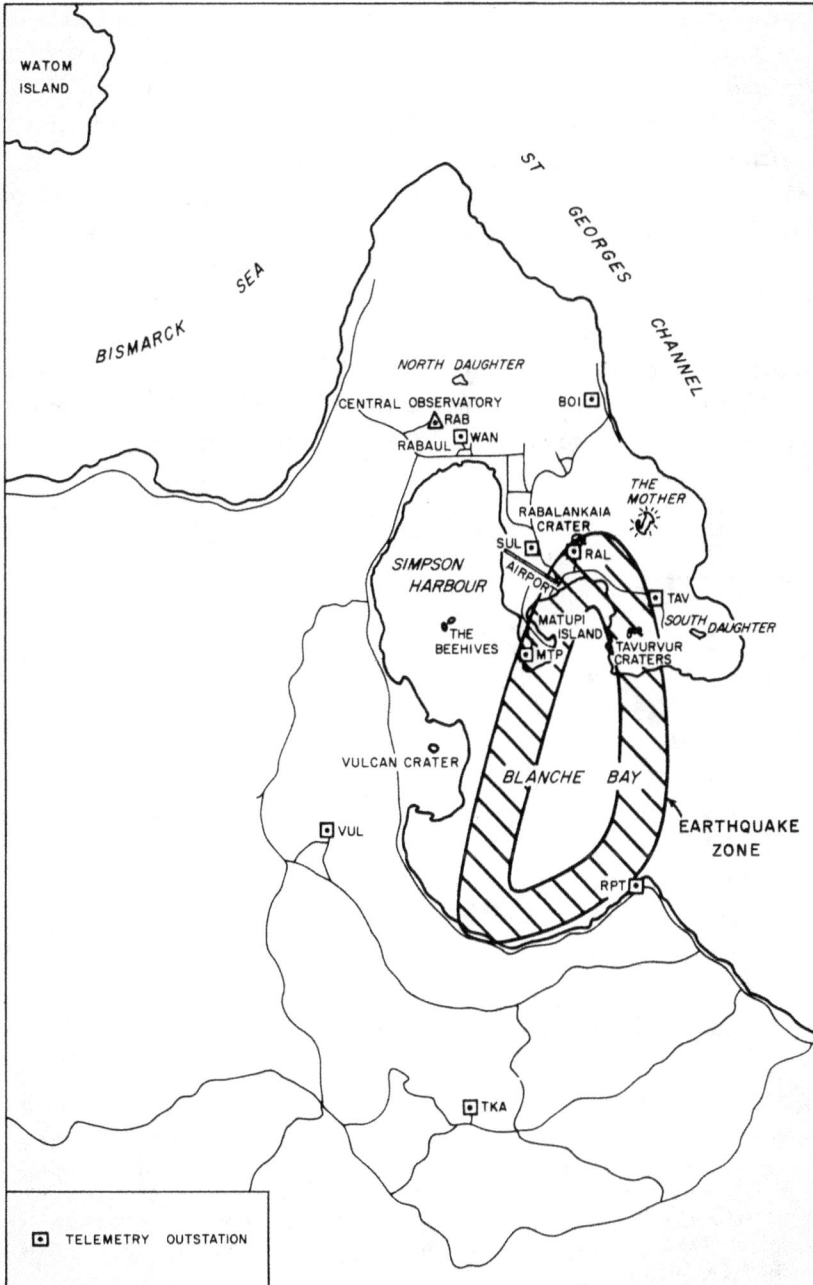

Figure 7.4. Early version of the 'seismic annulus' at Rabaul.

The 'seismic annulus' is seen here in its earliest published version (Myers 1976, Figure 1; see also Cooke 1977, Figure 1). Note that the hachured earthquake zone is 'flattened' on its western side, and that the annulus includes Tavurvur volcano but not Vulcan.

Another feature of the harbour earthquakes was that over the next 10 years or so they defined a series of time-clusters or seismic 'swarms' apparently triggered in some way by the two large earthquakes of July 1971 (Figure 7.5). Further, the largest number of monthly earthquakes in each swarm tended to increase over the same period (McKee et al. 1984). This increase was of concern, particularly as there was growing evidence too that the southern end of Matupit was slowly rising. This rise was in some ways welcome to the islanders who, one of us (N.A.T.) recalls, were gravely concerned about the loss of land caused by wave erosion at the southern end of the island. Some coastal scrub had died, apparently from flooding by seawater caused by subsidence of the south-western tip of the island during the 1971 earthquakes. This evidence for subsidence of Matupit has some similarities to that noted by George Brown in 1875 (see Section 1.2).

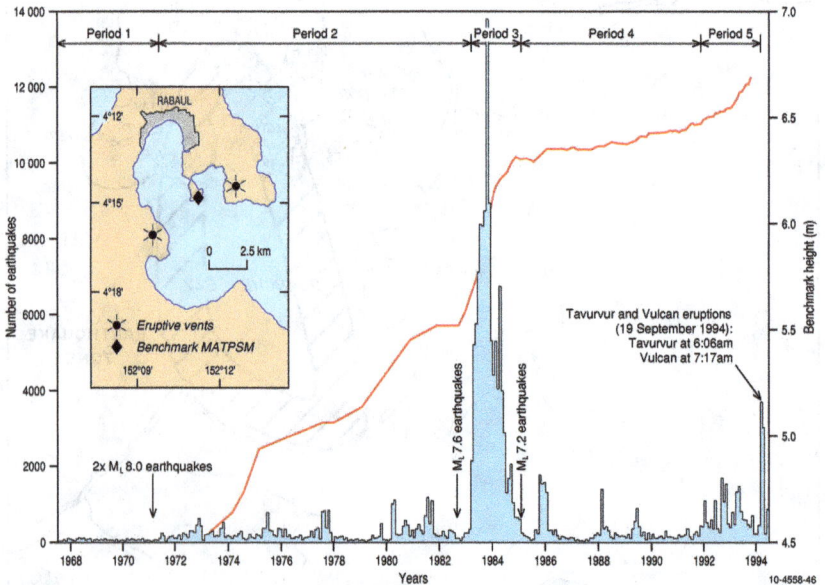

Figure 7.5. Graph of monthly earthquake counts between 1968 and 1994.
The monthly number of caldera earthquakes recorded between 1968 and 1994 on two or more stations of the Rabaul Harbour network are shown in this time series diagram adapted from the one presented by Itikarai (2008, Figure 2.9; see also Mori et al. 1989, Figure 2). Also shown is the progressive increase in height of the benchmark at the southern end of Matupit Island (see MATPSM in inset). Itikarai has here divided the 1968–94 time series into five periods.

Recognition of Matupit Island uplift led the RVO to instigate a new program of geodetic monitoring of height changes throughout the Blanche Bay area (de Saint Ours et al. 1991). The program began in 1973 using the normal optical-levelling methods deployed by surveyors, but other techniques were added, including portable tiltmeters ('dry tilt'); measurements of the changing gravity field; the monitoring of sea level using tide gauges and graduated rods (or 'tide sticks'); comparing bathymetric results from different marine surveys; and even measuring the heights of raised, stranded barnacles. Electronic (laser) horizontal-distance measurements were introduced in 1983 when Norm Banks of the United States Geological Survey (USGS) came with equipment donated to the RVO (Archbold et al. 1988). These geodetic data were used in conjunction with the earthquake datasets and ongoing temperature measurements to investigate the possible meaning of this growing geophysical unrest.

One question that emerged from the geodetic results collected throughout the 1970s and early 1980s concerned the nature of two centres, or foci, of sea floor uplift possibly caused by bodies of magma at shallow depths (McKee et al. 1984). In particular: what was pushing up and tilting Matupit Island towards the north and was there any additional evidence that might be obtained by studying the sea floor south and south-east of the island? A research vessel of the USGS, the RV *S.P. Lee*, was invited to undertake a bathymetric and seismic-reflection survey of Blanche Bay. The ship's survey in 1982 included tracks in and out of Greet Harbour and then south-westwards across the area south-east of Matupit (Greene, Tiffin and McKee 1986). A 'bulge' and associated active fault zones in the sea floor were detected (Figure 7.6), which

> are a result of emplacement of magma at a shallow depth. Contorted sediments and slumps adjacent to the bulge are probably the result of uplift and seismic activity. The pattern of activity appears to reflect increased magma pressure at depth beneath the caldera floor. This activity may eventually lead to an eruption. (Greene, Tiffin and McKee 1986, 327–9)

The possibility of such a submarine eruption taking place from a centre of uplift south-east of Matupit Island had already been considered, with concern, by RVO volcanologists: specifically, that a 'period of potentially very dangerous phreatomagmatic activity could occur in the early stage of the next eruption if a vent was established directly above the magma body' unless magma was channelled obliquely to one of the nearby subaerial volcanoes, such as Tavurvur (McKee et al. 1984, 408).

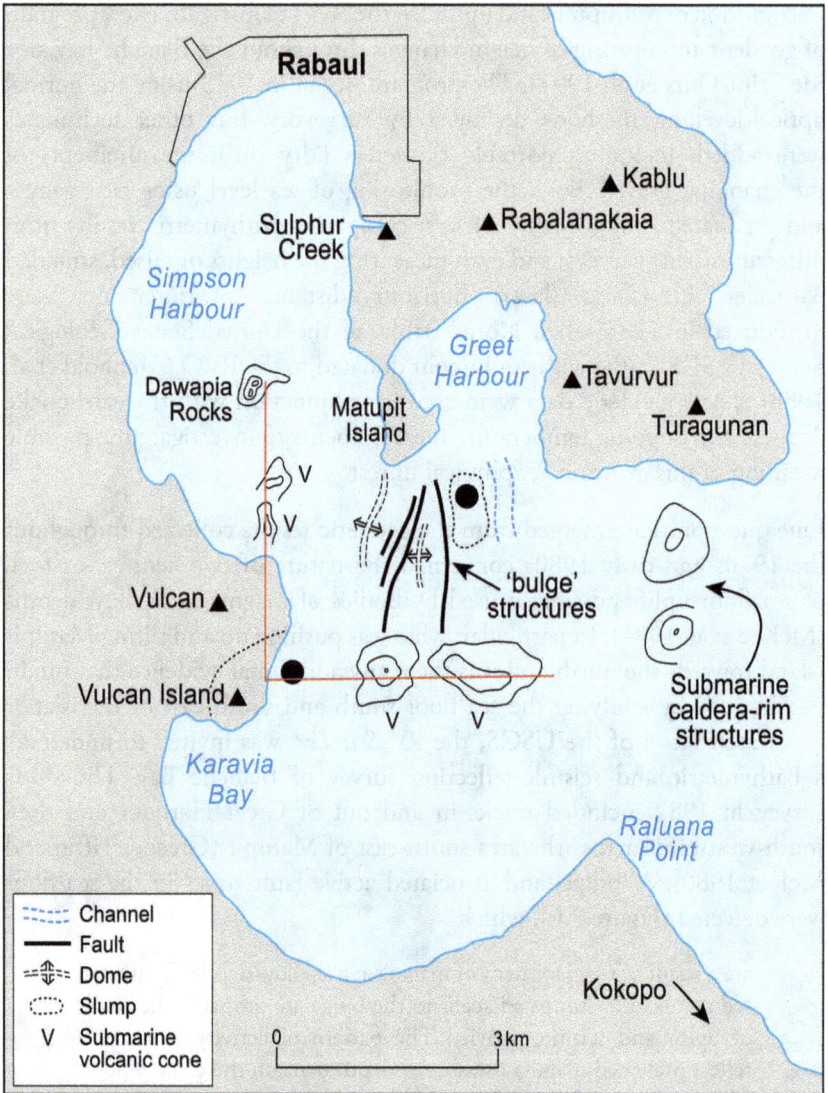

Figure 7.6. Centres of uplift and bulge structures in Blanche Bay.

This map of the 'bulge' structures south and south-east of Matupit Island is adapted from diagrams published by Greene, Tiffin and McKee (1986, Figure 16) and de Saint Ours et al. (1991, Figure 10). The filled circle south-east of Matupit Island represents the approximate position of the centre of uplift calculated from geodetic measurements (McKee et al. 1984). The other filled circle representing the centre of uplift near Vulcan Island to the south-west was mapped later using dry-tilt measurements (McKee et al. 1984). The RV S.P. Lee also mapped four sea floor cones of probable volcanic origin: two are in line with and east of Vulcan Island as well as with the centre of uplift there, and the other two are on a south–north line between Vulcan Island and Dawapia Rocks. The two lines intersect at right angles to each other.

7.2. A Time of Crisis, 1983–85

The seismic swarms continued into the early 1980s (Figure 7.5), as did the concerns of people throughout north-eastern Gazelle Peninsula, and a Volcanic Contingency Plan was completed for East New Britain Province by mid-1983. The content of the plan was guided by a disaster-management specialist, Brian Ward, from the United Nations Disaster Relief Organization, and was undertaken in consultation with national and provincial government authorities, the RVO and the private sector (McKee, Itikarai and Davies 2018). This contingency plan would be updated periodically in the years ahead and was part of an overall Provincial Disaster Plan managed by a Provincial Disaster Committee (PDC) and chaired by Nason Paulius, secretary of the Department of East New Britain (East New Britain PDC 1988).

The Provincial Disaster Plan was informed by the results of an assessment made by a volcanological team—mainly RVO staff—of volcanic hazards at Rabaul (McKee et al. 1985), thus updating the volcano-related hazard assessments made by N.H. Fisher 40 years previously (Fisher 1946b; Figure 6.1). McKee and his co-workers reported on the wide range of volcanic hazards accompanying different scales of eruption at Rabaul in the context of both eruption contingency planning and ongoing instrumental surveillance of the volcanoes at Rabaul. This study also included consideration, for the first time, of the impact of pyroclastic flows and involved a reinterpretation of photographs of pyroclastic flows from 1937 (Figures 3.11–3.13 and 3.17). The Provincial Disaster Plan assumed eruptions only of the scale of those in 1937 and 1878. Importantly, the plan also included a scheme of four stages of volcanic alert: Stage 1, in which a volcanic eruption was expected within years to months, up to Stage 4, in which one was expected within just hours to days.

The situation in Blanche Bay changed dramatically in September 1983 when many more earthquakes and a stronger period of uplift began to be detected (McKee et al. 1984; Mori et al. 1989). A magnitude 7.6 earthquake had taken place in March 1983 in the same region as the two large earthquakes of July 1971, possibly triggering the new period of enhanced unrest (Itikarai 2008; Figure 7.5). Monthly earthquake totals reached a maximum in April 1984 after which they declined gradually to July 1985, defining what became known as the 'Rabaul Seismo-Deformational Crisis'. This period of almost two years caused great concern and uncertainty among the wider community, including businesses, villages, crop-growing property owners, subsistence farmers and service providers—as well as insurance companies. It would demand virtually the sole attention of RVO staff, led by Englishman Dr Peter L. Lowenstein, who had taken over leadership of the observatory after Cooke's death in 1979 (Lowenstein 1988). A key question was: Are there now stronger precursory signs that a volcanic eruption would soon take place? If so, when would it occur, how big would it be and how long would it last?

The East New Britain PDC held its inaugural meeting on 13 April 1983 (Lowenstein 1988), and the first of several evacuation exercises was practised in late May and early June by government authorities. Hazardous and safe areas were shown on coloured posters that were displayed on public notice boards, together with the localities of key facilities such as evacuation pick-up points and care centres (McKee, Itikarai and Davies 2018). Film Australia in 1983 released a documentary by Bob Kingsbury appropriately entitled *Waiting for the Big Bang* (Kingsbury 1983). A range of other public awareness-raising activities were initiated, including—in the background in Canberra—preparation of the first edition of *Volcano Town* (Johnson and Threlfall 1985). Publication of the book was sponsored by the Insurance Underwriters' Association of Papua New Guinea and was aimed at illustrating how the previous eruptions had affected the Rabaul area.

A Stage 2 volcanic alert indicative of a possible volcanic eruption within only weeks to months was declared on 29 October 1983 following further intense earthquake swarms. The number of recorded earthquakes in April had reached a monthly maximum of more than 13,000, and the southern end of Matupit Island had risen a total of about 1.6 metres above its 1973 height. These events caused even greater concern among the communities of Rabaul town and the surrounding region. Peter Lowenstein was quoted on the front page of the *Sydney Morning Herald* newspaper for 26 January 1984 as saying that

> evidence is accumulating to suggest that the [Rabaul] volcano has embarked on an irreversible course towards the next eruption and that it is only a matter of time before this occurs … within the next few months. (Hastings 1974, 1)

The headline was: 'Volcano Set to Blow: 20,000 Plan Island Escape. Rabaul Fears Repeat of the 1937 Disaster'.

There was a partial and voluntary evacuation of the town itself, including businesses (Lowenstein 1988; Blong and Aislabie 1988; Neumann 1996). Many villagers moved to land outside the caldera area, including perhaps as many as 40 per cent of those people living in the perceived highest-risk areas south of the main business district. Blocks of government-owned land south of the Warangoi River were made available for settlement. Other people left the province altogether—for West New Britain, for example—and some expatriates departed for Australia.

An inevitable topic of debate in the community and media, as well as at different levels of government, concerned the suitability of Rabaul as a place for a town and provincial capital—a repetition of discussions that had taken place after the 1937–43 eruptions and after WWII. Prime Minister of Papua New Guinea Michael Somare entered the debate by announcing in mid-February 1984 that, should an eruption take place, Kokopo would be developed as a new administrative centre, and that Rabaul had insufficient land for expansion anyway (Darius 1984). The crisis was a local one, but it nevertheless prompted development of eight new national Acts of disaster-related legislation, which were passed by the National Parliament of Papua New Guinea in March 1984 (Davies 1995b). The national government also provided funds for an emergency airport at Tokua, 15 kilometres east of Kokopo, which was opened by Prime Minister Somare in November 1984. The airport for normal commercial aviation in the province, however, remained at Lakunai, south of Rabaul and north-west of nearby Tavurvur volcano.

There was increasing concern in the community about the limited flow of authoritative and up-to-date information from the provincial government and the RVO about the state of the volcanoes. The provincial government therefore established a Public Information Unit in early February 1984 to disseminate as much relevant information about the crisis as possible (McKee, Itikarai and Davies 2018). The unit was assisted for a time by the well-known Australian Government geologist Hugh L. Davies, who had many years of experience working in Papua New Guinea and who was based in Port Moresby. Town meetings and meetings with special-interest groups were arranged, and a local newspaper was produced that was supported by a local businessman (Davies 1995a). Its name was *The Rabaul Gourier*— a word combination of the English 'courier' and 'guria', which is Tok Pisin for earthquake (Davies 1984). Two other expatriates involved with the work of the Public Information Unit attempted to obtain information about community disaster preparedness through distribution of a questionnaire, but afterwards were dissatisfied by the way they had designed the questions (Kuester and Forsyth 1985).

A great deal of international interest was generated by the 1983–85 crisis at Rabaul, especially in Australia, the US, New Zealand and Japan. Overseas technical and professional assistance was provided to the RVO through international development-assistance agencies including, particularly, partnership collaborations involving staff of the USGS, Bureau of Mineral Resources, Geology and Geophysics (BMR), and RVO. The crisis also triggered a dramatic increase in the number of investigative geoscience publications on Rabaul volcano, a publication surge that continued well into the twenty-first century.

One reason for the overseas interest in Rabaul was because of geophysical unrest and concerns about volcanic outbreaks at caldera systems in two other countries. Alarming earthquake activity, ground deformation and new gas emissions (in one case) were also being reported in the early 1980s from the large active calderas at Long Valley, eastern California, US, and at Campi Flegrei ('burning fields') or Phlegrean Fields, west of Naples in Italy. This concurrent restlessness at three caldera systems in different parts of the world was quite coincidental, but the three were compared with considerable interest by volcanologists internationally. The three calderas also featured prominently in a major scientific literature review in the 1980s by two USGS volcanologists. These authors summarised almost 1,300 episodes of historical 'unrest' at 138 large calderas worldwide. Their conclusions included the following: 'The remarkable unrest at Rabaul Caldera that began in 1971 is perhaps the most threatening example in this compilation' (Newhall and Dzurisin 1988, 227).

The 1980s were also a time when there was global concern about aviation safety and dangerous encounters of aircraft—and jet-aircraft in particular—with volcanic ash clouds. High-rising eruptions at Galunggung volcano in Indonesia in 1982 caused the engines on a British Airways flight to cut out (Tootel 1985). The Australian Bureau of Meteorology in 1982 began using images from the Japanese geostationary meteorological satellite for the detection of ash clouds, and then issuing 'volcanic ash advisories' through an aeronautical telecommunications network. A Qantas aircraft sustained damage caused by high-rising ash clouds from Soputan in 1985.

International conferences were held on the subject (Casadevall 1994) and a worldwide network of Volcanic Ash Advisory Centers (VAACs) was set up in the early 1990s by the International Civil Aviation Organization, an agency of the United Nations, as part of the International Airways Volcano Watch. The Australian Bureau of Meteorology became responsible for running a VAAC in Darwin that covered the region encompassing Papua New Guinea, Indonesia and Australia.

Meanwhile, the 1983–85 crisis at Rabaul provided RVO scientists with even more seismological and geodetic data for interpretations of what might be happening geophysically beneath Blanche Bay. Attention was focused on the seismic annulus that, by this time, had become much better defined, although there was still considerable scatter in the map of epicentres caused by uncertainties in recording each individual earthquake (Mori et al. 1989). The annulus appeared to be mainly two long arcuate zones that faced each other but were linked in the north and south (Figure 7.7). Tavurvur and Vulcan volcanoes lay just outside the limits of the annulus. Another striking result was that the earthquakes were found to be mainly less than 4–5 kilometres deep—depths that apparently defined the top, or roof, of a large underlying reservoir of magma probably left over from the last period of caldera formation (Figure 7.8).

Further insights into the nature of the seismic annulus beneath Blanche Bay would be provided by seismologists in later years, using improved techniques and leading to a refinement of its likely structure (Jones and Stewart 1997; Itikarai 2008). The study by Jones and Stewart (1997), which recalculated or 'relocated' the original epicentres of 4,442 events recorded by the RVO between 1971 and early 1992, was particularly innovative. The method involved 'collapsing' two or more epicentres that were sufficiently close to one another and treating them as if they were just one event. This had the effect of reducing the number of epicentres *and* changing the previous scattering of epicentres (Figure 7.7) to a more focused configuration that appeared to define new structures within the zone of seismicity (Figure 7.9). In particular, a smaller inner ellipse of earthquakes in the north appeared to be younger than an incomplete but still seismically active and deeper one in the south.

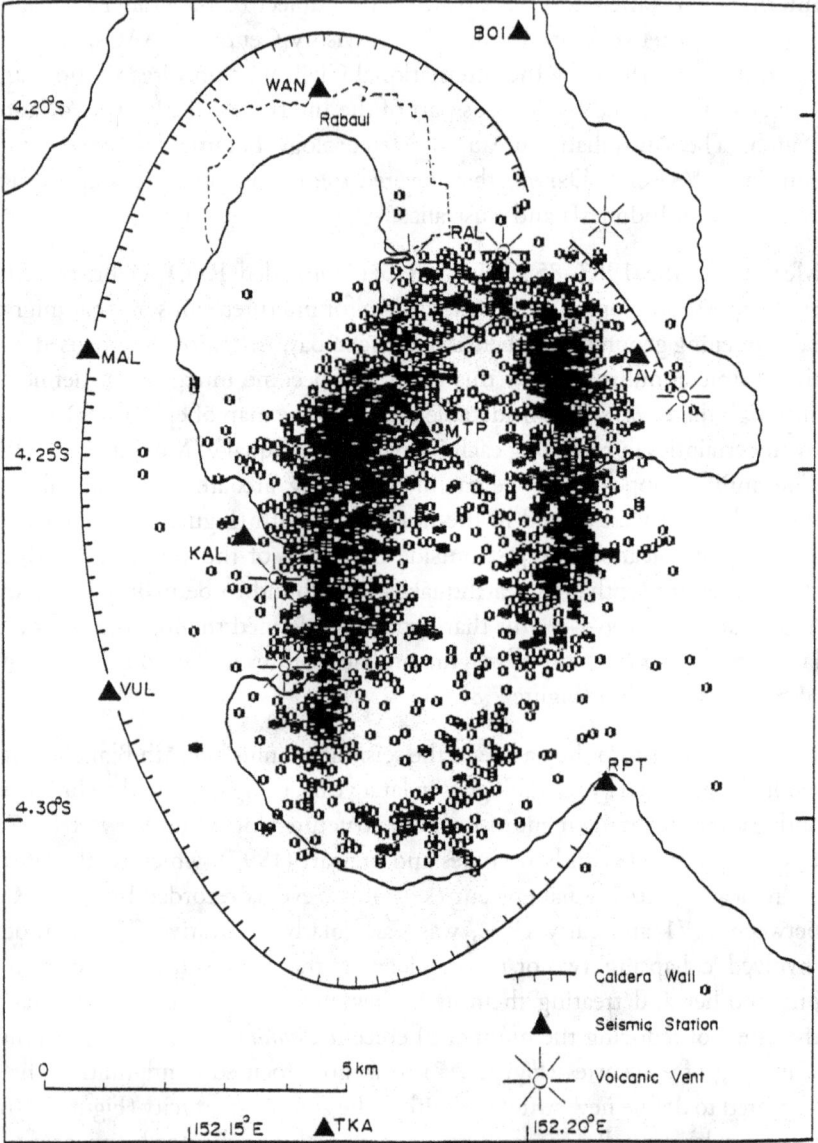

Figure 7.7. Map of earthquake epicentres defining the seismic annulus in 1983–85.

This map of earthquake epicentres is adapted from the one presented by Mori et al. (1989, Figure 7). These are the epicentres for over 2,500 earthquakes for the crisis period of September 1983–July 1985. All plotted events were recorded by seven or more stations and had horizontal errors of less than 1 kilometre.

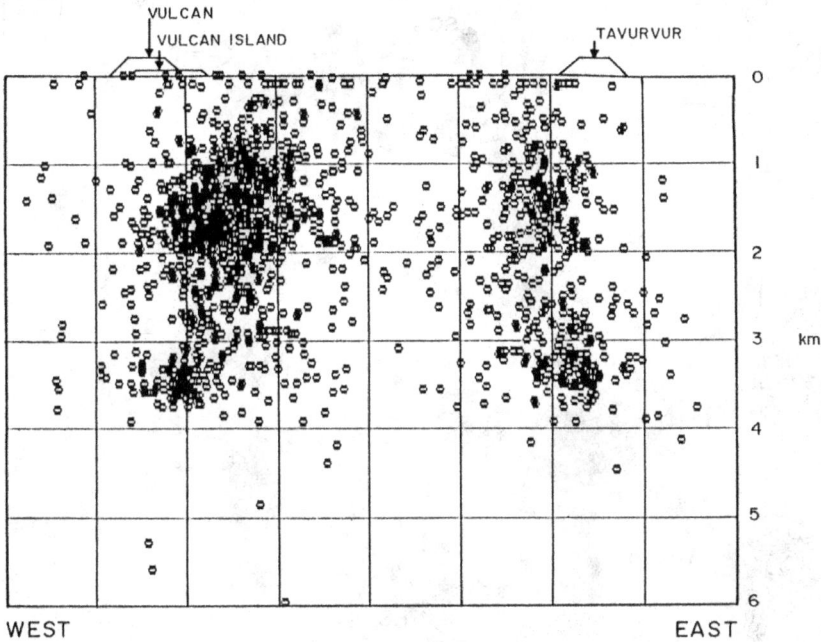

Figure 7.8. East–west cross-section through the seismic annulus shown in Figure 7.7.

This is a composite of three east–west cross-sections adapted from Mori et al. (1989, their Figures 9 and 10; see Figure 7.6). The plot includes all the well-located events between 4.24°S and 4.28°S.

The significance of the inner ellipse volcanologically is that the young eruptive centres within Blanche Bay generally have locations that are just outside the limits of the northern ellipse (McKee et al. 2020). An exception is the 1878 centre for Vulcan, which is close to the western margin of the older structure and in line with the two Karavia Bay centres, whereas its 1937 and 1994 centres are closer to the younger inner one. More importantly, however, an implication of the relationships shown in Figure 7.9 is that the distribution and growth of the young volcanoes appear to depend on the existence of the inner ellipse, meaning that there may have been earthquake activity on it prior to the 1937 and 1878 eruptions even though its existence at that time was unknown because of the absence of seismograph networks. Note also that the three Vulcan centres in Figure 7.9 may correspond to a northward progression between 1878 and 1994.

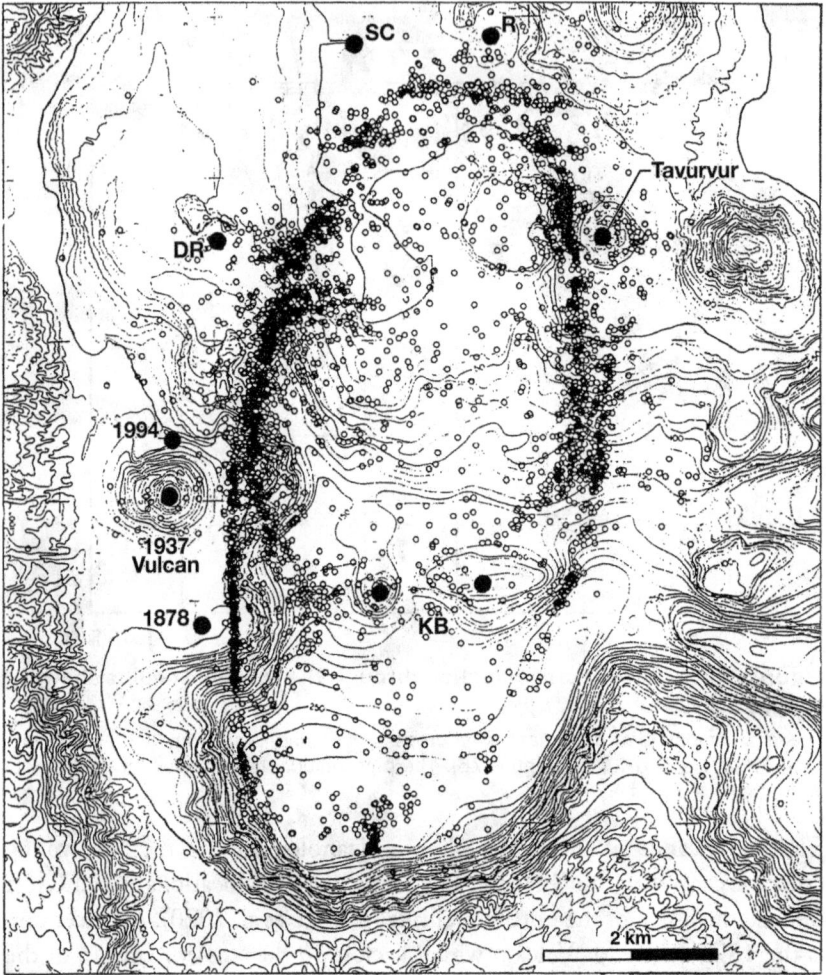

Figure 7.9. Ellipses formed by earthquake epicentres in Blanche Bay.

This map is adapted from ones published by Jones and Stewart (1997, Figure 7) and McKee et al. (2020, Figure 23). Abbreviations are as follows: DR, Dawapia Rocks; SC, Sulphur Creek; R, Rabalanakaia; KB, the two submarine cones in Karavia Bay. Only the initial outbreak point of the Vulcan 1994 eruption is shown here. Dawapia Rocks has not been in eruption historically and may have formed before creation of the northern ellipse of earthquakes (McKee et al. 2020). Itikarai (2008) showed that earthquakes in the northern earthquake ellipse are shallower than those in the south.

There was recognition by the early 1990s that the annulus earthquakes were not of the type expected from magma rising to the surface, and no volcanic tremor was recorded (McKee et al. 1984), meaning that there was uncertainty about whether magma was actually rising from below and how long the Stage 2 alert should be kept in place, or whether the level should be reduced or increased. The earthquakes were classified technically as being of 'volcano-tectonic' or 'V-T' type. An eruption may have been expected imminently during 1982–84, but the monthly totals of earthquakes dropped off after peak values in April 1984 and the Stage 2 alert was called off in November 1984. The monthly totals then declined further to pre-crisis levels (Figure 7.5). There was a feeling in the community that the crisis had been a volcanic 'false alarm', but this did not reduce overall safety concerns. The RVO continued its instrumental monitoring, and many more investigations would take place over the following years, some of which are highlighted below.

7.3. Ongoing Investigations and Uncertainty, 1986–94

Earth science staff members of UNESCO (Paris) promoted an international focus on volcanic disasters during the 1970s. Initially, interest focused on support for studies on the scientific prediction of volcanic eruptions (UNESCO 1972), but the emphasis changed during the decade towards how best the scientific results of volcano monitoring during volcanic crises could be integrated with social needs such as crisis management and evacuation decisions in a 'risk' context. This change was encapsulated by a simple but influential 'multiplication' formula (Fournier d'Albe 1979, 321):

Risk = Value or 'Exposure' (human lives, capital investment, agricultural land)

x Vulnerability (the proportion of the value likely to be lost)

x Hazard (the probability of different threats affecting an area).

There is no volcanic risk or disaster if any of these three factors is zero. Further, the expression 'natural disaster' becomes somewhat misleading as a disaster must also involve loss of, or in, a community caused by its own vulnerability, should a natural hazard strike. Therefore, a vulnerable community, such as the Rabaul one, is as much a cause of its own demise as any volcanic eruption that might overwhelm it. In this way, a community itself decides what level of risk it is prepared to accept, assuming it has a clear idea of what is at risk in the first place.

Another important clarification of terminology at this time centred on the meaning of eruption 'prediction'. Fournier d'Albe, for example, wrote:

> [W]e must face the fact that we do not have at present, nor are we likely to have in the foreseeable future, sufficient data or knowledge of the eruptive process to make predictions on a deterministic basis. (Fournier d'Albe 1979, 323)

The best that could be achieved would be to identify reliable precursors of eruptions empirically—such as earthquakes and ground-tilt—with the aim of making general forecasts rather than accurate predictions. This matter was highlighted by reflections on the prediction-versus-forecast problem in the case of other eruptions worldwide at this time (Opinion 1983). Scientific uncertainty and, therefore, lack of accurate predictions were unavoidable for volcanologists at the RVO, even though Rabaul people might have wanted otherwise. The best that could be done, using the accepted four-stage alert system, was to make general forecasts in the hope that a Stage 4 alert could be announced in good time.

Three studies on Rabaul were influenced by the overriding concept of disaster-risk reduction and were undertaken in response to the 1983–85 crisis. Two of these were sponsored by the Insurance Underwriters' Association of Papua New Guinea and were completed by consultants from New Zealand and Australia. The first of three, following Fournier d'Albe (1979), was produced by former RVO seismologist John Latter, who, with co-author A.W. Hurst, considered earthquake and tsunami threats as well as volcanic hazards (Latter and Hurst 1987). Both investigators were geophysicists from the New Zealand Department of Scientific and Industrial Research. Latter and Hurst (1987) concentrated on the hazards and the risk to human life.

The second, even more comprehensive, study of the Rabaul area, its volcanic hazards, vulnerabilities and likely damage impacts used more or less the same 'risk' principles as Latter and Hurst. The investigators in this study were Australian academics, physical geographer Russell J. Blong and economist Colin Aislabie. They considered much more 'exposure' information than did Latter and Hurst—on the built environment, agricultural investment and human life—as well as describing the contemporary disaster-management activities of the 1983–85 crisis (Blong and Aislabie 1988). Three eruption 'scenarios' were presented, together with assessments of their respective impacts. The first two cases were based on eruptions of the same scale as those of 1937, and for two different wind regimes corresponding to the changes of season in any one year. Most attention was directed towards the impact of a damaging eruption from a new volcano, named 'New Matupit', at the submarine centre of uplift south-east of Matupit and starting with sea floor phreatomagmatic eruptions. A much smaller but simultaneous eruption was considered for 'Liklik Vulcan' near the site of the 1878 eruption. Pyroclastic flows from New Matupit would inundate Matupit Island in either season and would create a new volcanic island to the south-east. The third scenario considered by Blong and Aislabie was a major eruption similar in scale to the one that took place in 1400 CE. They concentrated on the devastating and widespread effects of major pyroclastic flows out to distances of more than 50 kilometres.

A third risk-related and groundbreaking study dealing with disaster management at Rabaul was begun in 1983 by another Australian geographer, K.J. 'Ken' Granger. This research led to a master's thesis that was submitted to The Australian National University under the title, 'The Rabaul Volcanoes: An Application of Geographical Information Systems [GIS] to Crisis Management' (Granger 1988). This contribution, perhaps more than any other up to that time, provided a reminder that the 'Digital Age', including personal computers, had arrived and that computer-based processing of digital images taken by aircraft and from space could be coupled with spatial modelling and relational databases to produce a valuable information tool for use by crisis managers at Rabaul. The 1983–85 crisis was over, but volcanic eruptions could still be expected, meaning that 'the present period of remission from volcanic activity' could be used to good effect (Granger 1988, 135). Granger also wrote, somewhat famously:

> There is nothing more certain in the disaster management business than the fact that once a disaster starts to unfold, it is too late to start looking for the information to manage it. (Granger 2000, 20)

These results do not represent the only studies of value and relevance to emerge in the late 1980s. The surge of interest in Rabaul from researchers eager to work with RVO staff was ongoing, even though the current crisis had ended. A major contribution of lasting significance resulting from field work in 1985–86 was the production of a new geological map of the Rabaul area and an accompanying report by two geologists from the New Zealand Geological Survey and two staff from the RVO (Nairn et al. 1989, 1995). This work added greatly to the earlier geological achievements of Heming (1973, 1974) and Walker et al. (1981). The research accomplished on the many ignimbrites in the area, and how the major explosive eruptions that produced them were related to different phases of caldera formation at Rabaul, were among the highlights (Figure 7.10; see also Section 4.2). These eruptions were many times more powerful and voluminous than the eruptions of 1937 and 1878.

Another conclusion reached from this more modern geological mapping was that Matupit Island was different from the other young volcanic centres in Blanche Bay (Nairn et al. 1989, 1995). The raised, flat-topped island, which has no visible crater, was found to consist of water-sorted sediments of sea-rafted pumice and ash of unknown thickness comprising the 'Matupit Pumice Sediments'. These pumice and ash sediments were overlaid by sands, gravels and coral, and shell fragments that were radiocarbon-dated at about 750 BP. The Matupit Pumice Sediments were laid down shortly before this time and were then raised above sea level, possibly episodically. They may have been produced from a sea floor volcanic centre to the west—that is, by an eruption similar to the one at Vulcan in 1878. The deposits of Dawapia Rocks are thought to have a similar age to those of the Matupit Island sediments, so this post-caldera volcano may have been the eruption source for the Matupit pumice and ash (Nairn et al. 1989, 1995).

Figure 7.10. Pyroclastic deposits in a road cut near Kabakada.
The volcanic deposits exposed in this new road cut near Kabakada, Rabaul, and photographed in 1969 by R.F. Heming, consist of thick grey ignimbrite deposits (the Rabaul Ignimbrite and the Kulau Ignimbrite) overlying the eroded surface or 'unconformity' of the near-horizontal brown and yellow pyroclastic deposits of the older Kabakada Subgroup in the lower centre and right (Nairn et al. 1989, Figure 4). This road cut soon became shrouded in vegetation and the geology concealed.

Scientific interest continued in trying to understand the nature of the 'bulge' south-east of Matupit Island, and postwar technologies were used to investigate if the sea floor in that area was any warmer than elsewhere in Blanche Bay. First, a remotely operated vehicle equipped with a video camera and temperature sensor undertook dives at 11 different places on or near the sea floor bulge. No evidence of elevated water temperatures, hot springs, upward streaming of gas bubbles or submarine craters was found (Johnson 1986). These results led to major heat-flow surveys being undertaken in 1990 and 1992 when 4-metre-long probes equipped with thermistors down their lengths were used to map what turned out to be a complex pattern of heat discharge from the floor of Blanche Bay (Graham et al. 1993). The highest heat-flow values were found in Greet Harbour—and, surprisingly, at one point north-east of Dawapia Rocks—but the area south of Greet Harbour, including the uplift area south of Matupit Island, had much lower values than those to the north. The earlier interpretation that a new volcano might be created at this sea floor–uplift area looked less certain. Neither were high heat-flow values found in the Vulcan area.

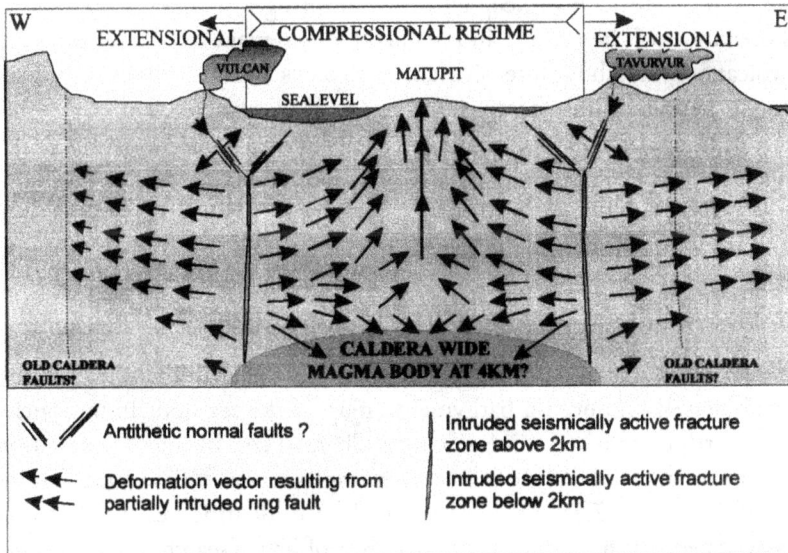

Figure 7.11. Intrusion of magma up ring fault and its resulting deformation of the central block.

Magma is shown rising in two stages from the large magma reservoir below (Saunders 2001, Figure 11) resulting in the eruptions at Vulcan and Tavurvur in 1994 (see also Roggensack et al. 1996, Figure 4). An 'antithetic' fault in geology refers to a minor secondary fault, commonly one of a set, whose sense of displacement is opposite to that of its associated major normal faults. Saunders has used the concept here as a way of explaining the slight outward displacement of the sites of the two volcanoes away from the seismic annulus (see Figure 7.9).

RVO geodesist Steve Saunders published the cross-section in Figure 7.11, showing both Vulcan and Tavurvur in eruption and the magma rising in underlying ring fractures. The injected magma is shown squeezing the central caldera block such that the area south-east of Matupit is up-bowed, thus explaining the doming up of the sea floor and the creation of the 'bulge' (Figure 7.6). Robertson and Kilburn (2016), however, promoted a different interpretation in which the surface up-doming was produced by the 4–5-kilometre-deep magma body, but was followed by a change in stress state from elastic to inelastic after 1991–92. In this interpretation, ascending magma does not utilise the fractures of the seismic annulus until one to two days before eruption.

The amount of financial support that could be provided to the RVO by the national government declined following the end of the 1983–85 crisis. Peter Lowenstein resigned from the RVO in January 1989 largely because of this reduced support. An Australian, C.O. 'Chris' McKee, who had joined the RVO as a geophysicist/geologist in 1973, took over as head of the Rabaul observatory, assisted by Ben Talai as deputy. By the late 1980s, the observatory was finding it difficult to keep all the monitoring instruments up-to-date and operational. The routine and manual measurement of temperatures on the volcanoes was labour intensive, and such work declined during and after the crisis period, mainly because of other work demands. Neither was the monitoring of changes in volcanic gas emissions at Tavurvur possible. Six villagers were in fact killed by gas asphyxiation in a crater on the southern side of the volcano (Figure 6.8) on 24 June 1990 when three people who had been collecting megapode eggs in the crater were overcome by carbon dioxide, and another three villagers were killed the day after when they tried to rescue them (RVO 1990).

The year 1990 was also when the scientific crew of the *Sonne*, a German research vessel, produced a bathymetric map of the sea floor immediately off the north-eastern coast in St Georges Channel confirming the existence of a large, mainly submarine caldera that impinged on the coastline (Tiffin et al. 1990; Figure 7.12). The caldera had been identified earlier, but not fully mapped, on a marine-research cruise of the research vessel *Moana Wave* in 1985–86, and its on-land, curved south-western side had been mapped previously as the Tavui Fault (Nairn et al. 1989). Tavui caldera is a major volcanic centre and the likely source of some of the volcanic deposits mapped on the land within the Rabaul volcanic complex. These deposits probably include the voluminous, 6,900-year-old Raluan Ignimbrite, which has been described as the 'penultimate major eruption deposit in the Rabaul area' (McKee 2015, 1).

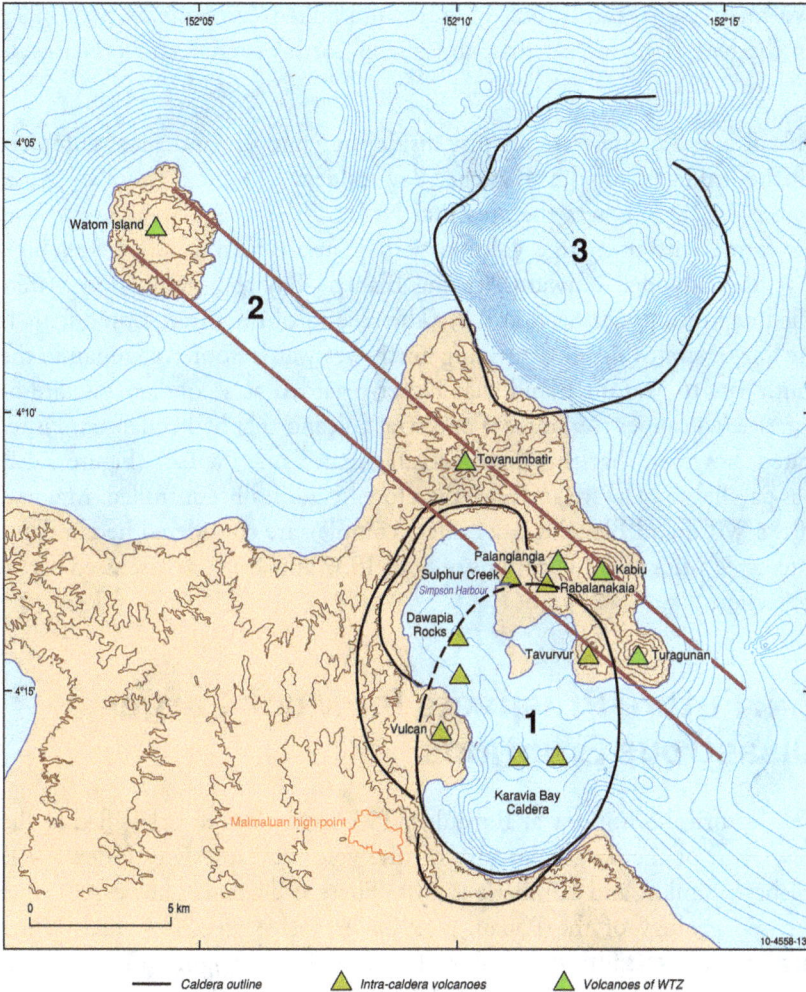

Figure 7.12. Map of proposed volcanic systems of the Rabaul area.

Three volcanic systems of the north-eastern Gazelle Peninsula are shown in this map: (1) Blanche Bay caldera complex; (2) Watom–Turagunan Zone; and (3) Tavui caldera (Johnson et al. 2010, Figure 13).

Tavui caldera is generally regarded as a neighbouring volcano to the main Rabaul volcanic complex and, accordingly, they have been considered separately by the Smithsonian Institution (Siebert, Simkin and Kimberly 2010; see Figure 0.2). There remains the possibility, however, that both centres are interrelated and are fed by common, albeit complex, conduit systems at depth—that is, they can be regarded as a single Rabaul–Tavui complex. Further, the 'Rabaul volcanic complex' itself may consist of two main parts, and a total of three interlinked components may make up the overall geological structure of the north-eastern Gazelle Peninsula (Figure 7.12).

Vulcan and Tavurvur are set just inside the youngest and most complete of the calderas of the Blanche Bay caldera complex (number 1 in Figure 7.12). This was the area that Norm Fisher (1939a) suggested was the former site of a large volcano perhaps more than 2,700 metres high that had exploded outwards leaving Blanche Bay as a giant, now water-filled depression (see Section 4.2). The second of the three components is the line of stratovolcanoes running north-westwards from Turagunan in the south-east to Watom Island that Fisher regarded as secondary satellite or parasitic volcanoes to the main ancestral volcano that had collapsed. Fisher stated, however, that this Watom–Turagunan Zone (WTZ) (Figure 7.12) was parallel to New Ireland and, therefore, tectonically controlled. Magmas of the WTZ and Blanche Bay caldera complex are considered here to have interacted prior to the eruptions of both 1937 and 1994, as discussed below in Section 9.4.

7.4. Future Prospects for Much Larger Disastrous Eruptions

The historical eruptions at Rabaul in 1937 and 1994 are classified at the lower end of the 'large' category on the volcanic explosivity index (VEI) (Siebert, Simkin and Kimberly 2010). There is evidence, however, in the exposed geology of the Rabaul area for explosive eruptions many times larger than this, as introduced above (in Section 6.4 on 'ignimbrites') and, in particular, with reference to the geological work of Ian Nairn and colleagues, who presented mainly radiocarbon dates for many of these deposits. At least five, and possibly as many as nine, ignimbrite-producing eruptions may have taken place in the past 20,000 years or so, each perhaps accompanied by a collapse forming a caldera (Nairn et al. 1989, 1995). Serial collapses are thought to have formed the nested caldera escarpments in the Blanche Bay caldera complex (Figure 7.12), the more recent collapses being the best preserved, and the older ones hardly preserved at all.

These results provided average return intervals for major explosive eruptions at Rabaul of between 2,600 and 6,000 years. A similar range of intervals was found later for the earlier part of the pyroclastic sequence at Rabaul using an 'argon–argon' (^{40}Ar/^{39}Ar) isotopic method on additional samples (McKee and Duncan 2016). These more recent results have emerged from a considerable amount of new geological and geochronological work that has been completed in the Rabaul area since the 1994 eruption, undertaken primarily by Chris McKee and RVO colleagues (McKee et al. 2020; McKee 2021). The tephra volumes for most of the larger explosive eruptions at Rabaul are between about 5 and more than 11 cubic kilometres, corresponding to the 'very large' (VEI 5–6) classification on the VEI scale.

The range of 2,600 to 6,000 years is wide, meaning that there is no reliable evidence for a constant eruption periodicity for the largest known eruptions at Rabaul. Further, there are many uncertainties in the geochronological record, including some imprecise radiometric dates. Perhaps the greatest uncertainty, however, is in mapping the stratigraphy of the tephra sequences. Exposures are generally poor or ephemeral and typically quite unlike the one illustrated in Figure 7.10. Also, tephras of different ages may look similar in the field, and making correlations between exposures over large distances can be challenging or not possible at all. In addition, the tephras expected at some exposures may have been removed by erosion from the sequence being examined.

Another complication is evidence of large eruptions at Rabaul that fall between the upper size limit of the small 1937–43 eruptions and the lower limit of the large caldera-forming, ignimbrite-producing ones. The best studied of these is the Talili Pyroclastics eruption sequence (Nairn et al. 1989, 1995; McKee and Fabbro 2018). These deposits are sandwiched stratigraphically between the two youngest of the larger caldera-forming ones. Their age range is given as 4200–1400 BP, and their total volume as 5 cubic kilometres or more (VEI 5). Part of the sequence consists of fine-grained pumice thought to have been derived from sea floor vents in Simpson and Greet harbours, whereas the remainder are the deposits of eruptions from WTZ volcanoes of Kabiu and Palangiangia. These prehistoric eruptions would have had an impact on Rabaul town far greater than those in either 1937 or 1994. Good outcrops of the Talili Pyroclastics are seen at Adelaide Street on the south-eastern edge of the now-destroyed town, and in road cuts on Namanula Hill, which was used as a major evacuation route for townspeople and others in 1937.

The best studied of the large, caldera-forming eruptions at Rabaul is the most recent one (Heming 1973, 1974; Walker et al. 1981; Nairn et al. 1989; McKee 2021). These deposits are referred to as the 'Rabaul Pyroclastics'.

They have an estimated volume of more than 11 cubic kilometres, meaning a VEI value of 6, and they consist of an upper ignimbrite layer (at least 8 cubic kilometres) and a lower plinian ash fall layer. The eruption led to formation of the most recent and best-preserved caldera escarpment in the Blanche Bay area, located mainly in Karavia Bay (Figure 7.12). The eruption is known, conveniently, as the '1400 BP' event, based on conventional radiocarbon-dating methods (Heming 1973, 1974), but a more recent age of 667–699 CE has been determined using a so-called wiggle-match radiocarbon-dating technique on a large charcoalised log from the ignimbrite (McKee, Baillie and Reimer 2015). Note that a seventh-century eruption is only a little more than 1,300 years ago. This value is less than the 2,600–6,000-year range of intervals between the earlier dated large eruptions and, on face value, could be taken as an indication that another major eruption is not due yet for quite some time. Such prognostications, however, need to be treated with considerable caution, bearing in mind the uncertainties in the record of geochronological results.

The seventh-century ignimbrite is found widely throughout the north-eastern Gazelle Peninsula. For example, it is known to be more than 40 centimetres thick more than 40 kilometres south of Blanche Bay (McKee 2021, Figure 10). Kokopo and other present-day settlements in the region are built on the deposit. The ignimbrite has been described as having a 'low-aspect ratio', meaning the pyroclastic flow that deposited it was energetic enough to roll over wide tracts of countryside, depositing ash thinly relative to the distance travelled, but filling some valleys during its passage (Walker et al. 1981). The eruption destroyed much of the environment and presumably was a major disruption to human life and settlement throughout the north-eastern Gazelle Peninsula. Afterwards, however, the revegetated area would have provided an opportunity for reoccupation for those migrants quick enough to take advantage of any newly recovered volcanic land.

An illustration of the hazard impact of larger eruptions in the north-eastern Gazelle Peninsula, including Rabaul and Kokopo, is presented in Figure 7.13 where the dashed line represents the area expected to be affected by a large ignimbrite-producing, pyroclastic-flow eruption. Note that this area includes Watom Island where the seventh-century ignimbrite can be identified but excludes the Duke of York Islands where it cannot.

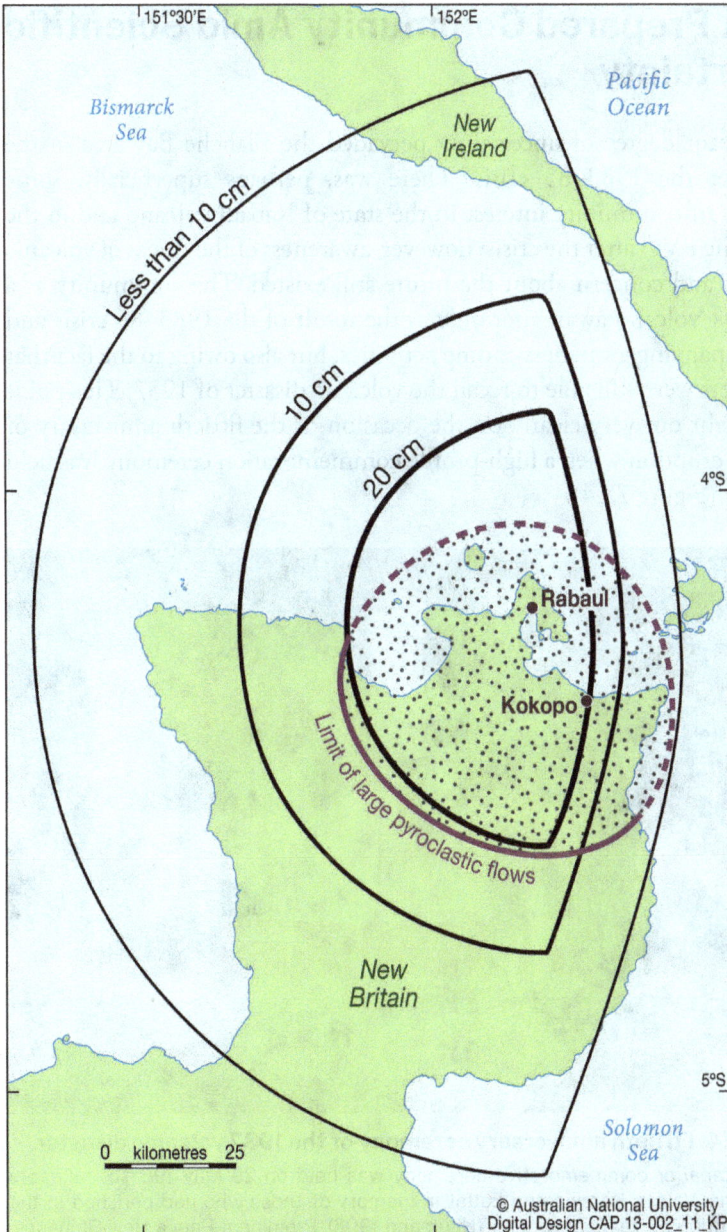

Figure 7.13. Anticipated extents of large pyroclastic flows and heavy ash falls in Gazelle Peninsula.

The dashed and solid line and the stippling (all in red) represent the area expected to be affected by large ignimbrite-producing pyroclastic flows across the north-eastern Gazelle Peninsula (adapted from McKee et al. 1985, Figure 13). The solid black lines signify the anticipated thickness of air fall ash. Bear-Crozier et al. (2013) provide a more detailed and recent assessment of the ash fall hazard in this area.

7.5. A Prepared Community Amid Scientific Uncertainty

A significant degree of uncertainty pervaded the Blanche Bay area in the years after the 1983–85 crisis. There was, perhaps superficially, some reduction of community interest in the state of Rabaul volcano and in the work of the RVO after the crisis; however, awareness of the threat of volcanic eruptions and concern about the future still existed. The community as a whole was 'volcano aware' not only as the result of the 1983–85 crisis and its accompanying awareness-raising activities, but also owing to the fact that some elders were still able to recall the volcanic disaster of 1937. This point was brought out very clearly on the occasion of the fiftieth anniversary of the 1937 eruption when a high-profile commemoration ceremony was held at Tavana (Figure 7.14).

Figure 7.14. Fiftieth anniversary ceremony of the 1937 volcanic disaster.
A *matamatanai* or commemorative ceremony was held on 29 May 1987 by villagers from Tavana, Valaur, Karavia and Latlat in memory of those who had perished in the Vulcan eruption 50 years previously (Neumann 1996). Prominent Papua New Guineans, including the governor-general, gave speeches and the large assembled crowd listened to choirs. The old monument shown here was erected shortly after the 1937 disaster and was the focus of the 1987 ceremony. Death tolls for each of seven villages are inscribed on the monument, totalling 352 people. The words at the foot of the inscription are: 'The earth covered them up at 4 o'clock 29th May 1937.' Photograph by R.W.J.

RVO staff also had the advantage of being more knowledgeable about the nature of the Blanche Bay volcanic complex as the result of post-WWII scientific advances. They were clearly in a much better state of preparation for an eventuality than were people in May 1937 when no volcanological observatory existed at all. Nevertheless, the RVO had declared in 1983 that the volcano was on an 'irreversible course' to an eruption 'within the next few months'. They had had to call off the Stage 2 alert in 1984 and there had been no volcanic eruption in the 10 years since then. In fact, the RVO, like many other volcano observatories worldwide, had the capacity to make only general volcanic forecasts and, as considered above, living with 'false alarms' was part of the responsibility of managing uncertain risk in the area.

Monthly earthquake totals had declined to pre-crisis levels by July 1985, but two magnitude 7.2 earthquakes within about 100 kilometres of Rabaul were recorded in May and July 1985, possibly marking the recommencement of higher monthly values in the following months (Itikarai 2008). Earthquake swarms from the seismic annulus were again recorded, and uplift continued at the southern end of Matupit Island, although at lower rates than before the crisis period (Figure 7.5). May 1992 marked the beginning of an increase in seismicity that was maintained for about two years, and the rate of uplift increased somewhat. A new feature of the seismic activity during this time were earthquakes in a well-defined zone running north-east of the caldera and well away from the seismic annulus (Itikarai 2008).

Friday 16 September 1994 was a national holiday, celebrating the anniversary of Papua New Guinea's Independence. Then, early on Sunday 18 September, two strong caldera earthquakes were felt at 2.51 am: one was located in Greet Harbour near Tavurvur volcano and the other appeared to be beneath the southern part of the Vulcan headland (McKee, Itikarai and Davies 2018; McKee et al. 2018). These events triggered an intense period of ongoing earthquake activity throughout the day that naturally gave rise to fears in the community and to uncertainties at the RVO. Some seismographs at the RVO 'were a mass of almost unreadable black scribble and the recording pens were having difficulty holding up with the constant movement' (Lauer 1995, 19, including a photograph of the seismograms). This was unlike the more 'normal' earthquakes of the 'swarms' recorded previously from the seismic annulus. People from the Vulcan area who visited the RVO later that morning were told they should not be too concerned (Neumann 1996); however, at about 6 pm a recommendation was made by the RVO to the PDC, and passed onto the National Emergency Disaster

Centre in Port Moresby, that a Stage 2 alert—signifying an eruption within months or weeks—should be declared, as in 1983 (Davies 1995a, 1995b, 1995c; McKee, Itikarai and Davies 2018).

A spontaneous evacuation of villagers from the Vulcan area had already started in the late afternoon, and by dusk most people on Matupit Island and in the southern end of town were resolutely evacuating along the road northwards towards an assembly place at Queen Elizabeth Park sportsground in Rabaul. People did not wait for any official recommendation from the authorities to do so. At least some listened to the advice of older people, their *patuana*, who recalled the 1937 eruption (Neumann 1995, 1996). Other Matupit people remained on the island. The unexpected night-time departure of aircraft from Lakunai appears to have confirmed in people's minds that evacuations should be continued if not hastened. By the early hours of 19 September, people from Rabaul town and villages within the caldera had evacuated by land and sea to safer villages, towns and missions in the south. Temporary camps, soon to be called 'care centres', were erected. The self-evacuation of people on the western slopes of Vulcan cone, including Tavana and Valaur villages, contrasts sharply to May 1937 when hundreds did not evacuate and so perished in the eruption, particularly in those two villages.

Meanwhile, the RVO advised the PDC at around midnight on the Sunday night that 'Rabaul was on an irreversible course towards an eruption' and that a Stage 3 alert—meaning an outbreak within weeks or days—should be declared (McKee, Itikarai and Davies 2018, 221). The PDC held off announcing this for the time being because of their concern that undue anxiety, even panic, might be caused in the community, which was already evacuating steadily. However, the National Disaster Emergency Services in Port Moresby was informed and its director, Leith Anderson, passed on the information about the Stage 3 alert to the prime minister, Sir Julius Chan, as well as to aviation authorities and the media. News about the Stage 3 alert was broadcast by ABC and NBC radio nationally, including in Rabaul, even though the alert had not been officially declared by the PDC itself.

There had been remarkable periods of uplift and exposure of the sea floor during the night of 18–19 September that were not seen fully until daybreak. The eastern shore of the Vulcan headland had risen about 6 metres and the southern shoreline of Matupit Island had migrated about 70 metres southwards (McKee, Itikarai and Davies 2018; McKee et al. 2018). Tavurvur broke out in eruptive activity at 6.06 am on Monday 19 September and was joined by Vulcan at 7.17 am in another 'twin' eruption at Rabaul like those of both January 1878 and May 1937 (Itikarai 2008; McKee, Itikarai and Davies 2018; McKee et al. 2018). There was no eruption from the area of the 'bulge' south-east of Matupit Island—the anticipated 'New Matupit'.

The Stage 2 alert was still being broadcast by the local radio station early on the morning of 19 September advising people to stay home and remain calm. Stages 3 and 4 were declared, belatedly, after the Tavurvur eruption had started; Stage 4 meant that an eruption could be expected within hours to days. This anomaly clearly illustrated the unsuitability of the four-stage system of alert, which was based, at least in part, on the expected time before an eruption. The RVO later abandoned the system.

About 17,000–18,000 people lived in the Rabaul town area in September 1994 and a further 27,000 had homes in immediately adjacent villages, including those on the north-east coast (Davies 1995a, 1995b, 1995c; Lindley 1995). Many thousands of other people lived outside the limits of Blanche Bay, all of whom would be affected in one way or another by the unfolding eruption. Probably as many as 105,000 displaced people would require assistance, including the provision of rations by the provincial government. Evacuees were accommodated in more than 270 care centres, ranging from large camps like the one at the Kokopo Showground, to numerous smaller ones scattered throughout the region.

8

Eruptions of 1994–2014

8.1. 'Twin' Eruptions, 19 September – 2 October 1994

Rabaul Volcanological Observatory (RVO) volcanologists undertook a helicopter-borne inspection of the changes taking place in Simpson Harbour at dawn on Monday 19 September 1994. They were over the general area of the Greet Geothermal Field when a convoluting ash cloud began emerging slowly from the 1937 crater of Tavurvur (GVP 1994b; Lauer 1995; McKee et al. 2018). The ash was then blown by south-east winds towards Lakunai Airfield and near-deserted Rabaul town. An Air Niugini F28 managed to take off from the airstrip just as the eruption started.

The eruptions increased in intensity over the next several hours (Figure 8.1), and the eruption columns above Tavurvur grew in height over the next few days, eventually reaching a maximum height about 6 kilometres above sea level. These eruptions were vulcanian and they deposited large amounts of ash to the north-west, covering the airport, stranding aircraft and causing numerous roof collapses, especially in the southern part of Rabaul town including Malaytown (Blong and McKee 1995). The first-deposited tephras consisted of dark-brown ash as well as a wet, mud-like, blue-grey ash strongly reminiscent of the 'blue gummy mud' that had been produced by Tavurvur in May 1937. The Tavurvur eruptions were again disrupting the clay- and sulphide-rich, geothermally altered interior of the volcano, and were, therefore, at this stage, technically of 'hydrovolcanic' origin.

Figure 8.1. Double-layer ash cloud from Tavurvur on 19 September 1994.

Ash clouds from Tavurvur volcano are here seen spreading across the eastern part of Rabaul town on Monday 19 September 1994, as photographed from Observatory Ridge at 7.30 am. The lower of the two layers of volcanic cloud represents the earlier, less intensive phase of the initial eruption, whereas the upper one is from the higher, stronger eruption column seen rising from Tavurvur. Vulcan had by this time started its activity on the other side of the harbour (off the photograph to the right) but its ash had not yet drifted over Rabaul. This digitally enhanced copy was provided courtesy of the photographer, Nick Lauer (Lauer 1995, 15).

RVO volcanologists on Observatory Ridge also observed—and photographed—the beginning of the eruption at Vulcan at 7.17 am on 19 September (Lauer 1995). The eruptions began low at the northern foot of the 1937 cone and appeared for a while to be 'phreatomagmatic', caused by the interaction of new magma with seawater in the harbour, much as had happened at the start of the Vulcan eruption in 1937. Pyroclastic flows or surges were produced, some extending out over the harbour floor and later creating tsunamis (Nishimura et al. 2005). New craters opened further upslope and a full plinian eruption began to develop. A conduit high on the 1937 cone was inclined, as seen by the prominent slant of a jetting ash column (Figure 8.2).

Figure 8.2. Initial explosive eruptions from Vulcan as seen from the south on 19 September 1994.

Vulcan is seen from the south jetting pumice, ash and water vapour out from an inclined vent shortly after 7.30 am on 19 September 1994. Pyroclastic flows are being emitted, moving to the north-west and north, and to the north-east across the waters of Rabaul Harbour. The water surface in the foreground is calm but would soon become disturbed after further pyroclastic flows began crashing down onto the water, generating volcanic tsunamis. The green vegetation seen here would also soon be destroyed by further eruptions. Accreditation: M. Phillips, B. Alexander and S. McGrade, Rabaul.

The plinian column grew higher, dumping pumice and ash to the west and north-west, but for a while its pyroclastic materials were carried north-east over Rabaul, adding to the loads from Tavurvur, much as had happened in 1937. Visibility deteriorated on Observatory Ridge as a result of this north-west drift and RVO volcanologists could no longer make good observations—a reflection on the unsuitability of the observatory's location during actual eruptions, at least in the south-east season. Forked lightning was seen in the developing Vulcan cloud, much as had appeared in 1937. Four more plinian or strong explosive eruptions took place at Vulcan over the following two days, but their strength declined by 22 September (Figure 8.3). This short duration is similar to that for the plinian phase at Vulcan in May 1937, although the lesser eruptions at Vulcan in 1994 lasted until 2 October. Floating pumice covered Simpson Harbour extensively in both 1937 and 1994, hindering ships and boats, until it dispersed in the following north-west season.

Figure 8.3. Vulcan and Tavurvur in double eruption on 22 September 1994.
The 'twin' eruption at Rabaul is seen in this photograph taken by Torsten Blackwood (Agence France-Presse) from the south on Thursday 22 September 1994 (see also front cover). Pyroclastic flows are being emitted by a vulcanian eruption on Vulcan (left). The well-established vulcanian cloud at Tavurvur is seen on the right still depositing ash on Rabaul town.

News of the 'twin' eruptions and volcanic disaster at Rabaul soon spread beyond East New Britain. There was full daily coverage by the Port Moresby–based *Papua New Guinea Post-Courier*, *National* and *Times* newspapers during the first two weeks, beginning on Tuesday 20 September and including articles, photographs and commentary on many aspects of the disaster. These covered the immediate disaster-relief arrangements and—almost inevitably—discussion on the future of Rabaul. The news spread also via radio and television to Australia and globally as many other international media outlets carried the story. Reporters, photographers and disaster specialists all started arriving in the province.

Volcanologists arrived too from overseas, most of them (but not all) invited by the RVO to assist in re-establishing instrumental monitoring at Rabaul, assessing the ongoing eruptions at Tavurvur and Vulcan, and—in the longer term—undertaking new scientific research support projects. International volcanological research interest in Rabaul and in other 'restless' calderas globally was high again, after the crisis periods of the mid-1980s. 'Erupting Neighbours—At Last' was the title of one article in the popular international journal *Science* (Williams 1995). However, the much-anticipated Rabaul eruption at Vulcan and Tavurvur was not the large-magnitude, caldera-forming event that some had thought might take place—and no eruptions at all took place at Campi Flegrei (Italy) and Long Valley (US), the other two restless volcanic areas of concern in the 1980s.

Interest in Rabaul has lasted up to the present day. There has been extensive reporting of all kinds since 1994, ranging widely from personal accounts and collections of colour photographs of the eruptions and their effects, through to specialised disaster-management and scientific articles, including major contributions from RVO staff and presentations at international conferences. There is now much more information on the 1994–2014 period than there is for 1937–43—easily enough for a comprehensive volume of its own. Accounts have been provided for the 1994–95 period from a community perspective—including from schoolchildren and students—by ANU historian Klaus Neumann (Neumann 1995, 1996; see also Neumann 2014, 2017; Matupit 2003); from a disaster-management perspective (Tomblin and Chung 1995; Davies 1995a, 1995b, 1995c; Lindley 1995; McKee, Itikarai and Davies 2018); and from a mainly volcanological standpoint by RVO staff led by Chris McKee (McKee et al. 2018). The US-based Global Volcanism Program (GVP) at the Smithsonian Institution, Washington, DC, played an important scientific role internationally in compiling reports on the Rabaul eruptions of 1994–2014 in their globally distributed *Bulletin of the Global Volcanism Network* (see, in particular, GVP 1994a, 1994b, 1994c, 2006, 2014, 2017).

Summaries of most of the 1994–2014 period are also provided by us in our earlier books (Threlfall 2012; Johnson 2013) and we draw on these here, concentrating on how the 1937–43 events compare with those of 1994–2014. We have been highly selective in the topics we discuss in the remainder of this final chapter and in the bibliographic sources we quote.

8.2. Pyroclastic Deposits, Damage and Deaths

Space technologies were not available in 1937 but they certainly were in 1994. The Vulcan plinian column and eruption cloud were clearly visible on 19 September 1994 from the Space Shuttle *Discovery* (Figure 8.4). The Vulcan cloud was seen as a well-defined fan that had an east–west axis consistent with the known direction of high-level winds, its fan-shape having been caused by wind shearing. Much information on satellite imagery of Rabaul eruptions was provided to the GVP, including from US agencies such as the National Oceanic and Atmospheric Administration (NOAA) and the National Aeronautics and Space Administration (NASA). The *Bulletin of the Global Volcanism Network*, therefore, contains valuable images of the Rabaul eruptions as seen from space. Tracking clouds using satellite imagery and distributing the results promptly also assist aviation authorities. There were no encounters between international jets and the drifting ash clouds from Rabaul in 1994, but airline companies sustained additional fuel costs because of aircraft diversions on longer routes.

Figure 8.4. Space shuttle image of Vulcan and Tavurvur eruption clouds.

Eruption clouds from both Vulcan and Tavurvur were photographed on 19 September 1994 by space shuttle astronauts. The larger, higher Vulcan cloud is seen spreading, fan-like, westwards across the Bismarck Sea, caused by shearing of winds at different levels. The lower and less pronounced cloud mainly from Tavurvur — indicated by the two arrows — extends north-westwards. Normal atmospheric weather cloud covers much of New Ireland in the foreground. This image was published in the *Bulletin of the Global Volcanism Network* (GVP 1994a, Figure 17) and credited to NASA (reference code STS064-116-064).

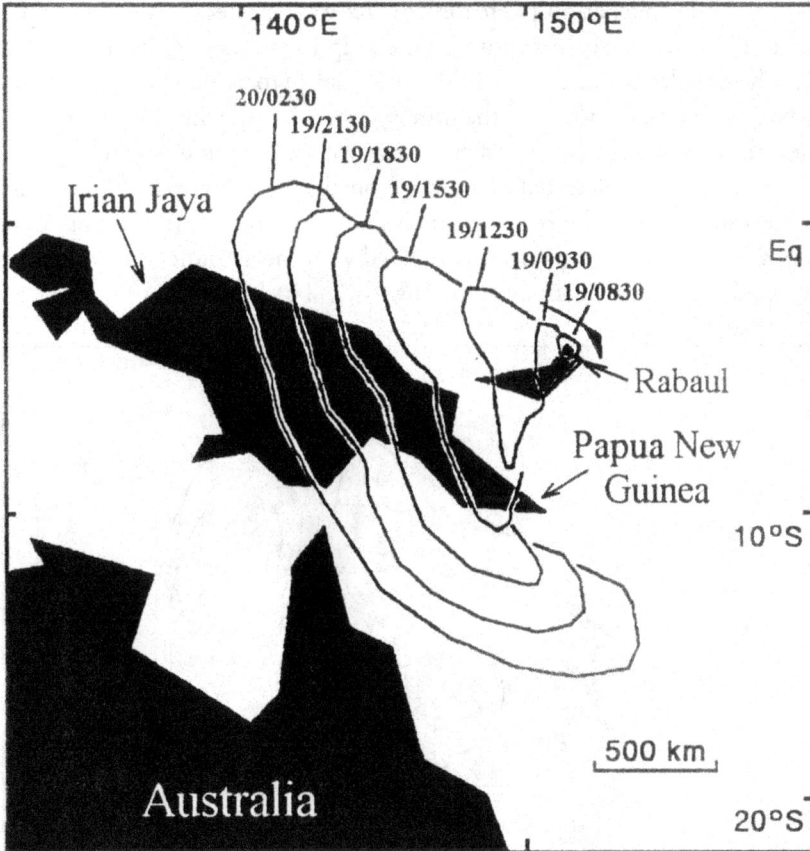

Figure 8.5. Vulcan ash cloud positions tracked by NOAA satellite at different times.

The westward progress of the Vulcan eruption cloud was tracked by NOAA using infrared imagery from the polar-orbiting NOAA-12 satellite. The notation '19/0830' refers to 8.30 am on 19 September 1994 and so on (GVP 1994b, Figure 19). The map was supplied by J. Lynch, NOAA.

The maximum height of the original plume from Vulcan was estimated to be 21–30 kilometres—that is, well into the stratosphere. This is much higher than the figure of about 8 kilometres for the Vulcan plume in 1937, which was estimated from the ground at an unspecified time and probably in limiting circumstances. The northern part of the space-imaged cloud in 1994 had stalled and was dissipating, leaving its more active south-western part to move across New Britain, then southwards and south-eastwards (Figure 8.5). This denser plume of ash and aerosols apparently did not remain in the upper atmosphere for long. Ice was detected in satellite images, the result of seawater thrown up by the eruptive activity evidently freezing on ash particles (Rose et al. 1995). The solid pyroclastic and ice particles may then have fallen out of the atmosphere fairly quickly, and mainly to the south-west just downwind from the volcano. Sulphur dioxide gas in the cloud was not detected from space on this occasion, possibly because it was caught up in the freezing process or, more likely, the amounts were very small to start with, at least compared with the amounts that would be detected later from Tavurvur eruptions.

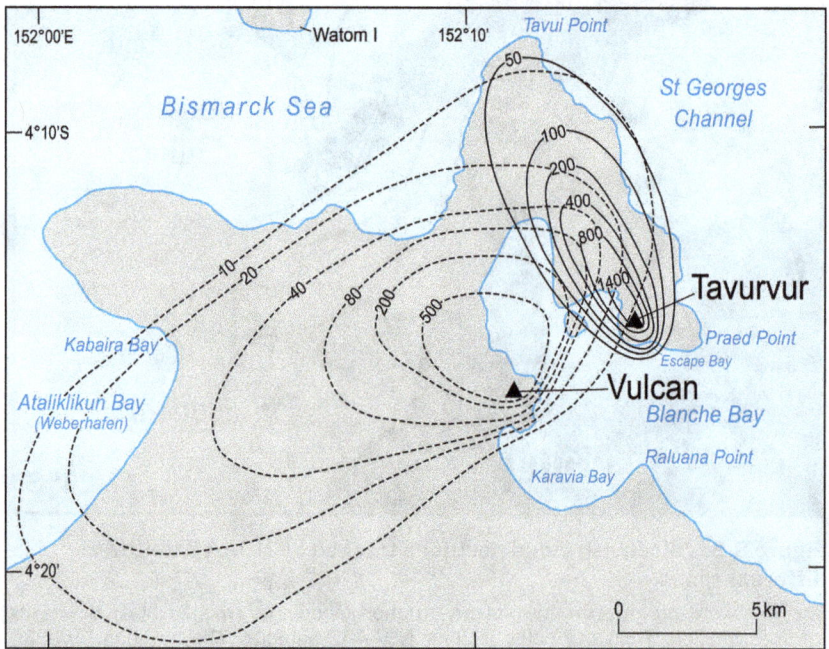

Figure 8.6. Thicknesses of 1994 ash from Vulcan and Tavurvur plotted as isopachs.

Distribution (isopachs) of air fall tephras (given in millimetres) for Vulcan and Tavurvur and deposited up to the end of September 1994 (Blong and McKee 1995, Figures 12 and 13; McKee et al. 2018, Figure 15). Thicknesses of pyroclastic-flow deposits near the volcanoes — especially those of Vulcan — are excluded. Compare these isopach patterns with those for 1937 in Figure 4.9 where pyroclastic-flow deposits must have been included in the thickness measurements.

The thicknesses of ash deposited from Vulcan and Tavurvur in 1994 were later mapped and 'isopachs' constructed for them (Figure 8.6) in much the same way as had been undertaken following the 1937 eruption (Figure 4.9). The patterns for Tavurvur in 1937 and 1994 are rather similar. Their respective isopachs trend roughly north-westwards, reflecting the low-level winds of the south-east seasons of both years, although the 1994 isopachs trend a little more northerly. The main difference, however, is in the much greater amount of ash deposited from Tavurvur in 1994—estimated to be one order of magnitude more voluminous (0.04 cubic kilometres) compared with the minimal Tavurvur deposits of 1937 (Blong and McKee 1995). Most of the main central business district of deserted Rabaul town was destroyed (Figures 8.7–8.8) and looting of the abandoned town began almost immediately until eventually brought under control by police and the military. The wet ash from Tavurvur flowed into buildings, probably enhanced by rain that fell on Rabaul that night (Blong and McKee 1995). Lakunai Airfield was made inoperable, stranded aircraft receiving heavy loads of ash, and Tokua Airport came into almost full use from thereon (Figure 8.9).

The Vulcan air fall–distribution patterns for 1937 and 1994 are different to each other (Figures 4.9 and 8.6). Those for 1994 are elongated north-east–south-west, whereas those for 1937 trend more to the north-north-west. The presence of a south-western limb for the 1994 isopachs is consistent with the satellite observations that the main part of the high-level eruption cloud trended in that same direction, but the explanation for the 1937 pattern is less clear. This apparent discrepancy may relate to the different heights reached by the two eruption columns: the 1937 pattern overall is the result of a lower column height (8 kilometres) and greater influence of the lower-level winds from the south-east. However, the north-eastern parts of both Vulcan patterns over Rabaul town are similar, apparently indicating that, in both 1937 and 1994, there were temporary, ephemeral changes in wind vectors in the early stages of column growth that permitted the ash clouds to drift over Rabaul for a relatively short time, adding to the ash load being produced by Tavurvur.

Figure 8.7. Aerial view of ash damage in northern part of Rabaul town.

The extent of damage to buildings and vegetation in the northern part of Rabaul town is seen in this aerial view in October 1994. The roofs of ash-covered houses have collapsed or buckled. Trees have been stripped of foliage, including those that once shaded the central strip of Malaguna Road, running left–right across the photograph. Graders by this time had cleared roadways and streets, leaving ash piled in ridges on either side of their tracks. Photograph by R.W.J.

Figure 8.8. Close-up aerial view of roof and tree damage in Rabaul town.

Roof collapses on houses and water tanks can be seen in this aerial photograph of damage in the southern part of Rabaul town. The leaves of coconut palm trees are folded back caused by ash loading. Photograph taken by R.W.J. in October 1994.

Figure 8.9. Lakunai Airfield damage including ash-covered aircraft.

The buildings at Lakunai Airfield were destroyed by the eruption and aircraft were left stranded. The rotor blades of helicopters were bowed down by the ash loading. Photograph courtesy of H.L. Davies.

The 1937 Vulcan cone grew in size as a result of the 1994 eruption. Much of that growth was caused by ash deposition from the pyroclastic flows that, as in 1937, were relatively limited in lateral extent. The maximum height of the cone grew only by about 20 metres, but the new craters greatly altered the cone's configuration (McKee et al. 2018). The eastern shore at Vulcan was extended into Simpson Harbour, and settlements and the Rabaul–Kokopo Road to the west were buried. These included the now deserted villages of Tavana and Valaur where there had been so many casualties in 1937, as well as the memorial at Tavana to those killed by the 1937 eruption.

People stranded by the twin eruptions of 19 September 1994 were rescued over the following few days by sea and land, including from coastal areas in the north and east outside the harbour area, such as the beach at Nodup where thousands of people had assembled and been evacuated in 1937. They joined evacuees, eventually numbering in the tens of thousands, who had already moved mainly southwards to safer places, including hastily erected camps. Nonga Hospital on the north coast was affected by ash fallout from the eruptions and about 400 patients had to be evacuated over the next two to three days to the safety of Vunapope Hospital, to other health centres or to the homes of relatives (Dent et al. 1995). Vunapope, therefore, undertook a role similar to the one it had adopted so successfully in 1937.

Ten fatalities were recorded by medical authorities, although there could have been more. Deaths were from several causes including traffic accidents, a lightning strike and entrapment in houses that had collapsed under ash loadings. This death total of about 10 is much lower than the total of more than 500 recorded in 1937, mainly in the villages near Vulcan.

How 'big' were the explosive eruptions at Rabaul in 1994 compared with those in 1937? Volcanologists measure the size or 'bigness' of such eruptions by estimating the total volume of expelled pyroclastic material (tephra) and by using the volcanic explosivity index (VEI) scheme devised by Newhall and Self (1982). The VEI is used in the Smithsonian Institution's definitive GVP database, and is a simple 0–8 index of increasing explosivity, each interval representing a tenfold increase in volume. The Rabaul eruptions of 1937 and 1994 are both indexed in the database as '4?' and represent the sum of the volumes for both Tavurvur and Vulcan combined (Siebert, Simkin and Kimberly 2010, 79). The question mark, presumably, acknowledges the difficulty of estimating the volumes at Rabaul, especially for Vulcan, where so much pyroclastic material is lost at sea or is deposited in distant places where any tephra would be poorly preserved in the tropical environment. VEI 4 is said to represent a 'large' eruption; however, it is considerably smaller than, say, the VEI 6 value for an eruption such as Krakatau in 1883. Estimates for the different volumes in cubic kilometres for both volcanoes in both years are as follows.

	Tavurvur	Vulcan	Total
1937 (Fisher 1939a)	0.003	0.3	0.30
1994 (Blong and McKee 1995)	0.04	(0.26)	0.30

Thus, the volumes for Vulcan in both 1937 and 1994 are much greater than those for Tavurvur. The two volumes for Tavurvur alone in 1937 and 1994 correspond to VEIs of much less than VEI 4 and, somewhat remarkably bearing in mind the uncertainties of measurements made at different times, the two final totals are the same. Further, the volume for Tavurvur in 1994 is much greater than it was in 1937. Therefore, the Australian administration in Rabaul in 1937 would have had a much greater recovery problem had Tavurvur produced an eruption of the same scale as it did in 1994.

8.3. Comparing the Timing of the May 1937 and September 1994 Outbreaks and Their Precursors

An initial and perhaps obvious point of emphasis concerns the quite remarkable similarities between the explosive eruptions that broke out in Blanche Bay in May 1937 and September 1994—as well as in January 1878. Vulcan and Tavurvur in these three years were the only two volcanoes in eruption, and they were in simultaneous activity—although for a fairly short time: less than a week in 1878, less than a day in 1937 and for two weeks in 1994. Both volcanoes in 1937 and 1994 produced what turned out to be immediately damaging explosive eruptions during the south-east season, meaning that Rabaul town did not escape the volcanic fallout, as might have been the case during a north-west or monsoon season—as in 1878, before Rabaul had been built.

A second similarity between the two eruptive periods of 1937–43 and 1994–2014 is that the eruptions at Tavurvur, in both cases, lasted much longer than those at Vulcan, although they later (in both cases) became interspersed with periods of inactivity lasting from weeks to years, while Vulcan remained totally inactive, as discussed further below.

A third notable point is the duration of the time gaps between the initial outbreaks of the three eruptions. The gap between January 1878 and May 1937 is 59 years and four months, and between May 1937 and September 1994 is 57 years and four months. This could suggest that the Rabaul volcanoes are active periodically—say, every 60 years or so—but this speculation would have be treated with caution as it is based on the durations of only two time gaps. The concept, therefore, cannot be used with confidence for any future disaster-planning purposes.

Perhaps the most striking similarity from an eruption-forecasting perspective is the 27-hour precursory lead-up time to eruptions in both May 1937 and September 1994. Strong harbour earthquakes were felt near the two volcanoes in both years, heralding 27 hours of near-constant and alarming ground-shaking that was considered different to the earthquakes of the 'swarms' recorded by the RVO in 1971–94.

1937 Earthquake at about 1.20 pm, Friday 28 May

1937 First eruption at about 4.15 pm, Saturday 29 May, **27 hours later**

1994 Earthquake at 2.51 am, Sunday 18 September

1994 First eruption at 6.06 am, Monday 19 September,
about 27 hours later

Earthquakes were not recorded instrumentally in the lead-up to the 1937 eruption so considerable care must be taken in suggesting that the earthquake activity in both 1937 and 1994 was similar before the start of these 27-hour periods. Further, there is no evidence that the 1878 outbreak was necessarily preceded by the same 27 hours of strong seismic activity as in 1937 and 1994. However, 27 hours is hardly much comfort to disaster managers who require longer times for early warnings and for the effective evacuation and relocation of displaced people. In addition, there is no available evidence that a seismic 'annulus' of the type recorded in the 1970s–90s existed before 1937, nor is there evidence of a clearly defined 'seismo-deformational crisis' between 1919 and 1937, like the one in 1983–85. These points are discussed further below.

An enduring theme in global volcanology, including in Papua New Guinea, is that some volcanic eruptions may be caused or 'triggered' by strong earlier earthquakes of tectonic origin that can take place either nearby or distant to the volcano being considered (e.g. Hill, Pollitz and Newhall 2002; Marzocchi 2002; Bebbington and Marzocchi 2011). This theme was dealt with in the 1950s by Tony Taylor and colleagues (e.g. Taylor 1955, 1958, 1960). Further, the two magnitude 7.9 earthquakes of July 1971 have been interpreted as 'destabilising' events for the start of the Rabaul Harbour seismic swarms four months later, leading up to the eruptions 23 years later (McKee et al. 2018). This is not the place to discuss the volcanological significance of the regional seismic-energy release hypothesis in detail—including statistical arguments and in the modern context of plate tectonics—but a few comments can be offered in the case of Rabaul.

The lead-in times of the six large-magnitude earthquakes that took place in the twentieth century in the New Britain area (Figure 6.9), and prior to the Rabaul eruptions in 1937 and 1994, are fairly similar to one another—about 20–30 years.

1937 minus 1906 = 31 years
1937 minus 1910 = 27 years
1937 minus 1916 = 21 years
1937 minus 1919 = 18 years
1994 minus 1971(2) = 23 years

Kizawa (1951) appears to have identified the 1906 earthquake as the tectonic precursor to the 1937 eruption—the one whose effects in Rabaul had been described by Pullen-Burry (1909) in German times. Further, Fisher (1939a) listed the strong earthquakes of 1910 (February), 1916 and especially 1919, just before his description of the immediate (late May) precursors to the May 1937 eruption. He seems to have been cautious, however, about the tectonic 'eruption-triggering' interpretation in general, and did not state specifically that the 1919 earthquake was the actual precursory 'trigger' for the 1937 eruption. Fisher (1941) was also critical of W.G. Woolnough, who suggested that the tectonic earthquake of January 1941 near Rabaul was a precursor to an outbreak of eruptions at Tavurvur (see Section 5.1).

Do the earthquake/eruption lead-in times of about 20–30 years have any geophysical significance? A key factor in answering this difficult but significant question may be understanding the timing of when magmas first form in the earth's mantle above the subduction zone beneath Rabaul, and how long they take to reach the surface and create a volcanic eruption. This complex process may be related in the first instance to a new period of northward subduction of the Solomon Sea plate and to dehydration or melting of the down-going plate, as referred to in Section 6 (see Figure 6.11). The newly formed 'primary' magmas in the deep mantle rise buoyantly towards the surface and begin crystallising. They may cool against and interact with solid conduit-wall rocks, and mix or mingle with any resident magmas that have preceded them. Their upward movement may be arrested at times, particularly when they reach the base of the lower-density earth's crust. Other factors are the volume and other physical properties (gas content, viscosity and so forth), and how effectively they are stored in reservoirs immediately beneath the volcano 'ready' for eruption.

Note also that the lead-in times to the eruptions in 1937 and 1994 reduce to 18–23 years where only the nearby 1916–19 and 1971 tectonic earthquakes are considered. Speculatively, therefore, values in the range 18–23 years could be indicative of the times taken for primary magmas, newly formed above the down-going subducted plate, to rise, evolve and reach the surface in eruption at Rabaul in both 1937 and 1994, as discussed further in Section 9.4.

Subduction, therefore, can be regarded as the primary kinetic driving force that produces *both* deep magma-formation and tectonic stress release. Occasions where a large nearby tectonic earthquake may trigger an eruption from an already 'charged-up' magma reservoir—by, say, breaking its roof rocks or opening old conduits—are possible but cannot always be established confidently. Furthermore, tectonic earthquakes caused by the same period of plate subduction might take place at any time during this overall magmatic process and at different depths and distances from Rabaul. In addition, there are many tectonic earthquakes that are not obviously triggers of volcanic eruptions and there are eruptions that seem to take place without tectonic triggering.

The 1937 and 1994 'twin' eruptions—and the 1878 one too—may have had similar geotectonic origins, but there were notable differences in the way that the 1937–43 and 1994–2014 eruptive periods developed, including their damaging effects on Rabaul town and the surrounding area.

8.4. Tavurvur Eruptive Activity, 1994–2014

Tavurvur ceased its eruptive activity in 1937 less than a day after the 'twin' outbreak of 30–31 May and it remained inactive for the next four years. While there were more vulcanian eruptions between June 1941 and into early 1942 when the Japanese invaded Rabaul, Takashi Kizawa witnessed only one eruption from Tavurvur during his four years there (1942–46)—the final one of the series in December 1943. This pattern has some similarities to the one at Tavurvur between 1994 and 2014—notably the intermittent nature of the eruptions, periods of non-eruption lasting up to months or years, and eruptions taking place in both the north-west (monsoon) and south-east (trade winds) seasons. Such a long period of intermittency, however, does not seem to apply in the case of the 1878 eruptions.

Strombolian eruptions took place at Tavurvur in eight separate phases between May 1996 and August 1997 (McKee et al. 2018; Figure 8.11). This type of explosive eruption tends to be less energetic than vulcanian explosions, producing less fine ash and tending to have lower VEI values, although distinguishing between the two is not always straightforward. The names 'strombolian' and 'vulcanian' both derive from those of historically active volcanic islands in the Mediterranean. The most distinctive feature of strombolian eruptions is the explosive ejection of incandescent lumps of lava on parabolic trajectories from magma that has risen into the active craters. The explosions produce spectacular sprays that are especially impressive on clear nights. Lava flows can also be produced by strombolian eruptions. Almost four months of quiescence followed the strombolian phases at Tavurvur, until a period of intensified vulcanian activity began again on 7 December 1997 lasting throughout 1998 (McKee et al. 2018). Intermittent vulcanian eruptions of different intensities continued into the new millennium, as reported in some of the monthly bulletins of the Global Volcanism Network (GVN) for 1999–2006.

Figure 8.10. Small Tavurvur eruption and general view along the Turagunan–Watom Island line.

A small vulcanian eruption from Tavurvur is seen near the left-hand edge of this aerial photograph taken on 2 or 3 January 1996 from the south-east. Greet Harbour, old Lakunai Airfield, Matupit Island and Dawapia Rocks are seen behind Tavurvur. Note especially the alignment of the four major stratovolcanoes of Turagunan (bottom left), through Kabiu and Tovanumbatir (overlooking Rabaul) to Watom Island in the distance. The photograph was taken on a commercial aerial survey and was supplied courtesy of the Gazelle Restoration Authority, East New Britain Province.

Figure 8.11. Night-time incandescent strombolian eruption at Tavurvur.
Strombolian activity at Tavurvur on 14 March 1997 is seen in this photograph taken by Steve Saunders (RVO) from Kaputin Point on Matupit Island. Note the expulsion of incandescent lava particles on parabolic trajectories. A new lava flow is moving down the southern slopes of the volcano to the right. A cloud of finer ash particles is being blown by the north-west winds of the monsoon season away from Rabaul.

This pattern for Tavurvur contrasts sharply with notably brief periods of eruption for Vulcan in 1878, 1937 and 1994. Further, there were significant differences between the 1937–43 and 1994–2014 periods, notably in two styles of explosive eruptive activity—'strombolian' and 'sub-plinian'—that had not been reported for Tavurvur for either 1878 or 1937–43. These two eruption styles were in addition to the 'normal', but intermittent, vulcanian eruptions that took place throughout 1994–2014 (Figure 8.10).

Local earthquakes continued to be recorded by the RVO after deployment of the Volcanic Disaster Assistance Program equipment in 1994 and were found to have a distribution quite unlike that of the 'seismic annulus' identified prior to the 1994 twin eruption. Instead, most local earthquakes after the twin eruption defined a broad linear zone that extended north-eastwards from the vicinity of Kabiu volcano into St Georges Channel (Itikarai 2008; Figure 8.12). Further, there appeared to be a relationship between the occurrence of some of these so-called north-east earthquakes and the timing of the vulcanian eruptions taking place at Tavurvur. The earthquakes apparently defined a major radial-fault zone running out from the caldera area, thus adding further complexity to what was known about the structure of the Rabaul and Tavui area. Some north-east earthquakes, in fact, had been recognised in May 1992, and even during the 1983–85 crisis period. However, they could not be well mapped owing to the limitations of the recording network at the time, meaning their possible significance remained unexplored. 'North-east' earthquakes, as expected, had not been recognised during or before 1937–43 owing to the total absence of seismographs in the area at that time.

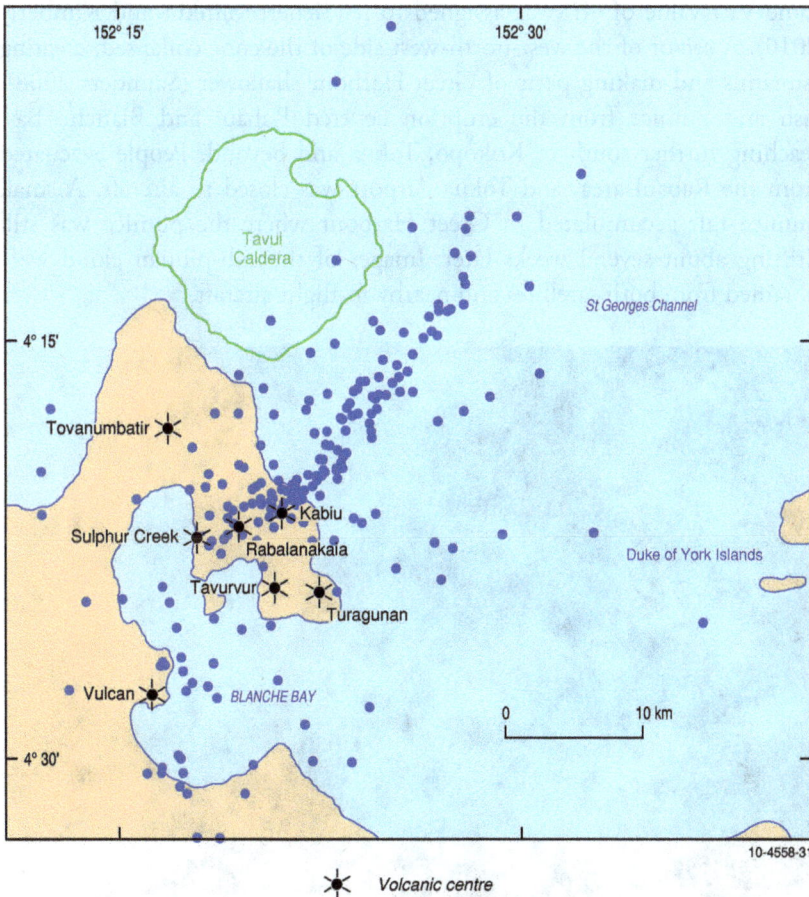

Figure 8.12. Plot of 'north-east' earthquake epicentres for 1994–98.

This map of earthquake epicentres for the period October 1994 – December 1998 is adapted from the map of 'relocated' earthquakes presented by Itikarai (2008, Figure 5.10a). The distribution is scattered except for the zone of earthquakes running north-east from the vicinity of Rabalanakaia and Kabiu volcanoes into St Georges Channel and parallel to the south-eastern margin of Tavui caldera. These form the North-East Earthquake Zone.

337

Unexpected explosions started at about 8.45 am on 7 October 2006, producing air blasts that caused doors to slam and windows to rattle in Rabaul town (GVP 2006; McCue 2007; Saunders 2008). These were the beginning of a sub-plinian eruption that continued into the early afternoon, the height of its column reaching 18 kilometres (Figure 8.13). This volcanic activity was the first time such an eruption had been recorded at Tavurvur, including in 1937–43. It had some similarities to the longer-lasting sub-plinian eruptions at Vulcan in both 1937 and 1994 but was shorter in duration and less voluminous (about 0.2 cubic kilometres) although the same VEI value of '4?' was assigned to it (Siebert, Simkin and Kimberly 2010). A sector of the west-north-west side of the cone collapsed, creating tsunamis and making parts of Greet Harbour shallower (Saunders 2008). Ash and pumice from the eruption covered Rabaul and Blanche Bay, reaching further south to Kokopo, Tokua and beyond. People evacuated from the Rabaul area, and Tokua Airport was closed to aircraft. A small pumice raft accumulated in Greet Harbour where the pumice was still drifting about several weeks later. Images of the sub-plinian cloud were obtained from both satellites and nearby in-flight aircraft.

Figure 8.13. October 2006 Tavurvur eruption as seen from the south.
This photograph was taken by Julie McLean at 9.57 am on 7 October 2006 from the south-east at Takubar, near Kokopo. The nearly vertical western side of the eruption column at Tavurvur can be seen clearly on the left, but the cloud has already risen to greater heights towards the west or south-west.

The cloud seen from space was roughly fan-shaped, as the Vulcan cloud had been in 1994, and most of it drifted south-westwards over central New Britain, as in 1994. The Tavurvur cloud in 2006, however, was seen to contain significant amounts of the volcanic gas sulphur dioxide, as deduced by results using a spectroradiometer on board NASA's Aqua satellite (GVP 2006). The nature of the eruption changed to strombolian at 2.25 pm but, by 5.30 pm, the eruption began to subside and smaller ash explosions resumed. An inspection on the following morning, 8 October, revealed that lava flows had been emplaced down the western and northern flanks of Tavurvur. Herman Patia noted that the eruption had declined by 28 October and that there were only occasional ash emissions accompanied by rare explosions (GVP 2006).

The last week of October 2006 was memorable for another reason: 97-year-old Dr N.H. 'Doc' Fisher visited RVO headquarters, almost 70 years after he began his studies of Rabaul volcano in 1937 (Saunders 2007; Figure 8.14).

Figure 8.14. Visitors to the RVO recording room in October 2006.

Dr N.H. and Mrs Molly Fisher are seen here on a visit to the RVO recording room on 25 October 2006. Herman Patia is on the left and Jonathan Kuduon, now retired, is behind him. Dr Fisher died in 2007 and Mr Patia in 2012. The photograph was supplied by Steve Saunders.

The October 2006 eruption at Tavurvur was short-lived but is of interest for other reasons. First, it appears to have been part of a 'time-cluster' of volcanoes in eruption in 2002–06, mainly in the eastern part of the Bismarck Volcanic Arc, as summarised elsewhere (Johnson 2013, Table 8). The cluster included volcanoes that had not been in eruption in historical time (Garbuna), or had unexpected gas emissions but no eruption (Likuruanga) or showed signs of magma having been intruded beneath the volcano but not leading to actual eruption (Sulu Range). In October 2006, Ritter Island had its first-reported eruptions since 1997, and both Ulawun and Langila had had mild ash eruptions during the 2002–06 period, the ones at Ulawun following an especially large eruption at the volcano on 28 September 2000.

Further, a remarkable sequence of tectonic earthquakes was recorded in the New Britain and New Ireland area in late 2000 (Anton and McKee 2005). A magnitude 6.8 earthquake along the New Britain submarine trench on 29 October 2000 started the sequence, followed on 16 November by a magnitude 8.2 earthquake along the Weitin Fault in southern New Ireland, and on 16 and 17 November by two more large earthquakes (magnitudes 7.8 and 8.2, respectively) along the New Britain trench (Park and Mori 2007). The earthquake on 16 November had an epicentre remarkably close to that of the major earthquake of 26 July 1971. The earthquake pattern in October–November 2000 seemed like a seismic chain of reaction of related releases of tectonic stress, and potentially may have been a 'trigger' for the 2002–06 volcano time-cluster. Note, however, that the large eruption at Ulawun in September 2000 preceded the earthquake sequence, leading to some questions. First, was the Ulawun eruption of late September 2000 an indication of crustal stress release and perhaps a precursor of the tectonic-earthquake sequence that would follow in October–November? Also, but even more uncertainly, was the large eruption at Ulawun in January 1970 a similar precursory signal of the two major earthquakes of July 1971? Both questions remain unanswered.

The 2006 eruption, and Rabaul volcanology in general, attracted research scientists from the Earth Observatory of Singapore to Rabaul in the new millennium. They and their co-workers published internationally on their petrological, geochemical and petrophysical findings, thus adding to the growing mountain of data and concepts that had arisen since 1994 (Bouvet de Maisonneuve et al. 2015; Bernard and Bouvet de Maisonneuve 2020; Fabbro et al. 2020; Bernard et al. 2022). Among their findings was confirmation that different magmas of basaltic and dacitic composition had mixed or mingled at least over the past few centuries, including in the pre-1937–43 period.

Tavurvur continued to produce intermittent vulcanian explosive eruptions after the October 2006 eruptions (Figure 8.15), as reported in the *Bulletin of the Global Volcanism Network*, but there were only four GVN reports between December 2009 and July 2014. Further, the southern end of Matupit Island had been subsiding since the 2006 eruption, as indicated by both local global positioning system (GPS) measurements undertaken by the RVO and by time series images from satellite-radar surveys (Garthwaite et al. 2015). Two phenomena were indicative of new changes taking place. The first was a resurgence of uplift of Matupit starting in about January 2010. The second was a series of tectonic earthquakes between 14 July and 1 August 2014 in a well-defined zone extending west-north-westwards from Cape Lambert on the north-western tip of the Gazelle Peninsula and in line with the north-western part of the Baining Fault west of Rabaul (Taranu and Herry 2015).

Figure 8.15. Distant incandescent explosive eruption at Tavurvur at night.

The lights of Rabaul town, shipping and wharves in the foreground, and of the Kokopo–Tokua area in the right background, are seen in this photograph taken on 16 April 2008 from Observatory Ridge (compare this view with that in Figure 6.7). Incandescent explosive activity is taking place at Tavurvur. Small, glowing pyroclastic flows appear to be moving down the upper northern flank of the volcano, and there is apparently some lava fountaining within the crater itself. The activity appears to be somewhere on the borderline between vulcanian and strombolian. Ash is drifting south-eastwards towards Tokua, caused by monsoonal winds. The photograph is reproduced here courtesy of Simon V. Hohl, who was working at the RVO at the time.

A large eruption from Tavurvur broke out on 29 August 2014 (GVP 2014, 2017). It included more strombolian activity, before another towering eruption cloud formed that is thought to have risen 18 kilometres and was imaged from space. The Darwin Volcanic Ash Advisory Centre issued warnings to aviation and a sulphur dioxide plume was again mapped by the spectroradiometer on board NASA's Aqua satellite. Analysed rocks from the eruption again retained evidence for magma mixing (Fabbro et al. 2020). The Cape Lambert earthquakes preceded the Tavurvur eruption, but the geodetic information for ground inflation at Matupit was an indication of magmatic unrest before both the earthquakes and the actual eruption of 29 August, which was maintained strongly only until the next day.

Tavurvur became inactive after August 2014, a condition that it has maintained up to the time of writing.

The amount of new scientific and disaster-management information that has emerged since 1994 is nothing less than prodigious, if not overwhelming, and we have had to be selective—and hopefully not too prejudicial—in the bibliographic sources we have chosen to reference so far and in the final Chapter 9.

9

Aftermath of the 1994 Twin Eruptions

9.1. Relief Measures, 1994–95

The disaster-management responses to the 1994 eruptions were very different to those in 1937 when Rabaul was still the capital of the Territory of New Guinea and the administration was accountable to the Australian Government in Canberra. The official responses in 1994 were considered initially by the national government in Port Moresby, the capital of independent Papua New Guinea (PNG), and a state of emergency in East New Britain Province was soon declared. Prime Minister Sir Julius Chan, on Monday 19 September, flew from Port Moresby to East New Britain and saw the eruptions and devastation for himself, as portrayed famously in the artwork of John Siune (Figure 9.1). Chan had a family home in Malaytown that was destroyed by the eruption.

Provincial and national government support for relief and then recovery assistance was soon provided, aided by donations from other parts of PNG. Australia was the first to respond from outside the country, when the Royal Australian Air Force began flying C120 Hercules aircraft into Tokua Airport with the first of several deliveries of relief supplies. Other major participants were the international development-assistance agencies of Australia, Japan, the European Union and the US; international development banks, such as the World Bank; agencies of the United Nations, including the World Health Organization and the Department of Humanitarian Affairs; and non-governmental organisations such as Red Cross and World Vision. The United Nations Development Program already had an office in Port

Figure 9.1. John Siune painting of Prime Minister Chan's helicopter visit over Rabaul.

'Dispela helekopta kisim Praim Minista bilong PNG igo lukim volkenu pairap long Rabaul': John Siune, 1996, acrylic on paper mounted on board. R.W.J. Collection. Intellectual property rights are held by the artist.

Moresby. This international participation is illustrative of a much more globalised and interconnected world than existed in 1937 and, thus, one more able to rally to a disaster-relief effort. The UN was evidently more effective in this regard than the League of Nations had been before WWII.

Disaster-relief activities in the province itself focused on a program called 'Operation Unity' that was managed from headquarters set up at Ralum Golf Club, Kokopo, where the controller was Brigadier-General Rochus I. Lokinap. The controller provided the national government in Port Moresby with a series of progress reports over the next several months that, together with local newspaper and other reports, provide a valuable source of disaster-management information about the relief phase. There were many disaster-relief issues to be addressed before the state of emergency could be called off and a 'Gazelle Restoration Authority' established to undertake and manage eventual resettlement in the province. Hugh Davies, professor of geology at the University of Papua New Guinea in Port Moresby, returned to the area to engage in liaison work between Operation Unity, the Rabaul Volcanological Observatory (RVO) and overseas visitors (Davies 1995a, 1995b, 1995c). Kokopo-based exploration geologist David Lindley spent four weeks assisting the Provincial Disaster Committee (PDC), briefing residents in care centres on aspects of the eruptions and hosting refugees in his own home. Lindley provided an insightful summary of his observations and experiences during the relief phase (Lindley 1995). In October, the PDC released an information newsletter for the community entitled *Tephra Tok: The Voice of the Rabaul Volcanoes*.

The priority for Operation Unity was ensuring that the tens of thousands of displaced people in the province were well cared for until resettlement could begin. Their health and safety were paramount. Law and order also had to be maintained, particularly as looters and armed gangs were roaming the Rabaul area. Non-Tolai Papua New Guineans, perhaps as many as 15,000 of them, were provided free passage back to their home provinces. A damage assessment was soon made of Rabaul town by provincial government staff, who prepared a coloured map (Lokinap 1995; Figure 9.2). The bright-green areas on the map refer to total damage (e.g. north of Lakunai Airfield, Malaytown, and most of the central business district of eastern Rabaul), whereas the maroon and light-blue colours signal areas that were relatively unharmed (e.g. Malaguna Road).

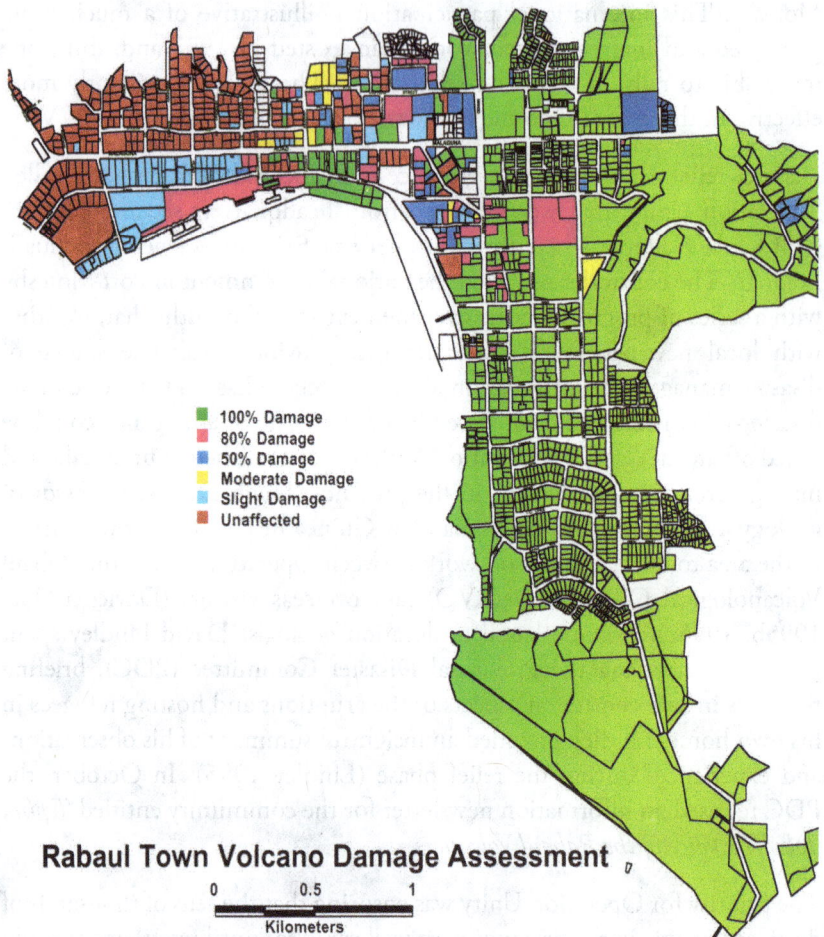

100% Damage
80% Damage
50% Damage
Moderate Damage
Slight Damage
Unaffected

Rabaul Town Volcano Damage Assessment

0   0.5   1
Kilometers

Figure 9.2. Map of damage assessment results for Rabaul town.

Officers of the East New Britain Provincial Administration produced this map of relative ash damage in Rabaul town. This is the front-cover illustration of a report by Lokinap (1995).

The Australian International Development Assistance Bureau (AIDAB, later AusAID) supported a 'Rabaul Volcanic Disaster Needs Assessment Mission' that was undertaken by external consultants (AIDAB 1994). Seventeen recommendations were listed, the first being that an urgent program of support be provided to upgrade the capacity of the RVO. Support was expressed also for the prevailing view that Kokopo should become the island's regional centre and the provincial capital for East New Britain, and what was left of Rabaul town in the north-west, including the wharves, should be partially redeveloped and remain as support for the region. The Insurance Underwriters Association of PNG funded a damage assessment from an insurance point of view, leading to a comprehensive and well-illustrated account of the 1994 eruption (Blong and McKee 1995).

These early damage assessments were undertaken towards the end of the dry season and so could not consider the impacts in the following wet seasons of flooding and mudflows in areas stripped of vegetation by the eruption as well as erosion and the redistribution of new ash and pumice, much as had happened in 1937. Hazard maps for mudflows and flooding were distributed in the community, including for Rabaul town, that showed places where mudflows and floods from the steep caldera walls to the north and east might impact on the remaining part of Rabaul town (Davies 1995a, Figures 17–18; Brown and Tutton c. 1994–95). Flooding and mudflow damage would lead to extra costs, and Department of Works staff were kept busy clearing debris from streets in Rabaul town, and from affected roads in general. Neumann (1996) concluded that the overall cost of the 1994–95 disaster was probably in excess of K300 million; this included the cost of caring for thousands of refugees, restoring electricity and telecommunications, short-term assistance to the education sector and looting, but did not include government-building losses (AIDAB 1994) and private losses (Blong and McKee 1995).

9.2. Recovery and Resettlement, 1995–2013

Widespread sadness was felt for old Rabaul as a result of the destructive eruption of 19 September 1994, as well as feelings of nostalgia—as articulated in the titles of both a song and a book, *Rabaul Yu Swit Moa Yet* or 'Rabaul you are so sweet' (Neumann 1996, 1997). Papua New Guinean author Grace Maribu wrote in 1994:

> The Premier and I were discussing Rabaul [at the Rabaul Disaster Control Centre at Ralum], the once beautiful, and peaceful town which was the pride of the Tolais in particular, and Papua New Guineans in general. The destruction of Rabaul and the surrounding villages was enormous, but the Premier was telling me how he was not going to abandon Rabaul, how his government was going to do everything in its power so that Rabaul would be back better than it was ever before …

> Rabaul was a special place, not just because of its picturesque setting
> and colourful history—it was a living monument of a people's pride.
> It was home to the East New Britain men and women—the Tolai,
> the Baining, the Sulka and the Tomoip, the Taulil and Mengen, the
> Kol, Makolkol, Nakanai and the Mumusi. (cited in Maribu 2021)

Self-organised resettlement took place over the next few years, even in
the knowledge that the places to which people were returning were at
risk of future volcanic eruptions. Major restoration and relocation were
the two main goals of the Gazelle Restoration Program created by the
PNG national government. The program was supported by the major
international funders AusAID and the World Bank, together with the
Japanese International Cooperation Agency (JICA), which concentrated on
construction works at Tokua Airport. The Gazelle Restoration Authority
(GRA) was established in early 1995. Its overall objectives were relocation of
urban services, infrastructure and population to the Kokopo area; relocation
of rural populations away from the high-risk eruption zone; and restoration
of economic and social linkages, including roads (Scales 2010). However,
given the importance for regional development of both the wharf area on
Simpson Harbour and the trunk-road link from the wharves to Kokopo,
what was left of Rabaul would not be totally abandoned.

A Development Rezoning Plan for Rabaul was released in 1997.
It described 'a town more limited in extent than previously ... [starting in
the western part] and extending sufficiently eastwards to include urgently
needed community facilities, such as market and sportsground' (Steering
Committee 1997, 3 and 2.10). People and businesses would return to the
now smaller town area. The provincial government headquarters, which was
previously located in Rabaul, would be re-established in the new capital,
Kokopo, but RVO headquarters would remain on Observatory Ridge. The
Rabaul–Kokopo Road was repaired and sealed, except for a dusty segment
immediately west of the new Vulcan cone. Other parts of the road would
remain susceptible in the years ahead to flooding and landslips from the
devastated inland areas.

Resettlement of displaced communities to safer places was encouraged by
the provincial government through the GRA. New land for resettlement
was made available towards the Warangoi River well south of Blanche Bay,
and services provided, including roads that connected the new settlements
to the slowly developing Kokopo area (Scales 2010). A new suburb,
Kenabot, was built at Kokopo, as well as a new satellite town at Baliora.

The resettlements at Sikut, Clifton, Warena and Gelagela did not evolve as planned, many displaced villagers facing practical difficulties, as described elsewhere (Neumann 1996). Many villagers were drawn back to the coastal areas of their traditional lands where their homes, gardens, cemeteries and fishing practices had long been established. Two Australian environmental researchers supplemented a growing body of information about the resettlement by conducting comparative botanical field surveys in 1997 and 1999, interviewing affected people and reporting on the 1994 eruptions and their impact (Lentfer and Boyd 2001).

Assessing future risk and achieving disaster-risk reduction were a key part of the challenge of resettlement in the north-eastern Gazelle Peninsula. This was addressed in two main studies funded by AusAID. Both studies examined risk throughout East New Britain Province for a range of geological hazards, not just volcanic threats. Their aim was to solve the simple but challenging 'risk equation' (exposure x vulnerability x hazard). These studies were driven in part by the 1990s being declared the International Decade of Natural Disaster Reduction, and in part by a devastating tsunami that killed at least 2,200 people at Aitape on the north coast of mainland New Guinea on 17 July 1998 (Davies 2017).

The first of the two studies was a community-based hazard and risk assessment conducted in the province in April–December 2001 in close association with the staff of the GRA, the provincial administration and communities. The leader and coordinator was Dr Isolde Macatol, a contracted disaster-risk advisor from AusAID's PNG Advisory Support Facility Project (Macatol 2002). Four PNG staff from the RVO were involved, each writing chapters for one of the final reports on geological hazards: Jonathan Kuduon considered volcanic hazards; Ima Itikarai, now head of the RVO, dealt with earthquakes; Felix Taranu discussed landslides; and tsunamis were considered by Kila Mulina. Luke Sendo from Kerevat National High School presented a chapter entitled 'Vulnerability Due to Geology'. A highlight of the compilation was a collection of 119 maps of geological-hazard impact throughout the province, together with population data for specific areas likely to be affected. Two other reports by Macatol and delivered to the GRA, the provincial administration and AusAID covered (1) an assessment of vulnerability and capability, and (2) community-risk perceptions.

The second study was even more ambitious. It aimed to be more quantitative than the Macatol study had been, concentrating on the collection of available geospatial digital data for the whole province. The data were then analysed mainly in a geographical information system (GIS) environment, much as had been pioneered by Ken Granger (1988, 2000) in Rabaul in the context of 'Risk-GIS', as referred to above. The work was part of a larger AusAID-funded activity called 'Strengthening Natural Hazard Risk Assessment Capacity in Papua New Guinea'. In 2010–13, in East New Britain, it involved the integration of natural-hazard and exposure information but only for the three hazards of earthquakes, tsunamis and volcanic ash. The volcanic-hazard methodology included a 'regional probabilistic assessment' based on computational modelling using a sophisticated volcanic ash dispersal model.

The final, highly technical report had 15 authors, 10 of them from Geoscience Australia in Canberra (Bear-Crozier et al. 2013). The authors were clear in identifying the limitations of their report, emphasising that it contributed only partly to the risk-analysis component of a full risk assessment. Spatial data needed to be accurate and constantly updated, something that might be achievable in some developed countries but not necessarily in provinces like East New Britain. Two of the authors, both spatial-data specialists, were clear on this point in one of the chapters:

> A plan for the ongoing maintenance and improvement of exposure information for East New Britain Province has not been developed, but future risk analysis activities would benefit from improvements to the input datasets that have been used to develop the exposure information, particularly for the themes identified [in this chapter]. (Jakab and Pirit 2013, 71)

Future planning in East New Britain can find value in continuing the approaches identified in both of these reports, even though many of the concepts were driven by outside expertise. Constantly updating digital data sources can be challenging and costly, and good GIS technical skills must remain and be valued in the province if the concepts are to survive. The concepts are an obvious reminder, too, that GIS technology—like earth-orbiting satellites—were not available before 1937–43.

9.3. Scientific Responses and New Insights, 1994–2022

There were also strong responses in 1994 from overseas geoscientists, most of them volcanologists who had been invited by the RVO to participate in assisting the ongoing post-disaster work of the observatory. Their number was well in excess of the three geologists who investigated the 1937 disaster—C.E. Stehn, W.G. Woolnough and N.H. Fisher. The most important immediate response came from the Volcanic Disaster Assistance Program (VDAP) of the United States Geological Survey (USGS), which arrived in the second week of the eruption, funded by the US Agency for International Development (USAID). VDAP was based at the Cascades Volcano Observatory in Vancouver, Washington State, near Mount St Helens, which had been in eruption in 1980. The volcanologists there had developed a suite of portable monitoring instruments that could be deployed quickly at volcanoes that seemed to be reawakening (Ewert et al. 1997). VDAP was created in response to the disastrous eruption at Nevado del Ruiz volcano, Colombia, in 1985 when more than 23,000 people were killed by fast-moving mudflows. Three VDAP volcanologists arrived at Rabaul at the end of September 1994 with about 40 crates of new monitoring equipment for the RVO—seismometers, tiltmeters and computers—which had lost monitoring capacity through eruption damage and vandalism. The earlier visit of USGS volcanologist Norm Banks in 1983, bringing in electronic distance-measuring equipment, had also been funded by USAID. All of this USGS technical capability was greatly in excess of that portrayed comically in Figure 3.65.

Two academics from the University of Arizona arrived at the same time. They brought a 'correlation spectrometer' to measure, remotely from the air, the amount of volcanic sulphur dioxide gas being emitted from Tavurvur, and they collected samples of the recently erupted ash and pumice for geochemical study, leading to early publications on the 1994 eruption in the international literature (Williams 1995; Roggensack et al. 1996).

Canberra-based staff from Geoscience Australia (GA), which in 1992 had changed its name from the Australian Geological Survey Organisation (AGSO, formerly BMR), arrived at Rabaul in October 1994. Their role was to provide initial support to the RVO in conjunction with VDAP staff and to organise additional technical-assistance support for the observatory. Rock samples were collected and brought back to Canberra, where petrologists and geochemists at The Australian National University provided mineral and whole-rock analyses. These early results were consistent with the view that the rocks were actually 'mixtures' of two different magma types that had mingled with one another: basalt and dacite (Johnson et al. 1995).

Discussions at the RVO in October 1994 led to a comprehensive AusAID initiative, the 'Papua New Guinea Volcanological Service Support (VSS) Project', which carried a total budget of A$6.8 million and ran from 1995 until 2000 as a joint project between RVO and GA staff. The VSS Project was multifaceted and its overall aim was to strengthen the technical and scientific capabilities of the national volcanological service based at the RVO. Ben Talai and Chris McKee exchanged roles at the end of 1994, meaning that Talai became the first Papua New Guinean to lead the RVO and thus to represent the RVO during the first years of the VSS Project. This VSS work also formed the basis for a subsequent, lower-budget 'RVO Twinning Program' between the RVO and GA that ran until 2015 (Nancarrow and Johnson 2015). AusAID during this time provided financial support for postgraduate studies by RVO volcanologists Herman Patia and Ima Itikarai at The Australian National University in Canberra, leading to the award of masters' degrees (Patia 2004; Itikarai 2008). Herman Patia's research was petrological and geochemical and included the study of 1937–43 samples among others.

There have been numerous laboratory studies of the volcanic rocks produced by the historical eruptions at Tavurvur and Vulcan. A dominant interpretation in these studies is that high-temperature basaltic magma has mixed or mingled with a cooler dacitic magma believed to occupy a Harbour low-velocity anomaly (LVA; see below). Petrologists who have analysed the constituent minerals of the rock samples, using electron-microprobes, repeatedly draw attention to the fact that the common rock-forming minerals of feldspar and pyroxene have 'bimodal' compositions—that is, one mineral mode corresponding to the basalt, the other to the dacite (Johnson et al. 1995; Patia 2004; Patia et al. 2017). The basalt component is more prevalent in the samples from Tavurvur than in those from Vulcan, particularly in the well-studied rocks produced during the 1994–2014 eruptive period. Herman Patia studied samples from the 1878 and 1937 eruptions of both volcanoes and confirmed the mixed/mingled character of some of them. There were also 'micropillows' of basaltic material with crenulate margins that had chilled against the cooler dacite (Figure 9.3).

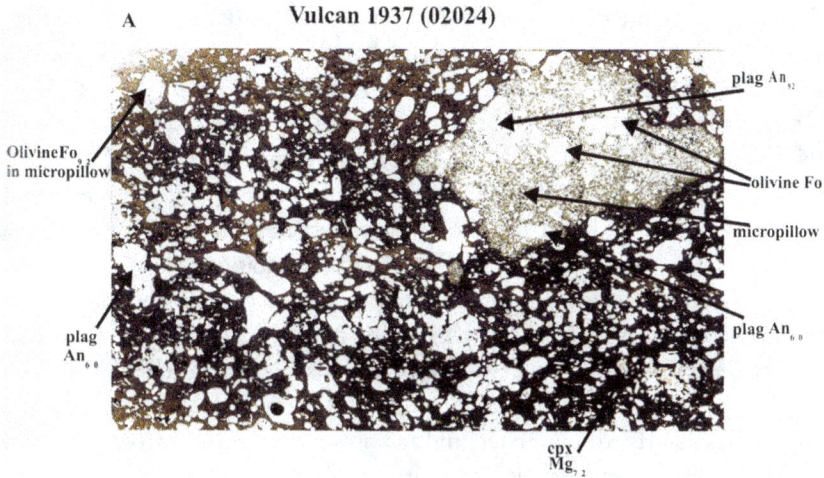

Figure 9.3. Vulcan 1937 sample as photographed through a microscope.

This photomicrograph is an image as seen through a petrographic microscope of a thin slice of a glassy sample taken from the 1937 deposits of Vulcan and described by Herman Patia in his master's thesis (Patia 2004, Figure 4.3a). The width of the image is about 20 millimetres, and the minerals are olivine, plagioclase feldspar (plag) and clinopyroxene (cpx). Note especially the two areas labelled 'micropillows'. These are of basaltic material containing olivine, a typical basaltic mineral.

New Zealand Government volcanologists Ian Nairn and B.J. 'Brad' Scott visited Rabaul in May 1995 as part of a fact-finding mission (Finnimore et al. 1995; Nairn and Scott 1995). Nairn knew Rabaul well from earlier visits when, with RVO staff, he produced a modern geological map and report on the area (Nairn et al. 1989). Geological mapping of the Rabaul area remained an important element of the RVO's work, much of it undertaken by Chris McKee, Herman Patia and Jonathan Kuduon. This work eventually led to major reviews being printed in locally produced reports (McKee et al. 2018, 2020; McKee 2021), the results of which place the 1937–43 eruptions in a strong geological context, as discussed further below. John Tomlin was another well-known visiting volcanologist who came to Rabaul in early 1995 accompanied by disaster-management specialist Joe Chung, both from the UN Department of Humanitarian Affairs (Tomblin and Chung 1995). Their visit was an investigative one aimed at identifying 'lessons learnt' as a basis for future disaster-mitigation planning internationally.

A major component of the VSS Project was a multinational geophysical survey of the Rabaul area, the 'Rabaul Earthquake Location and Caldera Structure' (RELACS) program (Gudmundsson et al. 1999). It was led by Doug Finlayson (AGSO) and Oli Gudmundsson (ANU) and was funded jointly by AusAID and JICA. RELACS was a modern rerun of a survey undertaken in the late 1960s in the Rabaul area that had seen artificial explosions detonated and the shock waves recorded at numerous stations (Brooks 1971; Finlayson 1972). The recording stations for RELACS in 1997–98 included not only instruments on land but also ocean-bottom seismometers provided by the University of Hokkaido, Japan. RELACS results form the basis for new interpretations of the nature of the deep interior of the Rabaul volcanic complex. Magma bodies were detected and mapped at depth by the technique known as 'seismic tomography' and results were reported in the international scientific literature (Finlayson et al. 2003; Gudmundsson et al. 2004; Bai and Greenhalgh 2005).

A major discovery of the RELACS geophysical survey was that parts of the crust beneath the Rabaul region were found to transmit earthquake waves so slowly that the transmitting material was inferred to be molten or at least partly so. Two such 'low-velocity anomalies or LVAs' were mapped by this 'seismic tomography' method, as seen in the 5-kilometre-deep horizontal 'slice' shown in Figure 9.4. The elongated 'Harbour LVA' in the south-west coincides with the northern seismic ellipse shown in Figure 7.9, whereas 'Rabuana LVA' is more circular and underlies both the shoreline between the Watom–Turagunan Zone and St Georges Channel, and the trend of the North-East Earthquake Zone. A third LVA was also detected below the Harbour LVA at depths of about 12–18 kilometres. It has been called the mid-crustal anomaly, but how it links with the two shallower anomalies is very unclear on current lines of evidence.

Satellite-borne radar imaging studies of the Rabaul area were also introduced to the RVO in the new millennium by GA through the RVO Twinning Program (Hutchinson and Dawson 2009; Romeyn and Garthwaite 2012; Garthwaite et al. 2015). These involved the precise measurement to the nearest centimetre of the distance between the satellite and the ground, comparing radar images of volcanic areas, including Rabaul, obtained from successive overpasses of the satellite and then constructing maps of changes to the elevation of the land surface. A multinational group of researchers combined satellite-radar results with RELACS data using geophysical finite-element modelling (Ronchin et al. 2017). Radar images can also be used in association with the results of ground-based stations of the GPS for accurate determination of locations and the measurements of relative plate motions (Tregoning et al. 2000). GPS data had been adopted widely in Papua New Guinea by this time by surveyors and geodesists in general, including at Rabaul, where four GPS stations had been installed in 1998 providing real-time geodetic monitoring of the harbour area by the RVO (Stanaway 2004). The sophisticated level of all these investigations was a measure of the strong local and international interest that had been generated in the nature of Rabaul volcano after 1994—something that certainly did not exist in Rabaul in 1937–43.

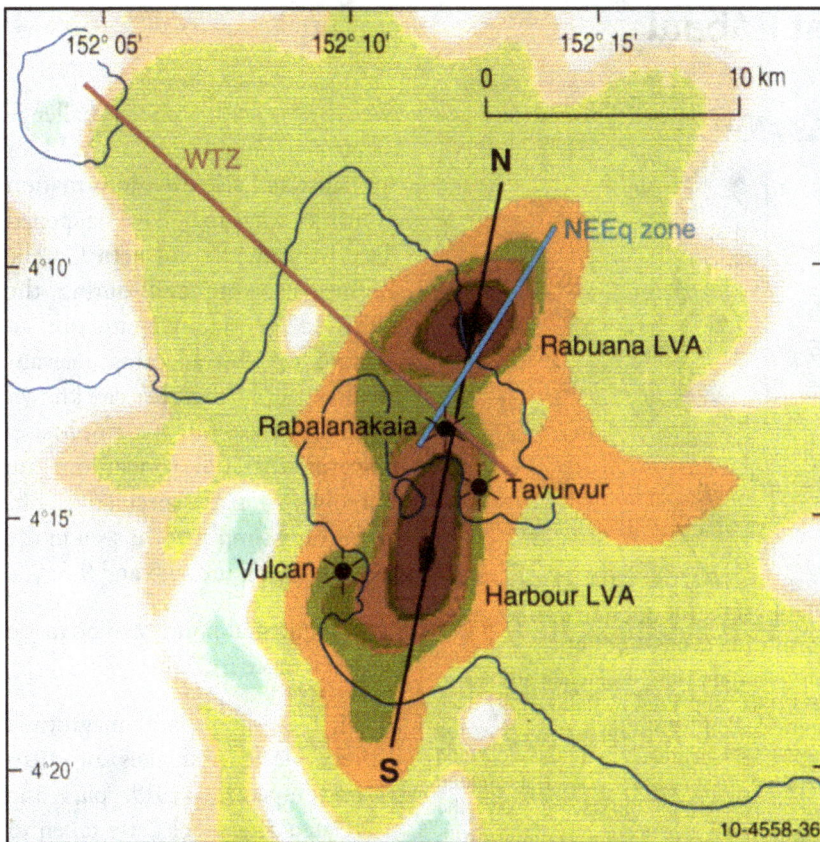

Figure 9.4. Seismic tomography of the north-eastern Gazelle Peninsula area.

Different seismic velocities (kilometres per second) are plotted here for a 5-kilometre-deep horizontal slice showing the Harbour LVA and the Rabuana LVA (adapted from Itikarai 2008, Figure 6.3a). The lowest velocity is shown in dark brown, grading up to olive green. WTZ signifies the Watom–Turagunan Zone, NEEq the general trend of the North-East Earthquake Zone shown in Figure 8.12, and N-S is the approximate line of the cross-section shown diagrammatically in Figure 9.6.

Ben Talai left the RVO in 1999 at the age of 51 and returned to his home in the Duke of York Islands. Ima Itikarai from Central Province, a physics graduate from the University of Papua New Guinea, took over as head of the RVO, a position he holds to the present day. Chris McKee transferred to Port Moresby, taking charge of the geophysical observatory there but maintaining a strong interest in Rabaul volcanology before his retirement and return to Australia.

9.4. Model for the 1937–43 Eruptions at Rabaul

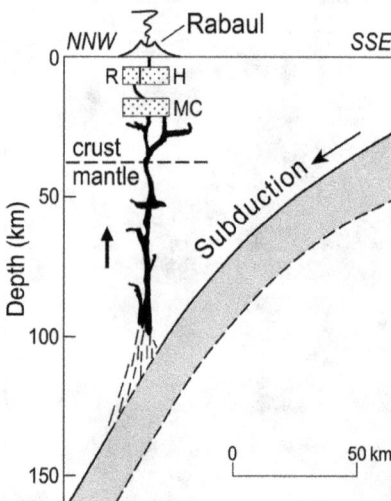

Figure 9.5. Subduction and magma-formation cross-section.

Magmas that were eventually erupted in both 1937 and 1994 at Rabaul may have had a similarly deep origin, as represented highly schematically in this cartoon adapted from Figure 7.2. The shaded subduction zone represents only the seismically active upper part of the down-going Solomon Sea plate as mapped by Everingham (1975) for the July 1971 earthquakes. R, H and MC refer to the Rabuana, Harbour and Mid-Crustal LVAs, respectively.

We are now faced with the challenge of summarising, from the wealth of specialist and scientific information available, what may have happened deep beneath the northern Gazelle Peninsula before and during the 1937–43 period. We attempt to achieve this by adapting relevant results from the better known 1994–2014 period. We emphasise, however, that our conclusions are not only, in part, speculative, but also greatly simplified, as seen in the cartoons in Figures 9.5 and 9.6.

Four sequential points can be made with reference to Figure 9.5:

1. First, the large-magnitude tectonic earthquakes of 1916 and especially 1919, plus any related aftershocks, are taken to correspond to a downward surge in subduction of the Solomon Sea plate. These earthquakes are equivalent to the two magnitude 7.9 earthquakes of July 1971 that preceded the 1994 eruption.

2. Fluids rise from the surface of the dehydrating and possibly melting subducting slab at depths of about 100–150 kilometres, into the overlying mantle causing partial melting and the creation of new 'primary' basaltic magma. The new magma is buoyant. It rises, recharging and rejuvenating the complex conduit and channel system used by earlier magmas that had been created by earlier subduction. The new, rising hot magma interacts with wall rocks and mixes or mingles with any stalled magma still residing in the conduit system.

3. Magma reaches the top of the earth's upper mantle, intruding on the 30–35-kilometre-thick crust beneath the Rabaul area (Finlayson et al. 1972). The structure of this crust is poorly known but the three LVAs referred to in the previous section are shown schematically in Figure 9.5 by the stippled boxes in the upper and middle crust. There may be more than three such LVA systems beneath Rabaul and how they are all connected remains unknown.

4. Much research has been accomplished globally on the nature of possible magma bodies directly beneath active volcanoes, and the notion of a shallow single crustal 'chamber' of magma—that is, completely molten material—has been sidelined to a large extent. A consensus now seems to have been reached that several systems may exist at different depths and that, as a whole, they are vertically extensive, unstable and some possibly connected (Cashman, Sparks and Blundel 2017). Further, the materials contained in such systems include magma, and also crystals of different compositions from different sources, some in such abundance that the name 'mush' is used for cases where the crystals form a continuous framework through which melt is distributed. Magma and crystals are stored during and after a long, and probably complex, process of magma ascent, magma and crystal mixing, and incorporation of wall rocks.

The Watom–Turagunan Zone is regarded as the primary, controlling geological feature of the three volcanic systems at Rabaul shown in Figure 7.12, at least during recent times encompassing the 1878, 1937 and 1994 eruptions. Basaltic magma derived from a long course of ascent and petrological evolution from the earth's upper mantle rises up into the Rabuana LVA but then moves obliquely into the Harbour LVA where it mixes with dacite magma. This magma mixing or mingling takes place beneath the general area of the Greet Geothermal Field, which includes the historically active volcanoes of Tavurvur, Rabalanakaia and Sulphur Creek. The sideways entry of the basaltic magma from the Rabuana LVA is regarded as the process by which the Harbour LVA has been replenished in modern times, not that other entry points can be excluded (see e.g. the question mark in Figure 9.6). The mixed magmas eventually also reach that part of the reservoir beneath Vulcan, although by this time they have lost much, but not all, of their basaltic component (Figure 9.3).

The Harbour and Rabuana LVAs are the two shallowest of the three LVAs detected beneath Rabaul. They are believed to be connected laterally, as shown diagrammatically in the cross-section in Figure 9.6. The section corresponds roughly to the N-S line in Figure 9.4, which runs through the two LVAs, both of which are seen to be roughly lenticular, based on the RELACS results. More exact shapes cannot be mapped for the two inferred magma- and mush-containing reservoirs, as the outer, cooler zones of both may contain more solid material than the hotter, more liquid central parts. One suggestion is that the Harbour LVA could correspond to a flat magma reservoir only 500 metres in thickness, or even less (Patia et al. 2017). Similarly, the width of the Harbour LVA shown in Figure 9.4 could correspond in reality to a much narrower reservoir of hotter magma, in which case the reservoir would be shaped more like a horizontal tube or cylinder.

Magmas in the Harbour LVA, comprising different mixed proportions of crystals, melt and mushes, rise up the ring fault or seismic annulus (Figure 9.6). This is a phenomenon that could not be recorded in either 1937 or 1994 but which could be modelled computationally, taking into consideration the geodetic and earthquake data that had been collected in the Rabaul area in the 23 years or so prior to the 1994 eruption (Saunders 2001; Robertson and Kilburn 2016; Figure 7.11). The notion that magma can be emplaced into a ring-like structure is well known from the geology of old volcanoes such as Glen Coe and Mull in Scotland, where 'ring dyke' intrusions had been mapped more than a century ago. The concept of dyke-formation had been discussed at the RVO, for example, when USGS volcanologist Norm Banks visited Rabaul in 1983. The fact that magma was erupted from Tavurvur and Vulcan in 1994, both of them close to the seismic annulus, is strong evidence that magma at some stage had indeed risen in conduits from the same 4–5-kilometre-deep magma reservoir identified beneath the northern ellipse and the Harbour LVA (Figure 7.9). More uncertain, however, was how much of other parts of the annulus were emplaced by magma; a complete ring dyke may not have formed. Also, when did the intrusions beneath Vulcan and Tavurvur actually start, and why did the other young volcanoes not form around the northern ellipse?

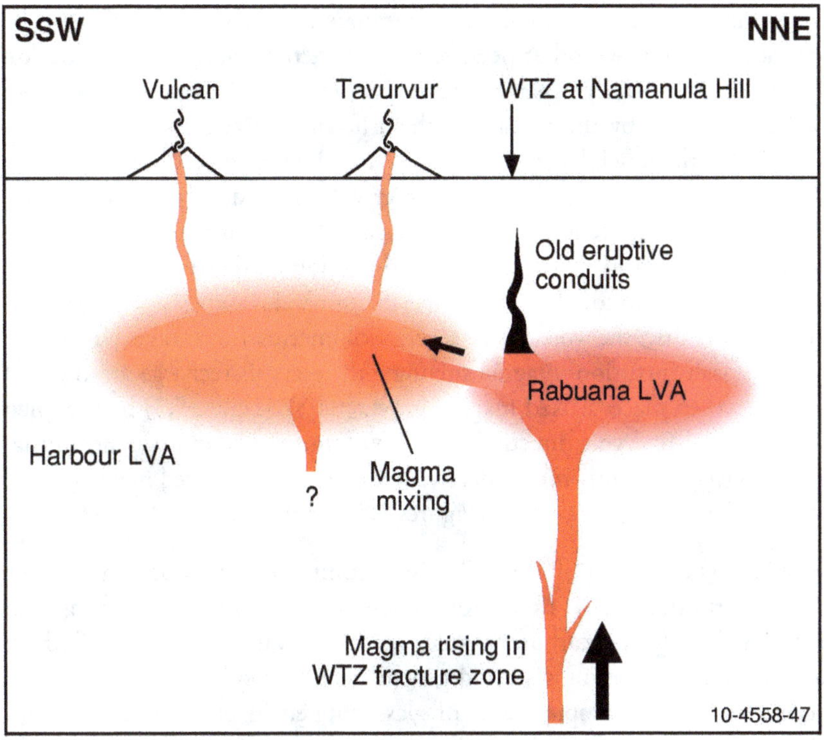

Figure 9.6. Magma mixing and low-velocity anomalies.

A zone of magma mixing beneath Tavurvur and the Greet Geothermal Field is shown in this schematic cross-section using RELACS seismic-velocity results for the two LVAs at Rabaul. Vulcan and Tavurvur are about 6 kilometres apart.

Eruptions at Tavurvur in 1878 and 1994–2014 were longer lasting than those at Vulcan in 1937–43, but even in 1941, after four years of inactivity, the recurrence of explosive eruptions at Tavurvur was consistent with the known longevity of Tavurvur eruptions—maintained within the Greet Geothermal Field—compared with those at Vulcan. Eruptions at Vulcan, including those in 1937, evidently require the underlying magma reservoir to be recharged relatively quickly, a process that may not be efficient or even possible as long as Tavurvur remains in eruptive activity.

The small size of the short-lived Tavurvur eruption in May 1937 can be considered from two other perspectives. First, the colonial administration in Rabaul and mainly expatriate townspeople were much more concerned, at least initially, by the impact of the ash fall on Rabaul town than they were about the much larger eruption at Vulcan where more than 500 village people had been killed. The much larger volume of deadly Vulcan pumice and ash did not become fully obvious until later, and particularly after volcanologist Norm Fisher calculated the amounts that had been erupted (Figure 4.9). Second, the dominance of the Vulcan eruption in 1937 is striking because the volcano is several kilometres from Tavurvur, below which basalt intrusion, magma mixing and reservoir recharge are thought to have taken place. Mixed magma in May 1937 (Figure 9.3) had reached places in the reservoir directly beneath Vulcan, and conduits beneath that volcano were evidently more effective in delivering the mixed magma to the surface than those at Tavurvur (Figures 9.5 and 9.6).

The basalts carried up in the WTZ have components that become volatile during eruptions and afterwards. These include the volcanic gases of sulphur dioxide and carbon dioxide, as well as water vapour, all of which are of interest to petrologists and geochemists as they can be detected and measured in microscopic pieces of glass trapped in pre-eruption crystals. Sulphur dioxide gas from the Tavurvur eruption of 2006 had been imaged in the large fan-like cloud seen from space (GVP 2006) and measured in 1994 by airborne spectrometers operating in Rabaul (Roggensack et al. 1996). Sulphur dioxide issuing from Tavurvur can be smelled locally, particularly during eruptions, as well as hydrogen sulphide, which is recognisable by its sulphuretted or 'bad-egg' smell when Tavurvur is quiescent. The magma between the zone of magma mixing and the roots of Tavurvur (Figure 9.6) is what keeps the Greet Geothermal Field active, its long-lived heat and volatiles chemically altering the volcanic rocks and deposits making up the cone of Tavurvur. This geothermally altered material was erupted hydro-explosively early in the eruptions of Tavurvur in both 1937 and 1994, as sticky blue-grey sulphide-bearing mud. Vulcan, by comparison, is different. It does not seem to be a major sulphur producer, or else the amounts are very small, and no geothermal activity survives for long after the explosive eruptions. The volatile element chlorine has been detected in Vulcan deposits from 1994; however, this is probably the result of seawater influence that is absent at Tavurvur.

The interpretations presented above are oversimplified, reflecting ongoing uncertainties over what actually happens beneath Blanche Bay in both a physical and a chemical sense. What are the amounts and rates at which the WTZ basalt reaches the Rabuana LVA and then the mixing zone and over what time frame? How much basalt or mixed magma remains in the Harbour LVA after any one eruption for involvement in later eruptions? How does the mixed magma beneath Tavurvur move laterally several kilometres to places beneath Vulcan? What happens in the deeper parts of crust and mantle below the levels shown in Figure 9.6? Also, can the interpretation be used to assess the chance of eruptions taking place that are substantially larger than those of 1878, 1937 and 1994?

Epilogue

Rabaul town still exists. It is much smaller than it was before the 1994 eruption but it continues to perform a strategically important role, situated alongside the shipping wharves that serve the economic needs of East New Britain Province. It is home to a vibrant local market and some development has taken place eastwards in recent years along Malaguna Road and elsewhere. The remnant town is situated at the foot of Tunnel Hill Road, which remains a convenient escape route to Talili Bay and the north coast, as it was in both 1937 and 1994, even during the south-east season. The administrative capital of the province has been returned to Kokopo, the place where the Germans had established their colonial centre of Herbertshöhe in the late nineteenth century before building their new capital of Rabaul in 1910. Papua New Guineans in the modern era have turned the tables.

The headquarters of the Rabaul Volcanological Observatory (RVO) is still on Observatory Ridge, on the same site chosen more than 80 years ago following the 1937 eruption, and overlooking the same spectacular scenery of Blanche Bay and its volcanic peaks. Instrumental and visual volcano monitoring continues there: maintaining and collecting earthquake data from the harbour seismometer network; undertaking GPS measurements at the southern end of Matupit Island; measuring ground temperatures, especially in the Greet Harbour area and on Tavurvur; making full use of a range of spatial datasets of the north-eastern Gazelle Peninsula provided from satellite-based imagery; and maintaining professional contact with overseas colleagues. The RVO continues to work in partnership with provincial authorities in the broad field of disaster-risk reduction, to supply its host department in Port Moresby with status reports on the condition of the volcanoes and is still in the business of eruption prediction.

Can a prognosis be made for the time and size of the next eruption in Blanche Bay? There should be more confidence in answering this question, bearing in mind the vastly increased post-WWII knowledge base relevant to understanding how the plumbing system of the Rabaul volcanic complex works, but considerable uncertainty remains. The reappearance of the seismic annulus, new uplift at Matupit Island, occurrence of 'north-east' earthquakes and increases in measured temperatures in the Greet Geothermal Field are all phenomena that need to be taken into consideration in forecasting the next eruption. Will it be a rerun of the twin eruptions of Vulcan and Tavurvur in 1878, 1937 and 1994, or will other young volcanoes of Blanche Bay, such as Rabalanakaia, break out in activity? Local communities will again be at risk, but at least they will have a stronger information and knowledge base than they did in 1937 when the eruptions killed hundreds.

References

ACSEE (Advisory Committee on Seismology and Earthquake Engineering). 1973. *Madang 1970 Earthquake*. Port Moresby: Department of Lands, Surveys and Mines. [118 pp.].

AGS (Allied Geographical Section). 1944. *Area Study of Gazelle Peninsula*. Terrain study. Allied Forces, South-West Pacific Area, no. 74.

Ahnon, W. 1953. Transcription from audio tapes recorded by Ray Sheridan in 1953 and used for the ABC radio program *Taim Bilong Masta*, 38–42. Interviewer Daniel Connell. Digital copy no. 2222 in RVO-IMS.

AIDAB (Australian International Development Assistance Bureau). 1994. *Rabaul Volcanic Disaster Needs Assessment Mission: Final Report*. Canberra: AIDAB. [131 pp.].

Anderhandt, J. 2012. *Eduard Hernsheim, die südsee und viel geld: biographie*. Two volumes. Münster, Germany: MV-Wissenschaft. [vol. 1: 581 pp., vol. 2: 620 pp.].

Anonymous. 1888a. [No title]. *Nachrichten über Kaiser Wilhelms-Land und den Bismarck-Archipel für 1888*, 76–9.

Anonymous. 1888b. 'Die Fluthwelle vom 13. Marz 1888'. *Nachrichten über Kaiser Wilhelms-Land und den Bismarck-Archipel für 1888*, 147–99.

Anonymous. 1914a. 'Earth Tremor'. *Government Gazette: British Administration— German New Guinea* 1 (3): 2.

Anonymous. 1914b. 'The Earth Tremor'. *Government Gazette: British Administration—German New Guinea* 1 (4): 2.

Anonymous. 1917a. 'Earth Tremor'. *Rabaul Record* 2 (2): 3–4.

Anonymous. 1917b. 'More Earth Tremors'. *Rabaul Record* 2 (3): 3.

Anonymous. 1932. 'No Scientific Records Kept of Rabaul's Quakes'. *Pacific Islands Monthly* 3 (5): 26.

Anonymous. 1937a. 'Vessel Struggles Out of Rabaul. Harbour Passage Found. Vulcan Crater 1600ft. Wide. Administrator Reports Easing of Tension'. *Sydney Morning Herald*, 3 June 1937, 11.

Anonymous. 1937b. 'Volcanoes. Less Danger in Rabaul. Reoccupation. Problem Causes Tension'. *Sydney Morning Herald*, 5 June 1937, 17.

Anonymous. 1937c. 'Future of Rabaul. Views Clash. Administration Returning'. *Sydney Morning Herald*, 9 June 1937, 17.

Anonymous. 1951. 'Routine Work of a Volcanologist: Inspecting Volcanic Craters and Hot Springs in Rabaul'. *Illustrated London News*, 3 February 1951, 168–9.

Anthony, A. 1981. Untitled account of Amy Anthony's experiences during the 1937 eruption and evacuation of Rabaul. R.W.J. Collection 30D, Folder 13, Sleeve 1. [8 pp.].

Anthony, M. 1941. Letter, 20 June 1941, from Rabaul, to Hon. Hubert L. Anthony, presumably in Australia. NAA: A518, AP836/4. Digital copy no. 902 in RVO-IMS.

Anton, L. and C.O. McKee. 2005. 'The Great Earthquake of 16 November 2000 and Associated Seismo-Tectonic Events Near the Pacific-Solomon-South Bismarck Plate Triple Junction in Papua New Guinea'. Geological Survey of Papua New Guinea, Department of Mining, Papua New Guinea, Report 2005/1. [74 pp.].

Aplin, D. 1980. *Rabaul 1942*. Melbourne: 2/22nd Battalion AIF Lark Force Association. [295 pp.].

Archbold, M.J., C.O. McKee, B. Talai, J. Mori and P. de Saint Ours. 1988. 'Electronic Distance Measuring Network Monitoring during the Rabaul Seismicity/Deformational Crisis of 1983–1985'. *Journal of Geophysical Research: Solid Earth* 93B10: 12, 123–12,136. doi.org/10.1029/JB093iB10p12123.

Arculus, A. (translator) and R.W. Johnson (compiler). 1981. '1937 Rabaul Eruptions, Papua New Guinea: Translations of Contemporary Accounts by German Missionaries'. Bureau of Mineral Resources, Report 229. [78 pp.].

Bai, C. and S. Greenhalgh. 2005. '3D Multi-Step Travel Time Tomography: Imaging the Local, Deep Velocity Structure of Rabaul Volcano, Papua New Guinea'. *Physics of the Earth and Planetary Interiors* 151: 259–75. doi.org/10.1016/j.pepi.2005.03.009.

Ball, W.B. 1937. 'Report on the Work of the Police during the Eruptions that Occurred in Rabaul on the 29th and 30th of May, 1937'. Unpublished typescript, Police Headquarters, Rabaul [8 pp.]. NAA: A518, V836/4. Digital copy no. 565 in RVO-IMS.

Bear-Crozier, A.N., G. Davies, M. Dunford, H. Ghasemi, N. Horspool, M. Jakab, L. Metz, V. Miller, M. Moihoi, K. Mulina, H. Patia, L. Pirit, L. Power, D. Robinson and F. Taranu. 2013. *Integrating Hazard and Exposure for East New Britain*. Professional Opinion 2013/07. Canberra: Geoscience Australia. [238 pp.].

Bebbington, M.S. and W. Marzocchi. 2011. 'Stochastic Models for Earthquake Triggering of Volcanic Eruptions'. *Journal of Geophysical Research: Solid Earth* 116, B05204. [16 pp.]. doi.org/10.1029/2010JB008114.

Bernard, O. and C. Bouvet de Maisonneuve. 2020. 'Controls on Eruption Style at Rabaul, Papua New Guinea—Insights from Microlites, Porosity and Permeability Measurements'. *Journal of Volcanology and Geothermal Research* 406: 107068. doi.org/10.1016/j.jvolgeores.2020.107068.

Bernard, O., W. Li, F. Costa, S. Saunders, I. Itikarai, M. Sindang and C. Bouvet de Maisonneuve. 2022. 'Explosive-Effusive-Explosive: The Role of Magma Ascent Rates and Paths in Modulating Caldera Eruptions'. *Geology* 50 (9): 1013–17. doi.org/10.1130/G50023.1.

Bignell, K.M. and M. Clarence. 1981–84. Correspondence between Mrs Clarence, R.W.J. and others. R.W.J. Collection 30D, Folder 2, Sleeves 19–22.

Biskup, P., ed. 1974. *The New Guinea Memoirs of Jean Baptist Octave Mouton*. Canberra: The Australian National University Press, Pacific History Series 7. [161 pp.].

Blong, R. and C. Aislabie. 1988. 'The Impact of Volcanic Hazards at Rabaul, Papua New Guinea'. Institute of National Affairs Discussion Paper 33, Port Moresby. [207 pp.].

Blong, R. and C. McKee. 1995. *The Rabaul Eruption 1994: Destruction of a Town*. [Sydney]: Natural Hazards Research Centre, Macquarie University. [51 pp.].

Boegershausen, G. 1906. 'Das Erdbeben auf der Insel Matupi und die Zerstoerung der Nonduper Kirche'. *Monatshefte zu Ehren Unserer Lieben Frau vom Heiligsten Herzen Jesu* 23: 111–15.

Boegershauser [sic], G. 1937. 'Eruption of a Volcano at Rabaul'. *Rabaul Times*, 6 August 1937, 15.

Border Watch. 1937. 'Rabaul Eruptions. King Bestows Honours. Administrator Knighted'. *Border Watch* (Mount Gambier, SA), 16 December 1937, 5.

Bouvet de Maisonneuve, C., F. Costa, H. Patia and H. Huber. 2015. 'Mafic Magma Replenishment, Unrest and Eruption in a Caldera Setting: Insights from the 2006 Eruption of Rabaul (Papua New Guinea)'. In *Chemical, Physical and Temporal Evolution of Magmatic Systems*, edited by L. Caricchi and J.D. Blundy, 17–39. London: Geological Society. doi.org/10.1144/SP422.2.

Braddock, R.D. 1973. 'The Solomon Sea Tsunamis of July 1971'. In *Oceanography of the South Pacific 1972*, compiled by R. Fraser, 9–14. Wellington: New Zealand National Commission for UNESCO.

Brooks, J.A. 1962. 'Seismic Wave Velocities in the New Guinea-Solomon Islands Region'. In *The Crust of the Pacific Basin*, edited by G.A. Macdonald and H. Kuno, 2–10. Washington, DC: American Geophysical Union of the National Academy of Sciences. doi.org/10.1029/GM006p0002.

Brooks, J.A. 1965. *Earthquake Activity and Seismic Risk in Papua New Guinea*. Canberra: Bureau of Mineral Resources Report 74.

Brooks, J.A., ed. 1971. *Investigations of Crustal Structure in the New Britain/New Ireland Region, 1969. Part 1: Geophysical and Geological Data*. Canberra: Bureau of Mineral Resources Record 1971/131. [45 pp.].

Brown, G. 1877. 'Notes on the Duke of York Group, New Britain, and New Ireland'. *Royal Geographical Society Journal* 47: 137–50. doi.org/10.2307/1798741.

Brown, G. 1878. *Journal of the Rev. G. Brown, 1860–1902*. 11 volumes. Sydney: Mitchell Library.

Brown, G. 1903. 'The Pacific, East and West'. In *Report of the Ninth Meeting of the Australasian Association for the Advancement of Science, held at Hobart, Tasmania, 1902*, edited by A. Morton, 458–79. Hobart: Australasian Association for the Advancement of Science.

Brown, G. 1908. *George Brown, D.D. Pioneer-Missionary and Explorer. An Autobiography. A Narrative of Forty-Eight Years Residence and Travel in Samoa, New Britain, New Ireland, New Guinea, and the Solomon Islands*. London: Hodder & Stoughton. [536 pp.].

Brown, T.J. and M.A. Tutton. c. 1994–95. 'Rabaul Town Mudflow/Flash Flood Assessment'. Unpublished map. Geological Survey Division, Department of Mining and Petroleum, Port Moresby.

Buckley, K. and K. Klugman. 1983. *The Australian Presence in the Pacific: Burns Philp 1914–1946*. Sydney: George Allen & Unwin. [392 pp.].

Bultitude, R.J. 1976. 'Eruptive History of Bagana Volcano, Papua New Guinea, between 1882 and 1975'. In *Volcanism in Australasia: A Collection of Papers in Honour of the Late G.A.M. Taylor GC*, edited by R.W. Johnson, 317–36. Amsterdam: Elsevier.

Burgess, H.E. 1937. Untitled article, 17 June 1937, on his experiences in Rabaul during the 1937 eruption. Photocopy of one page from the *Mittagong Star*. Incomplete citation details. Digital copy no. 613 in RVO-IMS.

Burns, Philp & Co. 1937. 'Rabaul Volcanic Disturbance'. Letter, 8 June 1937, to the secretary, Prime Minister's Department, Canberra. NAA: A518, V836/4. [2 pp.].

Cahill, P. 2012. *Needed—but Not Wanted: Chinese in Colonial Rabaul 1884–1960*. Brisbane: Copyright Publishing. [336 pp.].

Carey, S.W. 1938. 'The Morphology of New Guinea'. *Australian Geographer* 3 (5): 3–31. doi.org/10.1080/00049183808702186.

Carteret, P. (1767) 1965. 'Carteret's "Remarks" on the Navigation of St George's Channel, Log, 10–13 September 1767'. In *Carteret's Voyage Round the World 1766–1769*, edited by H. Wallis, 341–7. Cambridge: Hakluyt Society.

Casadevall, T.J., ed. 1994. *Volcanic Ash and Aviation Safety: Proceedings of the First International Symposium on Volcanic Ash and Aviation Safety*. Symposium held at Seattle, Washington, in July 1991. US Geological Survey Circular 1065. [450 pp.]. doi.org/10.3133/cir1065.

Cashman, K.V., R.S.J. Sparks and J.D. Blundel. 2017. 'Vertically Extensive and Unstable Magmatic Systems: A Unified View of Igneous Processes'. *Science* 355 (6331). doi.10.1126/science.aag3055.

Champion Hosking, H. 1937. 'Rabaul Sanitation'. Memorandum, 8 June 1937, to the administrator, Rabaul, from the Department of Public Health, Kokoko. NAA: A518, X836/4. Digital copy no. 553 in RVO-IMS.

Chinnery, E.W.F. and S. Chinnery. n.d. 'Eruption 29th May 1937: Rabaul'. Information (probably compiled by the Chinnerys) taken from E.W.F. Chinnery's report on eruption, and Sarah Chinnery's diaries. Digital copy available in RVO-IMS, reference no. 627. [4 pp.].

Chinnery, S. 1998. *Malaguna Road: The Papua and New Guinea Diaries of Sarah Chinnery*. Canberra: National Library of Australia. [246 pp.].

Cifali, G., G.W. d'Addario, E.J. Polak and W.A. Wiebenga. 1969. 'Rabaul Preliminary Crustal Seismic Test'. New Britain. Bureau of Mineral Resources, Canberra, Record 1969/125. [14 pp.].

Cilento, R. 1937a. 'Report on the Medical Significance of the Recent Eruption In Blanche Bay, Territory of New Guinea'. Unpublished typescript. NAA: A518, X836/4. Digital copy no. 629 in RVO-IMS. [33 pp.].

Cilento, R. 1937b. 'The Volcanic Eruption in Blanche Bay, Territory of New Guinea, May 1937'. *Journal of the Historical Society of Queensland* 2 (1): 37–49.

Clarence, M. 1982. *Yield Not to the Wind*. Sydney: Management Development Publishers. [154 pp.].

Clarke, G.A. 1960. 'Company History: Rabaul Volcanic Eruption—29th May 1927'. *Beep's Staff Bulletin*, Territory of Papua/New Guinea. A three-part article beginning in vol. 2, no. 1, pp. 1–2. R.W.J. Collection 30B, Folder 4, Sleeve 41. Digital copy no. 633 in RVO-IMS.

Clarke, G.[A]. 2001. 'The Rabaul Eruption of 1937—Another Perspective'. In *Tales of Papua New Guinea: Insights, Experiences, Reminiscences*, edited by S. Inder, 42–5. Roseville, NSW: Retired Officers' Association of Papua New Guinea Inc.

Cooke, R.J.S. 1977. 'Rabaul Volcanological Observatory and Geophysical Surveillance of the Rabaul Volcano'. *Australian Physicist* 14 (2): 27–30.

Cooke, R.J.S. 1981. 'Eruptive History of the Volcano at Ritter Island'. In *Cooke-Ravian Volume of Volcanological Papers*, edited by R.W. Johnson, 115–23. Port Moresby: Geological Survey of Papua New Guinea.

Cooke, R.J.S., C.O. McKee, V.F. Dent and D.A. Wallace. 1976. 'Striking Sequence of Volcanic Eruptions in The Bismarck Volcanic Arc, Papua New Guinea, in 1972–75'. In *Volcanism in Australasia: A Collection of Papers in Honour of the Late G.A.M. Taylor*, edited by R.W. Johnson, 149–72. Amsterdam: G.C. Elsevier.

Cummins, J.J. 1917. 'Early Experiences in New Britain'. *Rabaul Record* 2 (7): 9–10.

D'Addario, G.W., D.B. Dow and R. Swodoba. 1975. 'Geology of Papua New Guinea 1976'. 1:2,500,000-scale map. Canberra: Bureau of Mineral Resources.

Darius, W. 1984. 'Move to Kokopo Is On'. *Papua New Guinea Post-Courier*, 16 February 1984, 1.

Davies, H.L. 1973. 'Gazelle Peninsula, New Britain'. 1:250,000 Geological Series—Explanatory Notes Sheet SB/56-2 International Series. Canberra: Bureau of Mineral Resources. [22 pp.].

Davies, H.L. 1981. 'Robin John Seymour Cooke (1938–1939)'. In *Cooke-Ravian Volume of Volcanological Papers*, edited by R.W. Johnson, ix–xiii. Port Moresby: Geological Survey of Papua New Guinea.

Davies, H.L. 1984. 'The Rabaul Gourier (toktok long mauden paia)'. In *Volcanoes of the Sea, Rain Forest, and Ice: A Special Issue on Papua New Guinea*. Special issue *Volcano News* 19–20, edited and compiled by R.W. Johnson, 4–5.

Davies, H.L. 1987. 'Evan Richard Stanley, 1885–1924: Pioneer Geologist in Papua New Guinea'. *BMR Journal of Australian Geology and Geophysics* 10: 153–77.

Davies, H. 1995a. 'The 1994 Rabaul Eruption'. Unpublished Professorial Inaugural Lecture, University of Papua New Guinea, 21 February 1995. [40 pp.].

Davies, H. 1995b. 'The 1994 Eruption of Rabaul Volcano—A Case Study in Disaster Management'. Report prepared for the UNDP Office, Port Moresby, July 1995. Revised and re-issued by the University of Papua New Guinea, September 1995. [35 pp.].

Davies, H. 1995c. 'The 1994 Eruption of Rabaul Volcano: The Events of 18–19 September 1994—A Second Draft. (21 Nov 1994)'. In *Papua New Guinea Analysis of Lessons Learnt from Rabaul Eruption and Programming for Disaster Mitigation Activities in Other Parts of the Country*, by J. Tomblin and J. Chung, Annex 3, iv–vii. Report of Mission from 17–26 February 1995. Geneva: United Nations Department of Humanitarian Affairs. [8 pp., plus 4 annexes].

Davies, H.L. 2017. *Aitape Story: The Great New Guinea Tsunami of 1998*. Braddon, ACT: Halstead Press. [199 pp.].

Denham, D. 1969. 'Distribution of Earthquakes in The New Guinea—Solomon Islands Region'. *Journal of Geophysical Research* 74 (17): 4290–9. doi.org/10.1029/JB074i017p04290.

Dent, A.W., G. Davies, P. Barrett and P.J.A. de Saint Ours. 1995. 'The 1994 Eruption of the Rabaul Volcano, Papua New Guinea: Injuries Sustained and Medical Response'. *Medical Journal of Australia* 163 (4): 635–9. doi.org/10.5694/j.1326-5377.1995.tb124776.x.

D'Entrecasteaux, A.R.J. de B. 2001. *Bruny D'Entrecasteaux: Voyage to Australia and the Pacific, 1791–1793*. Edited by E. Duker and translated by M. Duker. Carlton, Vic.: Miegunyah Press, University of Melbourne Publishing. [This English translation is of D'Entrecasteaux's part of vol. 1 of the version of the voyages published in French by Rossel (1808)].

de Saint Ours, P.B. Talai, J. Mori, C. McKee and I. Itikarai. 1991. 'Coastal and Seafloor Changes at an Active Volcano: Example of Rabaul Caldera, Papua New Guinea'. Workshop on Coastal Processes in the South Pacific Island Nations, Lae, Papua New Guinea, 1–8 October 1987. *SOPAC Technical Bulletin* 7: 1–13.

Downs, I. 1980. *The Australian Trusteeship Papua New Guinea 1945–75*. Canberra: Department of Home Affairs, Australian Government Publishing Service. [587 pp.].

Downs, I. 1999. *The New Guinea Volunteer Rifles NGVR 1939–1943*. Broadbeach Waters, Qld: Pacific Press. [359 pp.].

Dunmore, J. 2005. *Storms and Dreams: Louis de Bougainville: Soldier, Explorer, Statesman*. Sydney: ABC Books. [296 pp.].

East New Britain PDC (Provincial Disaster Committee). 1988. *Rabaul Evacuation Plan 1987 Update*. Rabaul: East New Britain Provincial Government. [29 pp.].

Eggleston, F.W., H.L. Murray and H.O. Townsend. 1939. 'Site for the Capital of the Territory of New Guinea'. In *Australia Committee Appointed to Survey the Possibility of Establishing a Combined Administration of the Territories of Papua and New Guinea and to Make a Recommendation as to the Capital Site: Report*, 42–51. Canberra: Government Printer.

Epstein, A.L. 1969. *Matupit: Land, Politics, and Change among the Tolai of New Britain*. Canberra: The Australian National University Press. [335 pp.].

Epstein, A.L. 1992. *In the Midst of Life: Affect and Ideation in the World of the Tolai*. Berkeley, CA: University of California Press. [317 pp.]. doi.org/10.1525/california/9780520075627.001.0001.

Epstein, T.S. 1968. *Capitalism, Primitive and Modern: Some Aspects of Tolai Economic Growth*. Canberra: The Australian National University Press. [182 pp.].

Evans, H.W. 1937. Copy of handwritten letter, 6 June 1937, to Mr Evans' mother and sister, and provided by his brother Keith Evans to R.W.J. through Klaus Neumann in April 2021. R.W.J. Collection 37, Folder 4, Sleeves 89–92.

Everingham, I.B. 1974. 'Large Earthquakes in the New Guinea–Solomon Islands Area, 1873–1972'. *Tectonophysics* 23: 323–8. doi.org/10.1016/0040-1951(74)90068-7.

Everingham, I.B. 1975. 'Faulting Associated with the North Solomon Sea Earthquakes of 14 and 26 July 1971'. *Journal of the Geological Society of Australia* 22 (1): 61–9. doi.org/10.1080/00167617408728874.

Ewert, J.W. 2007. 'System for Ranking Relative Threats of US Volcanoes'. *Natural Hazards Review* 8 (4): 112–24. doi.org/10.1061/(ASCE)1527-6988(2007)8:4(112).

Ewert, J.W., C.D. Miller, J.W. Hendley II and P.H. Stuffer. 1997. 'Mobile Response Team Saves Lives in Volcano Crises'. US Geological Survey Fact Sheet 064-07. US Department of the Interior. doi.org/10.3133/fs06497.

Fabbro, G.N., C.O. McKee, M.E. Sindang, S. Eggins and C. Bouvet de Maisonneuve. 2020. 'Variable Mafic Recharge across a Caldera Cycle at Rabaul, Papua New Guinea'. *Journal of Volcanology and Geothermal Research* 393: 106810. doi.org/10.1016/j.jvolgeores.2020.106810.

Field, A. n.d. Unpublished account of the experiences of Mr Adrian Field during the 1937 Rabaul eruptions. R.W.J. Collection 30D, Folder 13, Sleeve 10. Digital copy no. 699 in RVO-IMS. [2 pp.].

Finlayson, D.M., ed. 1972. *Investigations of Crustal Structure in the Rabaul Region, New Britain 1967: Logistics and Data.* Canberra: Bureau of Mineral Resources Record 1972/45. [16 pp.].

Finlayson, D.M., J.P. Cull, W.A. Wiebenga, A.S. Furumoto and J.P. Webb. 1972. 'New Britain–New Ireland Crustal Seismic Refraction Investigations 1967 and 1969'. *Geophysical Journal of the Royal Astronomical Society* 29: 245–53. doi.org/10.1111/j.1365-246X.1972.tb06157.x.

Finlayson, D.M., A. Gudmundsson, I. Itikarai, Y. Nishimura and H. Shimamura. 2003. 'Rabaul Volcano, Papua New Guinea: Seismic Tomographic Imaging of an Active Caldera'. *Journal of Volcanology and Geothermal Research* 124: 153–71. doi.org/10.1016/S0377-0273(02)00472-9.

Finnimore, E.T., B.S. Low, R.J. Martin, P. Karam, I.A. Nairn and B.J. Scott. 1995. 'Contingency Planning for and Emergency Management of the 1994 Rabaul Volcanic Eruption, Papua New Guinea: Results of a Fact-Finding Mission'. New Zealand Ministry of Civil Defence. [19 pp., plus annexes: Annex C by I.A. Nairn and B.J. Scott].

Firth, S. 1978a. 'Albert Hahl: Governor of German New Guinea'. In *Papua New Guinea Portraits: The Expatriate Experience*, edited by J. Griffin, 28–47. Canberra: The Australian National University Press.

Firth, S. 1978b. 'Captain Hernsheim: Pacific Venturer, Merchant Prince'. In *More Pacific Islands Portraits*, edited by D. Scarr, 115–30. Canberra: The Australian National University Press.

Firth, S. 1983. *New Guinea under the Germans.* Carlton, Vic.: Melbourne University Press. [216 pp.].

Fisher, F.G. 1994. *Raphael Cilento: A Biography.* St Lucia, Qld: University of Queensland Press. [356 pp.].

Fisher, N.H. 1937. 'Geological Report on an Ascent of the Father, New Britain'. NAA: A518, AB836/4. Digital no. 707 in RVO-IMS.

Fisher, N.H. 1939a. 'Geology and Vulcanology of Blanche Bay, and the Surrounding Area, New Britain'. *Territory of New Guinea Geological Bulletin* 1. Digital no.181 in RVO-IMS. [68 pp.].

Fisher, N.H. 1939b. 'Rabaul's Volcanic Eruptions'. *Walkabout*, 1 March 1939, 13–20. Digital no. 714 in RVO-IMS.

Fisher, N.H. 1939c. 'Report on the Volcanoes of the Territory of New Guinea'. *Territory of New Guinea Geological Bulletin* 2. Digital no. 187 in RVO-IMS. [23 pp.].

Fisher, N.H. 1940a. 'Note on the Vulcanological Observatory at Rabaul'. *Bulletin Volcanologique* 4: 185–7. doi.org/10.1007/BF02994879.

Fisher, N.H. 1940b. 'Activity at Tavurvur Volcano, Sunday, 3rd March'. Unpublished report for government secretary, Rabaul, 4 March 1940. Digital copy no. 715 in RVO-IMS. Also in Rabaul Volcanological Reports, 1937–1942. [2 pp.].

Fisher, N.H. 1941. 'The Woolnough Article: Reviewed by Dr. Fisher'. *Rabaul Times*, 7 February 1941, 4.

Fisher, N.H. 1944. 'The Gazelle Peninsula, New Britain, Earthquake of January 14, 1941'. *Bulletin of the Seismological Society of America* 34 (a): 1–12. doi.org/10.1785/BSSA0340010001.

Fisher, N.H. 1945. 'The Fineness of Gold, with Special Reference to the Morobe Goldfield, New Guinea'. *Economic Geology* 40: 449–563. doi.org/10.2113/gsecongeo.40.7.449.

Fisher, N.H. 1946a. 'Administrative Centre for the Rabaul District'. Unpublished draft, 12 December 1946. Department of Supply and Shipping, Mineral Resources Survey Branch, Report No. 1946/32. [6 pp., plus Plan No. 1452].

Fisher, N.H. 1946b. 'Administrative Centre for the Rabaul District'. Record 1946/32. Canberra: Bureau of Mineral Resources. [6 pp.].

Fisher, N.H. 1957. *Catalogue of the Active Volcanoes and Solfatara Fields of Melanesia.* Catalogue of the Active Volcanoes of the World, Part 5. Naples: International Volcanological Association. [105 pp.].

Fisher, N.H. 1976a. 'Memorial—G.A.M. Taylor'. In *Volcanism in Australasia: A Collection of Papers in Honour of the Late G.A.M. Taylor*, edited by R.W. Johnson, ix–xiv. Amsterdam: G.C. Elsevier.

Fisher, N.H. 1976b. '1941–42 Eruption of Tavurvur Volcano, Rabaul, Papua New Guinea'. In *Volcanism in Australasia: A Collection of Papers in Honour of the Late G.A.M. Taylor*, edited by R.W. Johnson, 201–10. Amsterdam: G.C. Elsevier.

Fisher, N.H. and C.D. Branch. 1981. 'Late Cainozoic Volcanic Deposits of the Morobe Goldfield'. In *Cooke-Ravian Volume of Volcanological Papers*, edited by R.W. Johnson, 249–55. Port Moresby: Geological Survey of Papua New Guinea.

Fisher, N.H. and L.C. Noakes. 1942. 'Geological Reports on New Britain'. *Territory of New Guinea Geological Bulletin* 3. Digital no. 194 in RVO-IMS. [59 pp.].

Fooks, A.C.L. 1964. 'Preliminary Investigation of the Rabaul Geothermal Area for the Production of Electric Power'. Proceedings of the United Nations Conference on New Sources of Energy, vol. 2. *Geothermal Energy* 1: 230–6.

Forsyth, G. n.d. 'Rabaul—1937 Eruption'. Extract from unpublished memoirs, dealing with personal experiences during the 1937 eruption in Rabaul, and provided by N.A. Threlfall, 7–13. Digital copy no. 730 in RVO-IMS.

Fournier d'Albe, E.M. 1979. 'Objectives of Volcanic Monitoring and Prediction'. *Journal of the Geological Society of London* 136: 321–6. doi.org/10.1144/gsjgs. 136.3.0321.

Garnaut, R. 2010. 'Foreword'. In *Not a Poor Man's Field: The New Guinea Goldfields to 1942—An Australian Colonial History*, edited by M. Waterhouse, 4–5. Braddon, ACT: Halstead Press.

Garthwaite, M.C., S. Lawrie, S. Ampana, S.J. Saunders and M. Parkes. 2015. 'Pre- and Post-Eruptive Deformation at the Rabaul Caldera, Papua New Guinea Modelled Using PALSAR Time Series'. Paper presented at the Asia-Pacific Conference on Synthetic Aperture Radar, Singapore.

Graham, T.L., M.G. Swift, R.W. Johnson, J. Pittar, P. Musunamasi and I. Kari. 1993. 'Rabaul Harbour Heat Flow Project 1993 Papua New Guinea: Final Report'. Australian International Development Assistance Bureau. Report of surveys conducted by the Marine Geoscience and Petroleum Geology Program, Australian Geological Survey Organisation. [161 pp.].

Granger, K.J. 1988. 'The Rabaul Volcanoes: An Application of Geographical Information Systems to Crisis Management'. MA thesis, The Australian National University. [271 pp.].

Granger, K. 2000. 'An Information Infrastructure for Disaster Management in Pacific Island Countries'. *Australian Journal of Emergency Management* 15 (1): 20–32.

Green, R.C. and D. Anson. 1991. 'The Reber-Rakivai Lapita Site on Watom: Implications of the 1985 Excavations at the SAC and SDI Localities'. In *Report of the Lapita Homeland Project*, edited by J. Allen and C. Gosden, 170–81. Canberra: Department of Prehistory, The Australian National University, Occasional Papers in Prehistory, 20.

Greene, H.G., D.F. Tiffin and C.O. McKee. 1986. 'Structural Deformation and Sedimentation in an Active Caldera, Rabaul, Papua New Guinea'. *Journal of Volcanology and Geothermal Research* 30: 327–56. doi.org/10.1016/0377-0273 (86)90060-0.

Griffiths, T., W.C. Thomas and L. Thornton. 1938. 'Report of Committee Appointed to Investigate New Site for the Administrative Headquarters of the Territory of New Guinea'. Papers presented to Parliament. Volume 3, Parliamentary Paper 98 of 1938. [19 pp., plus appendices including the Stehn and Woolnough report of 1937 (Appendix A)].

Gudmundsson, O., D.M. Finlayson, I. Itikarai, Y. Nishimura and R.W. Johnson. 2004. 'Seismic Attenuation at Rabaul Volcano, Papua New Guinea'. *Journal of Volcanology and Geothermal Research* 130: 77–92. doi.org/10.1016/S0377-0273(03)00282-8.

Gudmundsson, O., R.W. Johnson, D.M. Finlayson, Y. Nishimura, H. Shimamura, A. Terashima, I. Itikarai and C. Thurber. 1999. 'Multinational Seismic Investigation Focuses on Rabaul Volcano'. *EOS, Transactions of the American Geophysical Union* 80 (24): 269, 273.

Guppy, H.B. 1887. *The Solomon Islands, Their Geology, General Features, and Suitability for Colonization*. London: Swan Sonnenschein, Lowrey & Co. [384 pp.].

Gutenberg, B. and C.F. Richter. 1954. *Seismicity of the Earth and Associated Phenomena*. Princeton, NJ: Princeton University Press. [310 pp.].

GVP (Global Volcanism Program). 1994a. 'Report on Rabaul (Papua New Guinea)', edited by R. Wunderman. *Bulletin of the Global Volcanism Network* 19 (8). [7 pp.]. doi.org/10.5479/si.GVP.BGVN199408-252140.

GVP (Global Volcanism Program). 1994b. 'Report on Rabaul (Papua New Guinea)', edited by E. Venzke. *Bulletin of the Global Volcanism Network* 19 (9). [6 pp.]. doi.org/10.5479/si.GVP.BGVN199409-252140.

GVP (Global Volcanism Program). 1994c. 'Report on Rabaul (Papua New Guinea)', edited by R. Wunderman. *Bulletin of the Global Volcanism Network* 19 (10). [4 pp.]. doi.org/10.5479/si.GVP.BGVN199410-252140.

GVP (Global Volcanism Program). 2006. 'Report on Rabaul (Papua New Guinea)', edited by R. Wunderman. *Bulletin of the Global Volcanism Network* 31 (9). [8 pp.]. doi.org/10.5479/si.GVP.BGVN200609-252140.

GVP (Global Volcanism Program). 2014. 'Report on Rabaul (Papua New Guinea)'. *Bulletin of the Global Volcanism Network* 39 (8). [9 pp.]. doi.org/10.5479/si.GVP.BGVN201408-252140.

GVP (Global Volcanism Program). 2017. 'Report on Rabaul (Papua New Guinea)', edited by A.E. Crafford and E. Venzke. *Bulletin of the Global Volcanism Network* 42 (2). [6 pp.]. doi.org/10.5479/si.GVP.BGVN201702-252140.

Habernecht, KorvettenKapitän. 1912. 'Erdbeben in der Blanche-Bucht (Neu-Pommern) am 8. September 1911'. *Annalen der Hydrographie und Maritimen Meterologie* 40: 167–8.

Hahl, A. 1980. *Albert Hahl: Governor in New Guinea.* Edited and translated by P.G. Sack and D. Clark. Canberra: The Australian National University Press. [164 pp.].

Hall, T. 1981. *New Guinea 1942–44.* Sydney: Methuen. [224 pp.].

Hammer, K.L. 1907. *Die geographische Verbreitung der vulkanischen Gebilde und Erscheinungen im Bismarckarchipel und auf den Salomonen.* Giessen Germany: Muenchow'sche Hof-und Universitaets-Druckerei.

Hantke, G. 1939. 'Uebersicht ueber die vulkanische Taetigkeit vom Januar 1937 bis Maerz 1938'. *Zeitschrift der Deutschen Geologischen Gesellschaft* 91 (2): 160–8.

Harrington, Captain. 1878. 'Unterseeische vulkanischer Ausbruch bei den Salomo-Inseln im Maerz 1878'. In *Unterseeische vulkanische Ausbrueche und Fluthwellen im suedlichen Stillen Ocean. Annalen der Hydrographie und Maritimen Meterologie* 6: 373–4.

Hasluck, P. 1976. *A Time for Building: Australian Administration in Papua and New Guinea 1951–1963.* Carlton, Vic.: Melbourne University Press. [462 pp.].

Hastings, P. 1974. 'Volcano Set to Blow: 20,000 Plan Island Escape'. *Sydney Morning Herald*, 26 January 1984, 1, 4.

Hawkesworth, J. 1773. *An Account of the Voyages Undertaken by the Order of His Present Majesty, for Making Discoveries in the Southern Hemisphere, and Successfully Performed by Commodore Byron, Captain Wallis, Captain Carteret and Captain Cook, in the Dolphin, the Swallow, and the Endeavour; Drawn up from the Journals Which Were Kept by the Several Commanders and from the Papers of Joseph Banks, Esq.* Vol. 1. London: W. Strahan and T. Cadell. [456 pp.].

Heming, R.F. 1967. 'The Kokopo (New Britain) Earthquakes of 14th August, 1967'. Territory of Papua and New Guinea, Department of Lands, Surveys and Mines, Geological and Volcanological Branch. Notes on Investigation No. 67503. [11 pp.]. See also BMR Record No. 1969/24.

Heming, R.F. 1973. 'Geology and Petrology of Rabaul Caldera: An Active Volcano in New Britain, Papua, New Guinea'. PhD thesis, University of California. [181 pp.].

Heming, R.F. 1974. 'Geology and Petrology of Rabaul Caldera, Papua New Guinea'. *Bulletin of the Geological Society of America* 85: 1253–64. doi.org/10.1130/0016-7606(1974)85<1253:GAPORC>2.0.CO;2.

Heming, R.F. and I.S.E. Carmichael. 1973. 'High-Temperature Pumice Flows from the Rabaul Caldera Papua, New Guinea'. *Contributions to Mineralogy and Petrology* 38: 1–20. doi.org/10.1007/BF00371723.

Hempenstall, P.J. 1978. *Pacific Islanders under German Rule: A Study in the Meaning of Colonial Resistance*. Canberra: The Australian National University. [264 pp.].

Hernsheim, E. (1878a) 2015. 'E. Hernsheim an Korvettenkapitän Mensing, SM Korvette Albatross, Hongkong, 14 April 1878'. In *Südseekaufmann: Gesammelte Schriften/Eduard Hernsheim; bearbeitet und herausgegeben von Jakob Anderhandt, Monenstein und Vannerdat OHG, Münster (Germany)*, by E. Hernsheim, 640–4. An unpublished English translation of Hernsheim's letter to Captain Mensing was kindly provided to R.W.J. by Hilary Howes.

Hernsheim, [E]. 1878b. 'Unterseeische vulkanische Ausbrueche und Fluthwellen im suedlichen Stillen Ocean. 4. Unterseeischer vulkanischer Ausbruch bei Neu-Britannien im Februar 1878'. *Annalen der Hydrographie und Maritimen Meteorologie* 6: 372–3.

Hernsheim, E. 1983. *Eduard Hernsheim: South Sea Merchant*. Edited and translated by P. Sack and D. Clark. Boroko, PNG: Institute of Papua New Guinea Studies. [230 pp.].

Hiery, H.J. 1995. *The Neglected War: The German South Pacific and the Influence of World War I*. Honolulu: University of Hawai'i Press. [387 pp.].

Hilder, B. 1937. Unpublished extract from the journal of Lieutenant Brett Hilder, RANR(S). Includes three drawings by Hilder. Digital copy available in RVO-IMS, reference no. 785. [6 pp.].

Hilder, B. 1961. *Navigator in the South Seas*. London: Percival Marshall & Co. [232 pp.].

Hilder, B. 1980. Transcription of an audio interview used for the ABC radio program *Taim Bilong Masta*, 26–34. Interviewer Daniel Connell. Digital copy no. 788 in RVO-IMS.

Hilder, B. 1980–81. Unpublished letters to W.D. Palfreyman (12 November 1980) and R.W. Johnson (6 April 1981). Digital copy no. 787 in RVO-IMS. [3 pp.].

Hill, D.P., F. Pollitz and C. Newhall. 2002. 'Earthquake–Volcano Interactions'. *Physics Today* 55 (11): 41–7. doi.org/10.1063/1.1535006.

Hiromi, T. 2004. 'Japanese Forces in Post-Surrender Rabaul'. In *From a Hostile Shore: Australia and Japan at War in New Guinea*, edited by S. Bullard and T. Keiko, 137–53. Canberra: Australian War Memorial. [In both Japanese and English].

Hoogerwerff, J. 1937. Copy of typed letter to Mr Mouton, 8 June 1937, written from Vunapope. Transcription of extracts from the personal and business papers of J.B.O. Mouton (1911–1948) held on microfilm at the Pacific Manuscripts Bureau, The Australian National University, reference code PMB 603. Digital copy no. 793 in RVO-IMS.

Hoogerwerff, J. 1941. Copies of handwritten diary entries. Transcription of extracts from the personal and business papers of J.B.O. Mouton (1911–1948) held on microfilm at the Pacific Manuscripts Bureau, The Australian National University, reference code PMB 603. Digital copy no. 793 in RVO-IMS.

Hopkins, E. 1937. Copies of letters written by Eric Hopkins from Rabaul to his parents including two CDs of photographs of the 1937 eruption, all supplied by Vince Neall in 2012. R.W.J. Collection 30B, Folder 6, Sleeve 1.

Hopper, P.W. 1986. 'John Charles Mullaly (1895–1973)'. *Australian Dictionary of Biography*. Vol. 10. Carlton, Vic.: Melbourne University Press. [2 pp.].

Hosking, J.S. 1938. 'Some Recent Volcanic Deposits and Volcanic Soils from the Island of New Britain in the Territory of New Guinea'. *Transactions of the Royal Society of South Australia* 62 (2): 366–77.

Hugo, H. 1980. Transcription of an audio interview used for the ABC radio program *Taim Bilong Masta*, 3–10. Interviewer Daniel Connell. Digital copy no. 2222 in RVO-IMS.

Hunter, J. (1793) 1968. *An Historical Journal of Events at Sydney and at Sea 1787–1792*. Sydney: Angus & Robertson. [452 pp.].

Hutchinson, D. and J. Dawson. 2009. *Satellite Radar Interferometry: Application to Rabaul Caldera, Papua New Guinea*. Canberra: Geoscience Australia, Record 2009/39. [15 pp.].

I.M. 1937. 'Rakaia'. *A Nilai ra Dowot* 332: 2.

Intemann, K. 2017. 'Deutsches Vermessungsschiff S.M.S. PLANET von 1905'. In *Schiffe und Mehr. Zusammengestellt für*. www.schiffe-und-mehr.com. [2 pp.].

Isacks, B., J. Oliver and L.R. Sykes. 1968. 'Seismology and the New Global Tectonics'. *Journal of Geophysical Research* 73 (18): 5855–99. doi.org/10.1029/JB073i018p05855.

Itikarai, I. 2008. 'The 3-D Structure and Earthquake Locations at Rabaul Caldera, Papua New Guinea'. PhD thesis, The Australian National University. [137 pp.].

Jakab, M. and L. Pirit. 2013. 'Spatial Data and Exposure Information'. In *Integrating Hazard and Exposure for East New Britain*, edited by A.N. Bear-Crozier et al., 55–71. Professional Opinion 2013/07. Canberra: Geoscience Australia.

Johnson [sic Johnston], G.J. 1919. 'Memorandum Addressed to the Secretary for Defence, Melbourne'. *Proceedings of the Geological Society (London)* 76 (1): v–vi.

Johnson, R.W. 1979. 'Geotectonics and Volcanism in Papua New Guinea: A Review of the Late Cainozoic'. *BMR Journal of Australian Geology and Geophysics* 4 (3): 181–207.

Johnson, R.W. 1986. 'Underwater Video Survey of the Volcanic Bulge on the Floor of Rabaul Harbour, Papua New Guinea, December 1985'. CCOP/SOPAC Proceedings of the 15th Session, Rarotonga, Cook Islands, 138–9. See IMS No. 1377 for the typescript of this report together with three draft illustrations.

Johnson, R.W. 1987. 'Large-Scale Volcanic Cone Collapses: The 1888 Slope Failure of Ritter Volcano, and Other Examples from Papua New Guinea'. *Bulletin of Volcanology* 49: 669–79. doi.org/10.1007/BF01080358.

Johnson, R.W. 2013. *Fire Mountains of the Islands: A History of Volcanic Eruptions and Disaster Management in Papua New Guinea and the Solomon Islands*. Canberra: ANU Press. [391 pp.]. doi.org/10.26530/OAPEN_462202.

Johnson, R.W. 2020. *Roars from the Mountain: Colonial Management of the 1951 Volcanic Disaster at Mount Lamington*. Canberra: ANU Press. [356 pp.]. doi.org/10.22459/RM.2020.

Johnson, R.W., I.B. Everingham and R.J.S. Cooke. 1981. 'Submarine Volcanic Eruptions in Papua New Guinea: 1878 Activity of Vulcan (Rabaul) and Other Examples'. In *Cooke-Ravian Volume of Volcanological Papers*, edited by R.W. Johnson, 167–79. Port Moresby: Geological Survey of Papua New Guinea.

Johnson, R.W., I. Itikarai, H. Patia and C.O. McKee. 2010. 'Volcanic Systems of the Northeastern Gazelle Peninsula, Papua New Guinea: Synopsis, Evaluation and a Model for Rabaul Volcano'. Rabaul Volcano Workshop Report. Papua New Guinea Department of Mineral Policy and Geohazards Management, and Australian Agency for International Development, Port Moresby. [84 pp.].

Johnson, R.W., C.O. McKee, S. Eggins, J. Woodhead, R.J. Arculus, B.W. Chappell and J. Sheraton (Rabaul Petrology Group). 1995. 'Taking Petrologic Pathways towards Understanding Rabaul's Restless Caldera'. *EOS, Transactions of American Geophysical Union* 76 (17): 171, 180. doi.org/10.1029/95EO00093.

Johnson, R.W. and N.A. Threlfall. 1985. *Volcano Town: The 1937–43 Rabaul Eruptions*. Bathurst, NSW: Robert Brown and Associates. [151 pp.].

Johnson, T. and P. Molnar. 1972. 'Focal Mechanisms and Plate Tectonics of the Southwest Pacific'. *Journal of Geophysical Research* 77 (26): 5000–32. doi.org/10.1029/JB077i026p05000.

Jones, A.S. 1937a. Letter, 3 June 1937, from Kabakada to Rev. Jones's parents, Mr and Mrs H. Jones of Hamilton, Victoria. Digital copy no. 812 in RVO-IMS. [9 pp.].

Jones, A.S. 1937b. A two-part article entitled, respectively, 'After Eruption' and 'Stricken Country', based on a letter to his parents (see Jones, 1937a) and published in the *Newcastle Morning Herald and Miners' Advocate*, 21 June (p. 8) and 22 June (p. 5) 1937.

Jones, H.L. 1937. An untitled account (probably part of a letter) of Mrs Jones's experience of the 1937 eruption at Rabaul and of the wedding of the Trevitts. R.W.J. Collection 30D, Folder 14, Sleeve 24. Digital copy no. 814 in RVO-IMS. [6 pp.].

Jones, R.H. and R.C. Stewart. 1997. 'A Method for Determining Significant Structures in a Cloud of Earthquakes'. *Journal of Geophysical Research* 102 (B4): 8245–54. doi.org/10.1029/96JB03739.

Joycey, D. c. 1937. 'The Rabaul Volcanic Eruption of 1937'. *Bulletin for Society Members*. Printed for the Historical Society of New Britain by Pacific Publications Pty Ltd, Sydney. [4 pp.]. The date of publication is not given in the bulletin.

Joycey, D.C. 1981. Correspondence between Mr Joycey and R.W.J. R.W.J. Collection 30B, Folder 6, Sleeve 10.

Kalua, K. 2022. 'The Kina Meets the *Tabu*: Ancient *Kastom* and a Safe Passage to the Afterlife'. *PNG Kundu* (Papua New Guinea Association of Australia Inc.) 3 (11): 36–9. Originally published in the *PNG Air Magazine*.

King, V.C.S. 1937. 'Banking under Extreme Difficulties'. *Bank Notes* (Staff Magazine of the Commonwealth Bank of Australia) 19, June, 4–10.

Kingsbury, R. 1983. *Waiting for the Big Bang*. Documentary by Film Australia, 45 minutes. National Film and Sound Archives of Australia, title no. 1350930.

Kizawa, T. c. 1943. 'Death Line'. Photocopy and English translation of part of an article in Japanese probably written in 1943 before Kizawa returned to Rabaul for a second time. R.W.J. Collection 1, Folder 5, Sleeve 40. Digital copy no. 30 in RVO-IMS.

Kizawa, T. 1951. 'Volcanic Tremor and Tilting of the Ground'. English translation of Japanese original. *Kenshin Jihô* [Earthquake testing review] 15 (2): 18–34. Five pages of the translation. R.W.J. Collection 1, Folder 5, Sleeve 41. Digital copy no. 823 in RVO-IMS.

Kizawa, T. 1961. Letter, 23 August 1961, to G.A. Taylor at the Vulcanological Observatory, Rabaul, written from the Meteorological Research Institute, Tokyo. R.W.J. Collection 1, Folder 1, Sleeves 1–3. Digital copy no. 824 in RVO-IMS. [5 pp.].

Kuester, I. and S. Forsyth. 1985. 'Rabaul Eruption Risk: Population Awareness and Preparedness Survey'. *Disasters* 9 (3): 179–82. doi.org/10.1111/j.1467-7717.1985.tb00935.x.

Kusaka, J. 1976. 'Rabaul Sensen Ijyo [All Quiet on the Rabaul Front]'. Tokyo: Kowa-do. R.W.J. Collection 1, Folder 5, Sleeve 43. [In Japanese]. [291 pp.].

Lacroix, A. 1904. *La Montagne Pelée et ses eruptions*. Paris: Masson et Cie. [602 pp.].

Lambert, S.M. 1941. *A Doctor in Paradise*. London: J.M. Dent & Sons Ltd. [421 pp.].

Langdon, R. 1973. 'That Day in 1937 When Vulcan Island Blew Its Top'. *Pacific Islands Monthly*, November: 47–8.

Latter, J.H. 1966. 'Notes On Near Earthquakes, Interference, and the Problem of Volcanic Tremors at Rabaul, New Britain'. Bureau of Mineral Resources, Canberra, Record 1966/19. [89 pp.].

Latter, J.H. and A.W. Hurst. 1987. 'An Assessment of Volcanic, Seismic, and Tsunami Hazard at Rabaul and Neighbouring Areas of New Britain, Papua New Guinea'. New Zealand Department of Scientific and Industrial Research, Contract Report 25. Prepared for the Insurance Underwriters' Association of Papua New Guinea, Port Moresby. [186 pp.].

Lauer, S.E. 1995. *Pumice and Ash: An Account of the 1994 Rabaul Volcanic Eruptions.* Lismore, NSW: CPD Resources. [80 pp.].

Laufer, C. 1956. 'Die Verwandtshaftsverhältnisse innerhalb des Gunantuna-Stammes (Südsee)'. *Anthropos* 51: 994–1028. See footnote on p. 1001.

Lentfer, C. and B. Boyd. 2001. *Maunten Paia: Volcanoes, People and Environment: The 1994 Rabaul Volcanic Eruptions.* Lismore, NSW: Southern Cross University Press. [85 pp.].

Le Pichon, X. 1968. 'Sea-Floor Spreading and Continental Drift'. *Journal of Geophysical Research* 73 (12): 3661–97. doi.org/10.1029/JB073i012p03661.

Lewis, F.G. 1937a. 'The Rabaul Disaster'. *Missionary Review* 46 (1): 3–10.

Lewis, F.G. 1937b. 'Volcano and Earthquake: The Rabaul Disaster'. *Methodist,* 31 July 1937, 1, 4.

Lindley, I.D. 1995. 'The 1994–1995 Rabaul Volcanic Eruptions: Human Aspects'. *Bulletin of the Royal Society of New South Wales.* Three parts: 188 (July), 7–9; 189 (August), 4–6; 190 (September), 4–6.

Lindsay, P. 2010. *The Coast Watchers.* Sydney: William Heinemann Australia. [416 pp.].

Linggood, W.L.I., H. Fellmann and R.H. Rickard. 1940. *The New Britain Dictionary.* Raluana, PNG: Methodist Mission Press. [143 pp.].

Lokinap, R.I. 1995. 'The Rabaul Volcanic Emergency: "Operation Unity" Report to the National Parliament by the Controller'. Report no. 4, 7 March 1995. [12 pp.].

Lowenstein, P.L. 1988. 'Rabaul Seismo-Deformational Crisis of 1983–85: Monitoring, Emergency Planning and Interaction with Authorities, the Media and the Public'. *Geological Survey of Papua New Guinea Report* 88 (32). [66 pp.].

Lowenstein, P.L. and B. Talai. 1984. 'Volcanoes and Volcanic Hazards in Papua New Guinea'. *Geological Survey of Japan Report* 263: 315–31.

Ludwig, E. 1927. *Bismarck: The Story of a Fighter.* Translated by E. and C. Paul. London: Allen & Unwin. [646 pp.].

Macatol, I.C., ed. 2002. *Disaster Risk Assessment Report for East New Britain Province.* Gazelle Restoration Authority, East New Britain Provincial Administration, and the PNG Advisory Support Facility Project, AusAID.

Macdonald, G.A. 1972. *Volcanoes.* Hoboken, NJ: Prentice Hall Inc. [510 pp.].

Mackenzie, D.P. and R.L. Parker. 1967. 'The North Pacific: An Example of Tectonics on a Sphere'. *Nature* 216: 1276–80. doi.org/10.1038/2161276a0.

Mackenzie, S.S. (1927) 1987. *The Australians at Rabaul: The Capture and Administration of the German Possessions in the Southern Pacific.* St Lucia, Qld: University of Queensland Press. [412 pp.].

Macnab, R.P. 1970. 'Geology of the Gazelle Peninsula, T.P.N.G'. Bureau of Mineral Resources, Canberra, Record 1970/63. [127 pp.].

Maribu, G. 2021. 'Rabaul … September 1994'. *PNG Kundu* (Papua New Guinea Association of Australia Inc.) 2 (8), inside front cover, including photographs by C. Read. Originally published as 'The Haven That Was Rabaul' in *Papua New Guinea Post-Courier*, 1994.

Marshall, P. 1935. 'Acid Rocks of the Taupo-Rotorua Volcanic District'. *Transactions of the Royal Society of New Zealand* 64: 1–44.

Martin, K. 2013. *The Death of the Big Men and the Rise of the Big Shots.* New York: Berghahn Books. [256 pp.].

Maru Special. 1984. 'Japanese Naval Operations in WWII Attack Wake Island Rabaul'. *Maru Special,* no. 94 IJN. [Japanese text and photograph captions].

Marzocchi, W. 2002. 'Remote Seismic Influence on Large Volcanic Eruptions'. *Journal of Geophysical Research* 107 (B1). [7 pp.]. doi.org/10.1029/2001JB000307.

Mason, E.C. 1937. 'Rabaul Eruption—May, 29th, 1937'. Typed transcription from diary of Miss Emmy Carol Coleman (married name E.C. Mason). R.W.J. Collection 30A, Folder 3, Sleeve 30. Digital copy no. 548 in RVO-IMS. [2 pp.].

Massey, Captain. 1918. 'Vulcanicity of Rabaul District'. *Rabaul Record* 3 (1): 8–10.

Matupit, C. 2003. 'Toboberatagul'. In *Legends of Papua New Guinea: A Collection of Traditional Stories from PNG Compiled by PNG Studies Diploma Two Students,* 7–8. Madang, PNG: Divine Word University.

McAulay, L. 1986. *Into the Dragon's Jaws.* Mesa, AZ: Champlin Fighter Museum Press. [148 pp.].

McBirney, A.R. and V. Lorenz. 2003. 'Karl Sapper: Geologist, Ethnologist, and Naturalist'. *Earth Sciences History* 22: 79–89. doi.org/10.17704/eshi.22.1.w78 q7v821t825922.

McCarthy, J.K. 1951. 'Appreciation and Plan—Volcanic Eruption, Rabaul'. Papua New Guinea National Archives, File GH1-5-7 (SN677, AN244, BN163).

McCarthy, J.K. 1963. *Patrol into Yesterday: My New Guinea Years*. Melbourne: Cheshire Pty Ltd. [252 pp.].

McCarthy, J.K. 1971a. 'Warning Flags were Flying in Rabaul but Nobody Took Heed'. *Papua New Guinea Post-Courier*, 19 July 1971, 5.

McCarthy, J.K. 1971b. 'When Matupit Blew It Was Time to Go'. *Papua New Guinea Post-Courier*, 20 July 1971, 5.

McCarthy, J. 1980. Transcription of an audio interview with Mrs Jean McCarthy used for the ABC radio program *Taim Bilong Masta*, 12–16. Interviewer Tim Bowden. Digital copy no. 2222 in RVO-IMS.

McCue, K. 2007. 'Three Days in the Life of a Rabaul PNG Volcano'. *Lava News* (Geological Society of Australia) 15: 3–5.

McCue, K. and H. Letz. 2019. 'The Last of the Great Earthquakes of 1906—Finisterre Ranges New Guinea'. Paper presented at the Australian Earthquake Engineering Society 2019 Conference, 29 November – 1 December, Newcastle (NSW). [21 pp.].

McKee, C.O. 2015. 'Tavui Volcano: Neighbour of Rabaul and Likely Source of the Middle Holocene Penultimate Major Eruption in the Rabaul Area'. *Bulletin of Volcanology* 77 (80). doi.org/10.1007/s00445-015-0968-1.

McKee, C.O. 2021. 'A Review of the Mid-Late Holocene Volcanism in the Rabaul Area of Papua New Guinea: Contrasting Eruptive Histories of Three Adjacent Volcanic Systems, Bimodal Sequences, and Magma-Mixing'. PNG Department of Mineral Policy and Geohazards Management, Geohazards Management Division Report 2021/02. [82 pp.].

McKee, C.O., M.J. Baillie and P. Reimer. 2015. 'A Revised Age of AD 667–669 for the Latest Major Eruption at Rabaul'. *Bulletin of Volcanology* 77 (65). [7 pp.]. doi.org/10.1007/s00445-015-0954-7.

McKee, C.O. and R.A. Duncan. 2016. 'Early Volcanic History of the Rabaul Area'. *Bulletin of Volcanology* 78 (24). [28 pp.]. doi.org/10.1007/s00445-016-1018-3.

McKee, C.O. and G.N. Fabbro. 2018. 'The Talili Pyroclastics Eruption Sequence: VEI 5 Precursor to the Seventh Century CE Caldera-Forming Event at Rabaul, Papua New Guinea'. *Bulletin of Volcanology* 80 (79). [28 pp.]. doi.org/10.1007/s00445-018-1255-8.

McKee, C., I. Itikarai and H. Davies. 2018. 'Instrumental Volcano Surveillance and Community Awareness in the Lead-Up to the 1994 Eruptions at Rabaul, Papua New Guinea'. In *Observing the Volcano World: Volcano Crisis Communication. Advances in Volcanology*, edited by C.J. Fearnley, K. Deanne, K. Haynes, W.J. McGuire and G. Jolly, 205–33. Cham, Switzerland: Springer. doi.org/10.1007/11157_2017_4.

McKee, C., I. Itikarai, J. Kuduon, N. Lauer, D. Lolok, H. Patia, P. de Saint Ours, S. Saunders, L. Sipison, R. Stewart and B. Talai. 2018. 'The 1994–1998 Eruptions at Rabaul: Main Features and Analysis'. PNG Department of Mineral Policy and Geohazards Management, Geohazards Management Division Report 2018/02. [135 pp.].

McKee, C.O., R.W. Johnson, P.L. Lowenstein, S.J. Reilly, R.J. Blong, P. de Saint Ours and B. Talai. 1985. 'Rabaul Caldera, Papua New Guinea: Volcanic Hazards, Surveillance, and Eruption Contingency Planning'. *Journal of Volcanology and Geothermal Research* 23: 195–237. doi.org/10.1016/0377-0273(85)90035-6.

McKee, C.O., J. Kuduon, H. Patia and F.B. Taranu. 2020. 'Volcanism at Rabaul since the 7th Century CE Caldera-Forming Eruption'. PNG Department of Mineral Policy and Geohazards Management, Geohazards Management Division Report 2020/01. [87 pp.].

McKee, C.O., P.L. Lowenstein, P. de Saint Ours, B. Talai, I. Itikarai and J.J. Mori. 1984. 'Seismic and Ground Deformation: Crises at Rabaul Caldera: Prelude to an Eruption?' *Bulletin Volcanologique* 47 (2): 397–411. doi.org/10.1007/BF01961569.

McNicoll, W.R. 1937a. 'Telegram to Prime Minister's Department, Canberra, from Wau, 9 am 31 May 1937'. NAA: A518, V836/4. Digital copy no. 549 in RVO-IMS.

McNicoll, W.R. 1937b. 'Telegram to Prime Minister's Department, Canberra, from the Montoro, 10 pm 31 May 1937'. NAA: A518, V836/4. Digital copy no. 549 in RVO-IMS.

McNicoll, W.R. 1937c. 'Circular Despatches [CD] 1–6, 1–15 June, Central Administration, Rabaul'. NAA: A518, V836/4, Part 2. Digital copy no. 880 in RVO-IMS.

McNicoll, W.R. 1937d. Decode of cablegram, 12 June 1937, to Sir George Pearce, Minister for External Affairs, Canberra, from Government House, Rabaul. NAA: A518, X836/4. Digital copy no. 549 in RVO-IMS.

McNicoll, W.R. 1937e. Personal letter, 14 June 1937, to Sir George Pearce, Minister for External Affairs, Canberra, from Government House, Rabaul. NAA: A518, X836/4. Digital copy no. 553 in RVO-IMS.

McNicoll, W.R. 1937f. Personal and confidential memorandum, 26 June, to Sir George Pearce in Canberra from Rabaul. NAA: A518, X836/4. Digital copy no. 549 in RVO-IMS.

McNicoll, W.R. 1937g. 'Volcanic Eruption, May 1937—Missing Natives'. Memorandum, 20 July, to the secretary, Prime Minister's Department, Canberra. NAA: A518, V836/4. Digital copy no. 881 in RVO-IMS.

Meares, C.D. 1968. 'Town in a Volcano'. *Enemelay: Review for the National Mutual Staff, Melbourne* 14 (4): 12–13. [First part only of the complete account given by Meares (n.d.)].

Meares, C.D. 1980. Transcription of an audio interview with Mr Meares used for the ABC radio program *Taim Bilong Masta*, 16–26. Interviewer Tim Bowden. Digital copy no. 2222 in RVO-IMS.

Meares, C.D. 1980–84. Correspondence between Mr Meares, R.W.J. and Mrs Rosemary Kennedy. R.W.J. Collection 30B, Folder 8, Sleeves 51–7.

Meares, C.D. n.d. 'Town in a Volcano'. Written after 1939 and probably in the late 1960s. Unpublished typescript. Digital copy no. 883 in RVO-IMS. The first part of the account (but apparently not the remainder) was published by Meares (1968).

Meier, J. 1908. 'A *kaja* oder Der Schlangenaberglaube bei den Eingebornen der Blanchebucht (Neupommeren)'. *Anthropos* 3: 1005–29.

Mennis, B. and M. Mennis. 2019. *The Babau of Rabaul: Tolai Fish-Traps of Papua New Guinea*. Port Moresby: University of Papua New Guinea. [112 pp.].

Mennis, M.R. 1972. *They Came to Matupit: The story of St. Michael's Church on Matupit Island*. Vunapope, PNG: Catholic Press. [119 pp.].

Mercalli, G. 1907. *I Vulcani Attiva della Terra*. Milan: Ulrico Hoepli. [421 pp.].

Michie, W. 1937. Copy of typed letter, 14 June 1937, from on board the SS *Montoro*, to the manager of the Island Department of Burns, Philp & Co. Ltd, Sydney. NAA: A518, V836/4. [2 pp.].

Mill Valley Record. 1937. 'Edgar Pedersen Relates Experiences in Noted South Sea Volcano Rescue: Mill Valley Youth Cabin Boy on Golden Bear of Disaster Fame'. *Mill Valley Record*, (California), 30 July 1937, 3.

Miller, L. 1980. Transcription of an audio interview used for the ABC radio program *Taim Bilong Masta*, 34–6. Interviewer Daniel Connell. Digital copy no. 2222 in RVO-IMS.

Miyake, Y. and Y. Sugiura. 1953. 'On the Chemical Compositions of the Volcanic Eruptives in New Britain Island, Pacific Ocean'. Proceedings of the 7th Pacific Congress, New Zealand, 361–3.

Mori, J., C. McKee, I. Itikarai, P. Lowenstein, P. de Saint Ours and B. Talai. 1989. 'Earthquakes of the Rabaul Seismo-Deformational Crisis September 1983 to July 1985: Seismicity on a Caldera Ring Fault'. In *Volcanic Hazards: Assessment and Monitoring, IAVCEI Proceedings in Volcanology*, edited by J.H. Latter, 429–62. Berlin: Springer-Verlag. doi.org/10.1007/978-3-642-73759-6_25.

Morobe News. 1941. 'Arrival of Sir Walter and Lady Nicholl—To Take Up Residence in the New Capital'. Newspaper cutting, *Morobe News*, 28 November 1941. NAA: A518, AK800/1/3/Part 4.

MSC (Missionarii Sacratissimi Cordis). 1937. English translations of numerous articles by Sacred Heart missionaries on the 1937 volcanic eruption at Rabaul, originally published in German in the mission magazine *Hiltruper Monatshefte* (1937). See Arculus and Johnson (1981) for the quoted page numbers of the English translations.

MSC (Missionarii Sacratissimi Cordis). 1982. *Yumi Pipel Bilong God 1882–1982: 100 Yia Lotu Katolik long Papua Nu Gini*. East New Britain Province, PNG: Vunapope Catholic Mission. [n.p.].

Murray Administrator. 1951. Typescript of international telegram from Popondetta for the minister, Department of External Territories, Canberra, Wednesday 14 February at 1410 hours. NAA: A518/1, AV918/1/Part 3.

Myers, N.O. 1976. 'Seismic Surveillance of Volcanoes in Papua New Guinea'. In *Volcanism in Australasia: A Collection of Papers in Honour of the Late G.A.M. Taylor*, edited by R.W. Johnson, 91–9. Amsterdam: G.C. Elsevier.

NAA (National Archives of Australia). 1937–39. 'New Guinea—Transfer of Administration Headquarters to Salamaua'. NAA: A518, AK800/1/3/Parts 1–3.

NAA (National Archives of Australia). 1937–50. 'New Guinea Miscellaneous, Volcanic Eruption 1937 Claims for Assistance', 1937–38 and 1938–50. NAA: A518, Z836/4, Parts 1–2.

NAA (National Archives of Australia). 1941–46. 'Transfer of New Guinea Administration to Lae'. NAA: A518, AK800/1/3/Part 4.

Nairn, I.A., C.O. McKee, B. Talai and C.P. Wood. 1995. 'Geology and Eruptive History of the Rabaul Caldera Area, Papua New Guinea'. *Journal of Volcanology and Geothermal Research* 69: 255–84. doi.org/10.1016/0377-0273(95)00035-6.

Nairn, I.A. and B.J. Scott. 1995. 'Scientific Management of the 1994 Rabaul Eruption: Lessons for New Zealand'. Institute of Geological and Nuclear Sciences, Science Report 95/26. [27 pp., plus appendices]. Annex C in E.T. Finnimore, B.S. Low, R.J. Martin, P. Karam, I.A. Nairn and B.J. Scott. 1995. 'Contingency Planning for and Emergency Management of the 1994 Rabaul Volcanic Eruption, Papua New Guinea: Results of a Fact-Finding Mission'. New Zealand Ministry of Civil Defence.

Nairn, I.A., B. Talai, C.P. Wood and C.O. McKee. 1989. 'Rabaul Caldera, Papua New Guinea—1:25 000 Reconnaissance Geological Map and Eruption History'. A report prepared for the Ministry of External Relations and Trade, New Zealand. New Zealand Geological Survey. [76 pp., plus coloured map and legend].

Nancarrow, S.N. and R.W. Johnson. 2015. 'Rabaul Volcanology Twinning Program: Completion Report'. Unpublished report, Geoscience Australia, Canberra, to AusAID, Canberra. [21 pp.].

Nelson, H. 1976. *Black, White and Gold: Goldmining in Papua New Guinea 1878–1930*. Canberra: The Australian National University Press. [298 pp.].

Nelson, H. 1982. *Taim Bilong Masta: The Australian Involvement with Papua New Guinea*. Sydney: Australian Broadcasting Commission. [224 pp.].

Nelson, H. 1992. 'The Troops, the Town and the Battle: Rabaul 1942'. *Journal of Pacific History* 27 (2): 198–216. doi.org/10.1080/00223349208572707.

Nelson, H. 1995. 'The Return to Rabaul 1945'. *Journal of Pacific History* 30 (2): 131–53. doi.org/10.1080/00223349508572791.

Neumann, K. 1992. *Not the Way It Really Was: Constructing the Tolai Past*. Honolulu: University of Hawai'i Press. [310 pp.]. doi.org/10.1515/9780824847098.

Neumann, K., ed. 1995. *Tavurvur I Puongo!: Students' Accounts of the 1994 Eruptions in East New Britain*. Vunadidir, PNG: Department of East New Britain. [65 pp.].

Neumann, K. 1996. *Rabaul Yu Swit Moa Yet: Surviving the 1994 Volcanic Eruption*. Port Moresby: Oxford University Press. [181 pp.].

Neumann, K. 1997. 'Nostalgia for Rabaul'. *Oceania* 67 (3): 177–93. doi.org/10.1002/j.1834-4461.1997.tb02603.x.

Neumann, K. 2014. 'A Volcano and Its People'. *Inside Story*, 19 September 2014. insidestory.org.au/a-volcano-and-its-people.

Neumann, K. 2017. Papers of Klaus Neumann, 1892–2017. National Library of Australia, MS10422.

Newhall, C.G. and D. Dzurisin. 1988. *Historical Unrest at Large Calderas of the World*. Vol. 1. United States Geological Survey Bulletin 1855. [598 pp.].

Newhall, C.G. and S. Self. 1982. 'The Volcanic Explosivity Index (VEI): An Estimate of Explosive Magnitude for Historical Volcanism'. *Journal of Geophysical Research* 87 (C2): 1231–8. doi.org/10.1029/JC087iC02p01231.

Newport, H. 1916. 'Report of the Earthquake of 1st January, 1916'. Report dated 15 January 1916. *Report on the Territory of New Guinea, 1921–22*, Appendix E, 2–3.

Newstead, G. 1968. 'Report on Visit to Rabaul, May 1968, for Inspection and Discussions on Rabaul Harbour Volcanic Warning System'. Unpublished report to the Administrator of Papua and New Guinea. The Australian National University. Digital copy no. 82 in RVO-IMS. [22 pp.].

Newstead, G. 1969. 'Keeping Watch on Volcanoes'. *Hemisphere* 13 (1): 32–7.

NFSAA (National Film and Sound Archive of Australia). 1937a. 'Visit by H.M.A.S. Moresby to Rabaul in Early June Following the May 1937 Eruptions'. 8 mm film of part of the ship's surveying season, attributed to Commander Crowther. NFSAA, Canberra, title number 16216.

NFSAA (National Film and Sound Archive of Australia). 1937b. 'Papua New Guinea, Volcano Eruption, 1937 Rabaul'. Documentary film on the effects of the May 1937 eruptions on the landscape, settlement and harbour at Rabaul, attributed to J. McColl. NFSAA, Canberra, title number 17719.

NFSAA (National Film and Sound Archive of Australia). c. 1939. 'Rabaul Devastated by Rain of Mud and Volcanic Ash'. *Movietone News*, vol. 9, no. 34. NFSAA, Canberra, title number 58089.

Nishimura, Y., M. Nakagawa, J. Kuduon and J. Wukara. 2005. 'Timing and Scale of Tsunamis Caused by the 1994 Rabaul Eruption, East New Britain, Papua New Guinea'. In *Tsunami Case Studies and Recent Developments. Advances in Natural and Technological Hazards Research*, edited by K. Satake, 43–56. Cham, Switzerland: Springer. doi.org/10.1007/1-4020-3331-1_3.

Nitta, J. 1980. 'Kazan Gun (Volcano Group)'. In *Hyogen. Hijyou no Burizaado* [Ice Field. Sad Blizzard], 55–116. Tokyo: Shincho-sha. [418 pp.]. R.W.J. Collection 1, Folder 5, Sleeve 42. [In Japanese.]

Official Handbook. 1937. *Official Handbook of the Territory of New Guinea Administered by the Commonwealth of Australia Under Mandate from the League of Nations*. Canberra: Commonwealth Government Printer. [551 pp.]. A reprint in 1943 includes a five-page supplement dealing in part with the 1937 eruptions.

Oliver, B. 2020. 'New Britain New Ireland Mission, South Pacific Division'. *Seventh-day Adventist Church Encyclopedia*, 12 July 2020. encyclopedia.adventist.org/article?id=B80Z.

Olsen, E.M. n.d. 'A Briefs Summary of Our Experiences during the Volcanic Eruption in the Harbour of Rabaul, Territory of New Guinea, on May 29th, 1937'. Report of SS *Golden Bear*, voyage no. 22, Matson Navigation Company. R.W.J. Collection 38A, Folder 4, Folio 48. [15 pp.].

Opinion. 1983. 'Articles by D.A. Swanson, "Forecasts and Predictions", and by J.L. Whitford-Stark and C.A. Wood, "Krafla Revisited"'. *EOS, American Geophysical Union* 64 (28): 452–3.

Overlack, P. 1972–73. 'German New Guinea: A Diplomatic, Economic and Political Survey'. *Royal Historical Society of Queensland* 9 (4): 128–51.

Pacific Islands Monthly. 1941. 'Eruption? Rabaul's Volcanoes Under Suspicion. From Our Own Correspondent. Rabaul, Jan. 3'. *Pacific Islands Monthly*, January 1941: 9.

Park, S.-C. and J. Mori. 2007. 'Triggering of Earthquakes during the 2000 Papua New Guinea Earthquake Sequence'. *Journal of Geophysical Research* 112 (B3). doi.org/10.1029/2006JB004480.

Patia, H. 2004. 'Petrology and Geochemistry of the Recent Eruptive History at Rabaul Caldera, Papua New Guinea: Implications for Magmatic Processes and Recurring Volcanic Activity'. MPhil thesis, The Australian National University. [111 pp.].

Patia, H., S.M. Eggins, R.J. Arculus, C.O. McKee, R.W. Johnson and A. Bradney. 2017. 'The 1994–2001 Eruptive Period at Rabaul, Papua New Guinea: Petrological and Geochemical Evidence for Basalt Injections into a Shallow Dacite Magma Reservoir, and Significant SO_2 Flux'. *Journal of Volcanology and Geothermal Research* 247: 200–17. doi.org/10.1016/j.jvolgeores.2017.08.011.

Pearce, G. 1937. Decode of cablegram, sent 11 June to the administrator, Rabaul. NAA: A518, X836/4. Digital copy no. 549 in RVO-IMS.

Perret, F.A. 1937. *The Eruption of Mount Pelée, 1929–1932*. Washington, DC: Carnegie Institution of Washington. [126 pp.].

Peterson, J. and C.R. Hutt. 2014. 'World-Wide Standardized Seismograph Network: A Data Users Guide'. USGS Open-File Report 2014-1218. [74 pp.]. doi.org/10.3133/ofr20141218.

Phillips, F.B. 1937a. 'Telegram to Prime Minister's Department, Canberra, 9 pm, Saturday 29 May 1937, from Rabaul'. NAA: A518, V836/4. Digital copy no. 549 in RVO-IMS.

Phillips, F.B. 1937b. 'Radio Message to Captain Michie on Board the *Montoro* at 11.30 pm, Saturday 29 May 1937, from Rabaul'. NAA: A518, V836/4. Digital copy no. 549 in RVO-IMS.

Phillips, F.B. 1937c. 'Situation in Rabaul'. Typescript of memorandum, 5 June 1937, from Rabaul, apparently for the attention of the administrator, W.R. McNicoll. NAA: A518, V836/4. Digital copy no. 549 in RVO-IMS.

Pigot, E.F. 1923. 'Brief Notes Regarding Earthquakes in the New Guinea Region'. Report dated 24 March 1923. *Report on the Territory of New Guinea, 1921–22*, Appendix E, 1–2.

Planters' Association. 1937, 1938. Correspondence with Planters' Association of New Guinea. Annual Reports for 1937 and 1938. United Nations Archives, Geneva, File R4130/6A/26678/4230.

PNGAA (Papua New Guinea Association of Australia). 2017. *When the War Came: New Guinea Islands 1942: Personal Stories of Those Who Faced WWII on Australian Territory and Our Greatest Maritime Disaster—The Sinking of Montevideo Maru.* Sydney: PNGAA. [514 pp.].

Powell, W. 1883. *Wanderings in a Wild Country: Or Three Years amongst the Cannibals of New Britain.* London: Sampson Low, Marston, Searle, & Rivington. [283 pp.].

Press Release. 1937. 'For the Press—8th July 1937'. NAA: A518, AB836/4. Digital copy no. 552 in RVO-IMS.

Pullen-Burry, B. 1909. *In a German Colony or Four Weeks in New Britain.* London: Methuen & Co. [238 pp.].

Purefoy Fitzgerald, K. 1937. Letter, 11 June 1937, from the rector of Rabaul to the administrator. NAA: A518, X836/4. Digital copy no. 553 in RVO-IMS.

Rabaul Times. 1937. 'Empire Day: Patriotic Programmes at Schools'. *Rabaul Times,* 28 May 1937, n.p.

Rabaul Times. 1940. 'Local and General News'. *Rabaul Times,* 28 June 1940, 5.

Radio Roundsman. 1937. Typescript of a broadcast made by Ben Sullivan on Monday 7 June at 7.40 pm on the radio station Sydney 2FC. NAA: A518, V836/4. Digital copy no. 549 in RVO-IMS. [4 pp.].

Rannie, D. 1912. *My Adventures among South Sea Cannibals: An Account of the Experiences and Adventures of a Government Official among the Natives of Oceania.* London: Seeley, Service & Co. [314 pp.].

Rechinger, L. and K. Rechinger. 1908. *Streifzüge in Deutsch-New-Guinea under auf den Salomons-Inseln.* Berlin: Deitrich Reimer (Ernst Vohsen). [108 pp.].

Reeves, L.C. 1915. *Australians in Action in New Guinea.* Sydney: Australasian News Company. [97 pp.].

Reynolds, M.A. 2005. 'Experiences in Volcanology and Life in the Territory of Papua and New Guinea 1953–57'. Unpublished memoirs. Digital copy no. 2267 in RVO-IMS. [51 pp.].

Ripley, S.D. 1947. *Trail of the Money Bird: 30,000 Miles of Adventure with a Naturalist.* London: Longmans, Green. [336 pp.].

Robertson, R.M. and C.R.J. Kilburn. 2016. 'Deformation Regime and Long-Term Precursors to Eruption at Large Calderas: Rabaul, Papua New Guinea'. *Earth and Planetary Science Letters* 438: 86–94. doi.org/10.1016/j.epsl.2016.01.003.

Robson, R.W., ed. 1937. 'Volcanic Eruptions Compel Evacuation of Rabaul: Residents Endure Many Days of Terror, Dirt and Discomfort—Followed by Period of Administrative Discord—Two Europeans and Several Natives Killed'. *Pacific Islands Monthly*, 23 June 1937: 9–14, 71–3.

Robson, R.W. 1965. *Queen Emma: The Samoan-American Girl Who Founded an Empire in 19th Century New Guinea.* Sydney: Pacific Publications. [239 pp.].

Roggensack, K., S.N. Williams, S.J. Schaefer and R.A. Parnell Jr. 1996. 'Volatiles from the 1994 Eruptions of Rabaul: Understanding Large Caldera Systems'. *Science* 273: 490–3. doi.org/10.1126/science.273.5274.490.

Romeyn, R.P. and M.C. Garthwaite. 2012. 'Broad-Scale Volcanic Monitoring in Papua New Guinea Using Satellite Interferometric Synthetic Aperture Radar'. Geoscience Australia, Canberra, Record 2012/49. [44 pp.].

Ronchin, E., T. Masterlark, J. Dawson, S. Saunders and J.M. Molist. 2017. 'Imaging the Complex Geometry of a Magmatic Reservoir Using FEM-Based Linear Inverse Modelling of Insar Data: Application to Rabaul Caldera, Papua New Guinea'. *Geophysical Journal International* 209: 1746–60. doi.org/10.1093/gji/ggx119.

Rose, W.L., D.J. Delene, D.J. Schneider, G.J.S. Bluth, A.J. Kreuger, I. Sprod, C. McKee, H.L. Davies and G.G.J. Ernst. 1995. 'Ice in the 1994 Rabaul Eruption Cloud: Implications for Volcano Hazard and Atmospheric Effects'. *Nature* 375 (8 June 1995): 477–9. doi.org/10.1038/375477a0.

Rossel, E.P.E. de, ed. 1808. *Voyage de D'Entrecasteaux, envoyé à la recherche de La Pérouse.* 2 vols. Paris: Imprimérie Impériale. [704 and 692 pp., plus atlas].

RVO (Rabaul Volcanological Observatory). 1937–42. Mainly weekly reports by N.H. Fisher, L.E. Clout and C.L. Knight, but including items by others. Volcanological Observatory, Rabaul, 2 December 1937 – 13 January 1942. Digital copy no. 718 in RVO-IMS. [250 pp.].

RVO (Rabaul Volcanological Observatory). 1990. 'CO$_2$ Kills 6 at Tavurvur; Seismicity Remains Low'. *Global Volcanism Network* 15 (6): 8–9.

RVO (Rabaul Volcanological Observatory). 2014. Nine unpublished lists, maps and guidelines resulting from the Papua New Guinea volcanoes threat-assessment workshop held in Rabaul, East New Britain. Note: The RVO workshop used the assessment methodology of Ewert (2007).

Ryan, W.B. 1980–84. Correspondence between Mr Ryan and R.W.J., including photographs. R.W.J. Collection 30A, Folder 4, Sleeves 49–55. Digital copy 1025 in RVO-IMS.

Sack, P.G. 1973. *Land between Two Laws: Early European Land Acquisition in New Guinea.* Canberra: The Australian National University Press. [197 pp.].

Sack, P. 1987. 'The Emergence and Settlement of Matupit Island: Vulcanological Evidence, Oral Tradition and "Objective" History in Papua New Guinea'. *Bikmaus, Institute of Papua New Guinea Studies, Boroko* 7 (1): 1–14.

Sack, P.G. and D. Clark, eds and trans. 1979. *German New Guinea: The Annual Reports.* Canberra: The Australian National University Press. [403 pp.].

Salisbury, R. 1970. *Vunamami: Economic Transformation in a Traditional Society.* Carlton, Vic.: Melbourne University Press. [389 pp.].

Sapper, K. 1910a. 'Wissenschaftliche Ergebnisse einer amtlichen Forschungsreise nach dem Bismarck-Archipel im Jahre 1908. 1'. *Beiträge zur Landeskunde von Neu-Mecklenburg und seinen Nachbarinseln. Mitteilungen aus den Deutschen Schutzgebieten. Ergänzungsheft*, no. 3. [130 pp.].

Sapper, K. 1910b. 'Eine Durchquerung von Bougainville'. *Mitteilungen aus den Deutschen Schutzgebieten* 23: 206–17.

Sapper, K. 1910c. 'Beiträge zur Kenntnis Neupommerns und des Kaiser-Wilhelms-Landes. Petermanns Mitteilungen aus Justus Perthes'. *Geographischer Anstalt* 56: 189–93, 255–6.

Sapper, K. 1927. *Vulkankunde.* Stuttgart: J. Engelhorns Nachf. [424 pp.].

Sapper, K. 1937. 'Vulkanausbrüche bei Rabaul (Neupommern)'. *Petermanns Mitteilungenaus Justus Perthes Geographischer Anstalt* 83: 279–80.

Saunders, S.J. 2001. 'The Shallow Plumbing System of Rabaul Caldera: A Partially Intruded Ring Fault?' *Bulletin of Volcanology* 63: 406–20. doi.org/10.1007/s004450100159.

Saunders, S.J. 2007. 'Dr Norman Fisher Visits Rabaul'. *Una Voce*, no. 2: 18–19.

Saunders, S. 2008. 'Proceed with Caution in Greet Harbour!' Unpublished note from the RVO and Rabaul Historical Survey. [5 pp., including bathymetric charts].

Saxton, A.E. 1937. 'Preliminary Report to Shareholders Re Volcanic Disturbance at Rabaul'. Rabaul Electricity Limited, Sydney. NAA: A518, V836/4. Digital copies of selected folios are available in RVO-IMS, reference no. 549. [2 pp.].

Scales, I. 2010. *Roads in Gazelle Peninsula Development: Impact of Roads in the Post-eruption Economic Landscape of East New Britain*. Canberra: Australian Agency for International Development. [117 pp.].

Schleinitz, Capitan von. 1876. 'Die Expedition S.M.S. Gazelle. X11. Hydrographische Beitraege fuer den westlichen Theil des suedlichen Stillen Oceans'. *Annalen der Hydrographie und Maritimen Meteorologie* 4: 399–404.

Schleinitz, Kapitan von. 1889. *Die Forschungsreise S.M.S. Gazelle in den Jahren 1874 bis 1876. I. Theil: Der Reisebericht*. Berlin: Ernst Siegfried Mittler & Sohn. See especially pp. 239–52.

Scrope, G.P. 1862. *Volcanos: The Character of Their Phenomena, Their Share in the Structure and Composition of the Surface of the Globe, and Their Relation to Internal Forces: With a Descriptive Catalogue of All Known Volcanos and Volcanic Formations*. London: Longman, Green, Longmans, and Roberts. [490 pp.].

Selby, D. (1956) 1971. *Hell and High Fever*. Sydney: Pacific Books. [198 pp.].

Selby, D.M. 1981. Letter, 16 February 1981, to R.W.J. R.W.J. Collection 30B, Folder 7, Sleeve 31.

Sieberg, A. 1910. 'Die Erdbebentätigkeit in Deutsch-Neuguinea (Kaiser-Wilhelms-Land u. Bismarckarchipel). Dr A. Petermanns Mitteilungen aus Justus Perthes'. *Geographischer Anstalt* 56, heft 2: 72–4, heft 3: 116–22.

Siebert, L., T. Simkin and P. Kimberly. 2010. *Volcanoes of the World*. 3rd ed. Washington, DC: Smithsonian Institution. [551 pp.].

Simet, J. 1991. 'Tabu: Analysis of a Tolai Ritual Object'. PhD thesis, The Australian National University. [453 pp.].

Simkin, T. and L. Siebert. 1994. *Volcanoes of the World.* 2nd ed. Washington, DC: Smithsonian Institution. [349 pp.].

Simpson, C.H. 1873. 'Hydrographical Extract from a Six Months' Cruise among the South Sea Islands'. Letter, 1 November 1872, HMS *Blanche*, Sydney, New South Wales, to the [British] hydrographer. Hydrographic Notice No. 1 of 1873 (Pacific Notice No. 23), 1–8.

Sloan, K. 2003. '"Aimed at Universality and Belonging to the Nation": The Enlightenment and the British Museum'. In *Enlightenment: Discovering the World in the Eighteenth Century*, edited by K. Sloan, 12–25. London: British Museum Press.

South Pacific Post. 1951a. Articles in the issue of 19 January: 'Mt Lamington Erupts' (p. 1); 'Administrator on Tour' (p. 2).

South Pacific Post. 1951b. Articles in the issue of 26 January: 'Plan Ready for Any New Emergency in Our Worst Disaster' (pp. 1–2); 'Tribute of a Native' (p. 1); 'Missing Man Took This Picture' (p. 1); 'Casualty List' (p. 4); 'Natives Have Burns Treated' (p. 4); 'Unheeded Warning!' (p. 8); 'The Legend of Mount Lamington' (p. 8; by L. Austen); 'Letters to the Editor: Reception of Survivors' (by E.L. Hand) (p. 9); 'Australia Volcano Proof' (p. 10).

South Pacific Post. 1951c. Articles in the issue of 16 February: 'Rabaul Volcanoes Unchanged' (p. 2); 'Dobell Landscape to Aid Fund' (p. 3); 'Donations Received' (p. 3); 'Letters to the Editor: The Lamington Disaster' (by T.W. Upson) (p. 9).

South Pacific Post. 1951d. 'Rabaul Party to See Mt Lamington Area—Move by Administrator to Stress Horrors of Eruption', 11 May 1951, 9.

Spate, O.H.K. 1979. *The Spanish Lake.* Canberra: The Australian National University Press. [372 pp.].

Special Correspondent. 1945. 'Horrors in Wrecked Rabaul: Australians Are Now Trying to Restore Order'. *Pacific Islands Monthly* 16 (3), 16 October: 59–60.

Spurling, K. 2017. *Abandoned and Sacrificed: The Tragedy of the* Montevideo Maru. London: New Holland Publishers. [303 pp.].

Stanaway, R.F. 2004. 'Implementation of a Dynamic Geodetic Datum in Papua New Guinea'. MPhil thesis, The Australian National University. [157 pp.]

Stanley, E.R. 1923. 'Report on the Salient Geological Features and Natural Resources of the New Guinea Territory Including Notes of Dialectics and Ethnology'. Report on the Territory of New Guinea, Commonwealth of Australia Parliamentary Paper 18 of 1923, Appendix B. [99 pp.].

Steering Committee. 1997. 'Rabaul Subject (Zoning) Development Plan'. East New Britain Provincial Physical Planning Board. [Five chapters individually paginated, plus 11 maps and 8 appendices].

Stehn, C.E. and W.G. Woolnough. 1937a. 'Report on Volcanological and Seismological Investigations at Rabaul'. Original typescript of report to Australian Government. Digital copy 1094 in RVO-IMS. [35 pp.].

Stehn, C.E. and W.G. Woolnough. 1937b. 'Report on Volcanological and Seismological Investigations at Rabaul'. Report to Council of the League of Nations on the Administration of the Territory of New Guinea for the year 1936–37. Papers presented to Parliament (and ordered to be printed). Vol. 3, Parliamentary Paper 84 of 1937, Appendix C, 147–58.

Stewart, D., 1937. 'Evacuation of Rabaul: How the "Montoro" transferred, without casualty, 200 Europeans and 5000 Natives to Kokopo'. *Pacific Islands Monthly*, 23 July, pages 16 and 72–73.

Stone, P. 1995. *Hostages to Freedom—The Fall of Rabaul*. Yarram, Vic.: Oceans Enterprises. [513 pp.].

Studt, F.E. 1961. 'Preliminary Survey of the Hydrothermal Field at Rabaul, New Britain'. *New Zealand Journal of Geology and Geophysics* 4: 274–82. doi.org/10.1080/00288306.1961.10423128.

Suarez, T. 2004. *Early Mapping of the Pacific: The Epic Story of Seafarers, Adventurers, and Cartographers Who Mapped the Earth's Greatest Ocean*. Singapore: Periplus Editions. [224 pp.].

Südsee-Handbuch. 1920. *Draft of a Guide to the South Seas Made by Direction of the German Imperial Admiralty*. Part IIA: New Britain and Adjacent Islands. English translation of the original German version, together with new footnotes covering later information from Australian Navy sources. Melbourne: Australian Government Printer. [127 pp.].

Sydney Morning Herald. 1904. 'Tragic End of TomaVI'. *Sydney Morning Herald*, 2 November 1904.

Symons, G.J., ed. 1888. *The Eruption of Krakatau, and Subsequent Phenomena*. Report of the Krakatau Committee of the Royal Society. London: Trubner. [494 pp.].

Talai, B. and L. Pue. 1981. 'Elias Ravian (1945–1979)'. In *Cooke-Ravian Volume of Volcanological Papers*, edited by R.W. Johnson, xiii–viv. Port Moresby: Geological Survey of Papua New Guinea.

Taranu, F. and M. Herry. 2015. 'The Cape Lambert Earthquakes of 2014: Evidence for NW Extension of the Baining Fault'. PNG Department of Mineral Policy and Geohazards Management, Geohazards Management Division Report 2015/01. [21 pp.].

Taylor, G.A. 1955. 'Tectonic Earthquakes and Recent Volcanic Activity'. Bureau of Mineral Resources Record, 1955/123. [5 pp.].

Taylor, G.A. 1956. 'Australian National Committee of Geodesy and Geophysics Report of the Sub-Committee on Vulcanology, 1953'. Review of volcanic activity in the Territory of Papua-New Guinea, the Solomon and New Hebrides Islands 1951–53. *Bulletin Volcanologique* 18: 35–7.

Taylor, G.A. 1958. *The 1951 Eruption of Mount Lamington, Papua.* Canberra: Bulletin of the Bureau of Mineral Resources. [117 pp.].

Taylor, G.A. 1960. 'An Experiment in Volcanic Prediction'. Bureau of Mineral Resources, Canberra, Record 1960/74. [17 pp.].

Termer, F. 1966. 'Karl Theodor Sapper 1866–1945: Leben und Wirken eines deutschen Geographen und Geologen'. *Deutsche Akademie der Naturforscher Leopoldina, Lebensdarstellungen deutscher Naturforscher* 12: 8–89.

Territories. 1938. Untitled, undated memorandum—probably a press release—from Minister W.M. Hughes's office in Canberra, containing details provided by Sir Walter McNicoll on the Rabaul earthquakes of 8 January 1938. A photocopy of the document was made by either W.D. Palfreyman or R.J.S. Cooke likely from an unidentified Territories Branch file, Canberra. Digital copy no. 1188 in RVO-IMS. See also Woolnough (1938).

Territories. 1952. International telegram, 14 June 1952, to the administration in Port Moresby for the attention of Cleland [Administrator D.M. Cleland] from Lambert [Secretary C.R. Lambert, Department of Territories] in Canberra. Papua New Guinea National Archives, File 44/4/4/4 (SN 677, AN247, BN353).

Thomas, G. 1937a. 'Our Volcanic Issue'. *Rabaul Times*, 4 June 1937. [7 pp.].

Thomas, G. 1937b. 'The Terror of the Eruption: Rabaul Journalist Describes His Experiences When "Black-Out" Occurred'. *Pacific Islands Monthly* 7 (12), 23 July 1937: 35–8.

Thomas, G. 1937c. 'Nearing Normal'. *Rabaul Times*, 23 July 1937, 8.

Thomas, G. 1941. 'And So to Lae. Editorial'. *Rabaul Times*, 29 August 1941, 6.

Thomas, G. 2012. *Prisoners in Rabaul: Civilians in Captivity: 1942–45.* Loftus, NSW: Australian Military History Publications. [266 pp.].

Thompson, J.E. and N.H. Fisher. 1965. 'Mineral Deposits of New Guinea and Papua and Their Tectonic Setting'. *Proceedings of the Eighth Commonwealth Mining and Metallurgical Congress* 6: 115–48.

Threlfall, N. 1975. *One Hundred Years in the Islands: The Methodist/United Church in the New Guinea Islands Region 1875–1975*. Rabaul, PNG: United Church, New Guinea Islands Region. [288 pp.].

Threlfall, N. 2012. *Mangroves, Coconuts and Frangipani: The Story of Rabaul*. Gosford, NSW: Rabaul Historical Society. [570 pp.].

Threlfall, N.A. 2021a. '1967 Earthquakes, East New Britain'. Unpublished typescript of personal experiences in the Rabaul area at the time of the earthquakes. R.W.J. Collection 37, Folder 4, Sleeves 122–3. [2 pp.].

Threlfall, N.A. 2021b. '1971 Earthquakes in East New Britain'. Unpublished typescript of personal experiences in the Rabaul area at the time of the earthquakes. R.W.J. Collection 37, Folder 4, Sleeve 115. [5 pp.].

Tiffin, D.L., B.D. Taylor, W. Tufar and I. Itikarai. 1990. 'A Seabeam snd Sampling Survey of the Newly Discovered Tavui Caldera, Near Rabaul, Papua New Guinea. Sonne Cruise SO-68, 7–9 May 1990'. South Pacific Applied Geoscience Commission Offshore Programme, Papua New Guinea Project PN.14. [18 pp.].

ToMaran. 1951. 'A Tale of '78'. *Pacific Island Monthly* 22 (4): 67.

ToMaran. n.d. 'Volcanic Eruption of 1878'. Unpublished typescript in the papers of Mr C.D. Meares, Rabaul. Account translated into English from the Tok Pisin and then transliterated. Digital copy no. 1137 in RVO-IMS.

Tomblin, J. and J. Chung. 1995. *Papua New Guinea Analysis of Lessons Learnt from Rabaul Eruption and Programming for Disaster Mitigation Activities in Other Parts of the Country*. Report of Mission from 17–26 February 1995. Geneva: United Nations Department of Humanitarian Affairs. [8 pp., plus four annexes].

Tootel, B. 1985. *All Four Engines Have Failed: The True and Triumphant Story of Flight BA 009 and the 'Jakarta Incident'*. [London]: André Deutsch Ltd. [178 pp.].

Tregoning, P., H. McQueen, K. Lambeck, R. Jackson, R. Little, S. Saunders and R. Rosa. 2000. 'Present-Day Crustal Motion in Papua New Guinea'. *Earth Planets Space* 52: 727–30. doi.org/10.1186/BF03352272.

Trevitt, M. (née Chaseling). 1937. Two typed versions of a letter to Miss Mavis (Mae) Betts in Australia, written initially from Vunairima on 4 July 1937, and a slightly different version edited by Mrs M. Walker (formerly Mrs Trevitt) in 1981. R.W.J. Collection 30D, Folder 14, Sleeves 21–6. [6 pp.].

UNESCO (United Nations Educational, Scientific and Cultural Organization). 1972. *The Surveillance and Prediction of Volcanic Activity: A Review of Methods and Techniques*. Paris: UNESCO Earth Science Series. [166 pp.].

United States Strategic Bombing Survey (Pacific). 1946. *The Allied Campaign Against Rabaul*. Washington, DC: Naval Analysis Division, Marshall-Gilberts-New Britain Party. [273 pp.].

Vagg, A.G. 1981. Correspondence between Mr Vagg and R.W.J. R.W.J. Collection 30B, Folder 7, Sleeve 33.

van Bemmelen, R.W. 1929. 'Het caldera problem'. *De Mijningenier, Vereeniging van Ingenieurs en Geologen, Bandoeng* 10 (5): 101–12. [In Dutch with English summary].

van Bemmelen, R.W. 1949. 'Charles Edgar Stehn (1884–1945)'. Nécrologie. *Bulletin Volcanologique* 8: 133–7.

Walker, G.P.L. 1981. 'Characteristics of Two Phreatoplinian Ashes, and Their Water-Flushed Origin'. *Journal of Volcanology and Geothermal Research* 9 (4): 395–407.

Walker, G.P.L., R.F. Heming, T.J. Sprod and H.R. Walker. 1981. 'Latest Major Eruptions of Rabaul Volcano'. In *Cooke-Ravian Volume of Volcanological Papers*, edited by R.W. Johnson, 181–93. Port Moresby: Geological Survey of Papua New Guinea. doi.org/10.1016/0377-0273(81)90046-9.

Wallace, J.I. 1938, 1948. Two handwritten letters (7 January 1938 and 19 April 1948) to the Australian Prime Minister concerning reimbursement for losses incurred during the May 1937 disaster. NAA: A518, Z836/4, Parts 1–2.

Wallis, H., ed. 1965. *Carteret's Voyage Round the World 1766–1769*. 2 vols. Cambridge: Hakluyt Society. [564 pp.].

Wanliss, D.S. 1937. 'Telegram to Prime Minister's Department, Canberra, from Rabaul, at 11 am, Saturday 29 May 1937'. NAA: A518, V836/4. Digital copies of selected folios are available in RVO-IMS, reference no. 549.

Waterhouse, M. 2010. *Not a Poor Man's Field: The New Guinea Goldfields to 1942—An Australian Colonial History*. Braddon, ACT: Halstead Press. [272 pp.].

Waugh, D. 1937. 'Telegram to Prime Minister's Department, Canberra, from the Montoro at Kokopo, 9.20 am, Monday 31 May 1937'. NAA: A518, V836/4. Digital copies of selected folios are available in RVO-IMS, reference no. 549.

Wayne, R. 1937. 'The Volcanic Eruptions at Rabaul'. Typescript of extract from his diary. The typescript has footnotes added by 'H.W.', who is almost certainly Mr Wayne's wife, Helen, and who may have arranged for the typing to be undertaken, probably in the late 1970s. Digital copies 1156 and 2337 in RVO-IMS. [20 pp., plus a map and list of photographs].

Weetman, C. 1937. 'Rabaul: A Study in Black and White'. *Walkabout Magazine*, 1 April: 19–23.

Wendt, H.W. 1879. 'Hydrographische und meteorologische Beitraege zur Kenntniss der Duke of York-Inseln und der Bai von Ratavul an der Nordkueste von Neu-Britannien'. *Annalen der Hydrographie und Maritimen Meteorologie* 7: 177–80.

West, F. 1968. *Hubert Murray: The Australian Pro-Consul*. Melbourne: Oxford University Press. [296 pp.].

Wharton, W.J.L. 1889. 'Volcanic Sea Wave'. *Nature* 39: 303–4. doi.org/10.1038/039303a0.

Whittaker, J.K. 1975. 'A French Naturalist Observes the New Irelanders, 1823'. Translated from the French. In *Documents and Readings in New Guinea History: Prehistory to 1889*, edited by J.K. Whittaker, N.G. Gash, J.F. Hookey and R.J. Lacey, 228–9. Milton, Qld: Jacaranda Press.

Whittaker, J.K., N.G. Gash, J.F. Hookey and R.J. Lacey, eds. 1975. *Documents and Readings in New Guinea History: Prehistory to 1889*. Milton, Qld: Jacaranda Press. [552 pp.].

Whitty, S.G. 1981. Undated, handwritten letter to R.W.J. R.W.J. Collection 30B, Folder 7, Sleeve 35.

Wichman, A. 1912. 'Entdeckungsgeschichte von Neu-Guinea 1885 bis 1902'. *Nova Guinea* 2: 371–1026.

Wiebenga, W.A. 1973. 'Crustal Structure of the New Britain–New Ireland Region'. In *The Western Pacific: Island Arcs, Marginal Sea, Geochemistry*, edited by P.J. Coleman, 163–77. Nedlands, WA: University of Western Australia Press.

Wiebenga, W.A. and E.J. Polak. 1962. 'Rabaul Geothermal Investigation, New Britain 1960'. Bureau of Mineral Resources, Canberra, Record 1962/9. [6 pp.].

Wigmore, L. 1957. *The Japanese Thrust. Australia in the War of 1939–1945*. Series 1 Army, Vol. IV. Canberra: Australian War Memorial. [715 pp.].

Wilkinson, R. 1996. *Rocks to Riches: The Story of Australia's National Geological Survey*. St Leonards, NSW: Allen & Unwin. [446 pp.].

Williams, H. 1941. 'Calderas and Their Origin'. *University of California Publications, Bulletin of the Department of Geological Sciences* 25: 239–346.

Williams, H. 1951. 'Volcanoes'. *Scientific American* 185 (5): 45–53. doi.org/10.1038/scientificamerican1251-45.

Williams, S.N. 1995. 'Erupting Neighbours—At Last'. *Science* 267, 20 January 1995: 340–1. doi.org/10.1126/science.267.5196.340.

Willson, A.C. 1937. 'Earthquake and Volcanic Eruption'. *Hydrographic Bulletin* 2500, 4 August 1937. Also referenced as A.G. Willson in 'Die Vulkanausbrueche bei Rabaul im Mai/Juni 1937'. *Der Seewart, Nautische Zeitschrift* 9: 330–4.

Woolnough, W.G. 1938. 'Rabaul Earthquakes'. Copy of memorandum dated 11 January 1938, presumably sent to Minister Hughes in Canberra. A photocopy of the document was made by either W.D. Palfreyman or R.J.S. Cooke, likely from an unidentified Territories Branch file, Canberra. Digital copy no. 1188 in RVO-IMS. See also Territories (1938).

Woolnough, W.G. 1941a. 'Volcanic Activity at Rabaul'. R.W.J. Collection 8, Folder 4, Folio 505.

Woolnough, W.G. 1941b. 'The Woolnough Article: Commenting on Recent Earthquake'. *Rabaul Times*, 31 January 1941, 10. Digital copy no. 1189 in RVO-IMS.

Appendix 1: Authors' Research Collections

R.W. Johnson: Collection of research notes and correspondence held at the National Library of Australia, Canberra.

- Collection 1: Takashi Kizawa correspondence and documents 1961–99.
- Collections 3–5: R.J. Cooke negatives, correspondence, manuscripts etc.
- Collection 8A: Rabaul 1937 eruption and aftermath: Rabaul eruption and town; National Archives of Australia folios, 1937–38.
- Collection 8B: Rabaul 1937 eruption and aftermath: Rabaul eruption and town 1937–54, plus volcanic areas other than Rabaul (1920–57); National Archives of Australia folios.
- Collection 9: M.A. Reynolds, J. Barrie and C.D. Branch, memoirs and photographs 1953–65.
- Collection 10: N.H. Fisher publications, correspondence and photographs 1939–2007.
- Collection 11: Rabaul volcano, 1983–85 crisis and pre-1994 surveys.
- Collection 12: Rabaul volcano and 1994–2002 visits: R.W.J. 35 mm slides and negatives.
- Collection 13: Rabaul Volcano and 1994–2006 eruptions: miscellaneous images and documents.
- Collection 14: Rabaul and RVO: images, documents and PAMBU microfilming 1991–2011.
- Collection 24: Correspondence on the history of colonisation of PNG volcanic areas.
- Collection 29: National Archives of PNG: copies of folios on TPNG files.
- Collection 30A: Rabaul Eruption 1937–41: A–He, photographs and correspondence.

- Collection 30B: Rabaul Eruption 1937–41: Ho–M, photographs and correspondence.
- Collection 30C: Rabaul Eruption 1937–41: N–S, photographs and correspondence.
- Collection 30D: Rabaul Eruption 1937–41: T–Z, photographs and correspondence.
- Collection 35: Photographic collections, mainly GA images of Lamington-51 and Rabaul-37.
- Collection 37: General PNG correspondence, November 2019 – May 2022.

N.A. Threlfall: Notes and research materials on the history of Rabaul and the Gazelle Peninsula, held at the Pacific Manuscripts Bureau, The Australian National University, Canberra. Identity area: AU PMB MS1305, 13 microfilm reels: asiapacific.anu.edu.au/pambu/catalogue/index.php/threlfall-rev-neville-1930.

Appendix 2: Acronyms and Abbreviations

ABC — Australian Broadcasting Corporation

ACSEE — Advisory Committee on Seismology and Earthquake Engineering

AGS — Allied Geographical Section

AGSO — Australian Geological Survey Organisation

AIDAB — Australian International Development Assistance Bureau (later AusAID)

ANU — The Australian National University

AusAID — Australian Agency for International Development

BMR — Bureau of Mineral Resources, Geology and Geophysics (now Geoscience Australia)

BP — Before Present

CD — circular despatches

CE — Common Era

GA — Geoscience Australia

GIS — geographical information system

GPS — global positioning system

GRA — Gazelle Restoration Authority

GVN — Global Volcanism Network

GVP — Global Volcanism Program

HMAS — Her or His Majesty's Australian Ship

HMS — Her or His Majesty's Ship

IMS — information management system

IAVCEI — International Association of Volcanology and Chemistry of the Earth's Interior

JICA	Japanese International Cooperation Agency
LVA	low-velocity anomaly
MATPSM	Matupit Public Survey Mark
MBE	Member of the Order of the British Empire
MSC	Missionarii Sacratissimi Cordis
NASA	National Aeronautics and Space Administration
NBC	National Broadcasting Company
NEEq	North-East Earthquake
NFSAA	National Film and Sound Archive of Australia
NGVR	New Guinea Volunteer Rifles
NOAA	National Oceanic and Atmospheric Administration
OBE	Order of the British Empire
PDC	Provincial Disaster Committee
PMGO	Port Moresby Geophysical Observatory
PNG	Papua New Guinea
POW	prisoner of war
RAAF	Royal Australian Air Force
RELACS	Rabaul Earthquake Location and Caldera Structure
RVO	Rabaul Volcanological Observatory
TPNG	Territory of Papua and New Guinea
UNDP	United Nations Development Programme
UNESCO	United Nations Educational, Scientific and Cultural Organization
USAID	United States Agency for International Development
USGS	United States Geological Survey
VAAC	Volcanic Ash Advisory Center
VDAP	Volcanic Disaster Assistance Program
VEI	volcanic explosivity index
VSS	Volcanological Service Support
WTZ	Watom–Turagunan Zone
WWSSN	World-Wide Standardized Seismograph Network
WWII	World War II

Appendix 3: Glossary

A mata	A hole or depression.
Andesite	A generally grey volcanic rock containing 53 up to less than 62 per cent silica (SiO_2) and forming a chemical series with basalt, dacite and rhyolite.
Ash	Pieces of generally glassy volcanic rock less than 4 millimetres in diameter.
Ash fall	A term commonly used to distinguish the gravitational fall of particles from eruption clouds, from the deposits of pyroclastic flows.
Basalt	A dark volcanic rock containing less than 53 per cent silica (SiO_2) and forming a chemical series with andesite, dacite and rhyolite.
Caldera	Large generally elliptical or sub-circular surface depressions at least 1–2 kilometres wide, formed by the collapse of the roofs of magma reservoirs.
Crater	The small depressions commonly seen on volcanoes, especially at their summits, and which are generally less than 1–2 kilometres in diameter (see also caldera).
Dacite	A generally light-greyish volcanic rock containing 62 up to less than 70 per cent silica (SiO_2) and forming a chemical series with basalt, andesite and rhyolite.
Doctor-boi	New Guinean medical assistant.
Dukduk	Masked ritual figure; secret society.
Geothermal	Relating to or produced by the internal heat of the earth, whether magmatic or produced by radioactivity.
Guria	Earthquake.

Hydrovolcanic eruption	Explosions caused by magma encountering subsurface water and producing intense fracturing and expulsion of fragmented rocks.
Ignimbrite	A pumice-rich deposit or rock commonly of dacitic or rhyolitic composition produced by deposition from a pyroclastic flow.
Intrusion	Rocks representing magmas that have filled and solidified in fissures, cracks, faults and other spaces beneath volcanoes but which have not erupted from them.
Kaia	Refers to different spirits that appear most commonly as giant snakes called *valvalir* or *kaliku*.
Kina	PNG national currency.
Kuanua	Tolai language.
Landeshauptmann	Administrator.
Lava flow	A ground-hugging stream of erupted magma and rock, commonly referred to simply as 'lava' and clearly distinguishable from pyroclastic flows.
Magma	Subsurface materials in the earth's crust and upper mantle that have become molten and that either erupt from volcanoes or form intrusions beneath them. Most magmas are technically alumina-silicate liquid and most contain at least some pre-eruption crystals.
Matamatanai	Commemorative ceremony.
Nuées ardente	The term means 'glowing cloud' and was introduced for the block-and-ash type pyroclastic flows observed at Mount Pelée on 8 May 1902.
Patuana	Older people.
Peléean eruption	The name derives from the explosive eruptions at Mount Pelée in 1902 when *nuées ardentes* were first described and named. Both terms are no longer volcanologically fashionable.
Phreato	A widely used prefix that corresponds commonly to the early 'hydroexplosive' phase that precedes some magmatic eruptions.

Plinian eruption	Paroxysmal ejections of large volumes of pyroclastic materials, at times accompanied by caldera formation. The eruptions form high-rising eruption columns and clouds. Large-volume pyroclastic flows may develop when the column collapses, including ignimbrites if pumice is abundant in the flows.
Pumice	Pieces of highly frothed volcanic rock commonly of dacitic or rhyolitic composition and which can float on water.
Pyroclastic	Literally meaning 'fire broken' and applied to fresh magma or hot volcanic rocks that have been broken up during volcanic explosions.
Pyroclastic flow	Fast-flowing hot emulsions or 'avalanches' of pumice, volcanic ash, dust, blocks and entrained air, which tend to follow the floors of valleys during their emplacement.
Ra-baul	The mangroves.
Scoria	A pyroclastic *rock* in fragments containing abundant round, bubble-like cavities (vesicles). Scoria is commonly basaltic or andesitic and ranges in colour from black or dark grey to deep rust-reddish depending on its state of oxidation. It is not pumiceous.
Strombolian eruption	Weak to violent ejection of pasty blebs of fluid lava, accompanied by spherical to fusiform 'bombs', cinders and ash. The activity can be spectacularly incandescent at night-time. Lava flows may be formed.
Subduction	The process whereby a tectonic plate is thrust under an adjoining plate and descends deep into the earth's interior.
Surge	Laterally propelled eruption clouds caused typically by the interaction of magma with lake water or in shallow-water coastal areas (see hydrovolcanic eruptions). They flow across the water surface away from the vent and the base of the main eruption cloud or column.
Tabu	Shell money.

Tephra	A commonly used synonym for pyroclastic materials or pyroclastic rocks in general. The term was used originally by Aristotle.
Tinata tuna	Our own language.
Tomography (seismic)	Mapping features beneath the earth's surface using the measured velocities of earthquakes.
Tsunami	A Japanese term referring to a long, high sea wave caused by an earthquake or some other disturbance such as a volcano collapsing into the sea. Tsunamis increase their heights where approaching coastlines and can be hazardous.
Tubuan	Masked ritual figure; secret society.
Vulcanian eruption	Violent ejections of solid or viscous hot volcanic fragments, at times in cauliflower- or mushroom-shaped clouds. Pyroclastic flows can take place, but lava flows are typically absent.
Vunatarai	Matrilineal descent lines.

www.ingramcontent.com/pod-product-compliance
Lightning Source LLC
Chambersburg PA
CBHW050623280326
41932CB00015B/2503